THE POPULATION OF
JAMAICA

GEORGE W. ROBERTS

THE POPULATION OF JAMAICA

WITH AN INTRODUCTION BY
KINGSLEY DAVIS

CAMBRIDGE
PUBLISHED FOR
THE CONSERVATION FOUNDATION
AT THE UNIVERSITY PRESS
1957

CAMBRIDGE UNIVERSITY PRESS
Cambridge, New York, Melbourne, Madrid, Cape Town,
Singapore, São Paulo, Delhi, Mexico City

Cambridge University Press
The Edinburgh Building, Cambridge CB2 8RU, UK

Published in the United States of America by Cambridge University Press, New York

www.cambridge.org
Information on this title: www.cambridge.org/9781107623422

First published 1957
First paperback edition 2013

A catalogue record for this publication is available from the British Library

ISBN 978-1-107-62342-2 Paperback

CONTENTS

APPENDICES

LIST OF FIGURES

MAP

LIST OF TABLES

PREFACE

This study, a formal demographic analysis of Jamaica, forms part of a broad research project sponsored by The Conservation Foundation. It is confined to the island of Jamaica; the dependencies—the Turks and Caicos Islands and the Cayman Islands—are not treated here because their vital statistics (which are not under the control of the Registrar General of Jamaica) are inferior to those of Jamaica.

As the last census of Jamaica was that of 1943 the analysis can in the main be carried no further than this date. However, in discussions of fertility and mortality full use is made of more recent data prepared by the Registrar General for years up to 1952. It remains an unfortunate but unavoidable consequence of the rapidly changing vital rates that any demographic study of the island tends to have dated somewhat by the time it is published.

Comparative material for other West Indian territories is drawn on where relevant historical and other factors suggest that the subject can be more satisfactorily developed in the wider context of the British Caribbean as a whole. Thus in Chapters 1, 4 and 7 material from other West Indian territories is used though in each case the focus of the analysis remains the population of Jamaica.

Some of the material used in this study has already appeared in papers published in *Population Studies*, and thanks are expressed to the editors of this journal for their permission to incorporate such material in the study.

The writer is most grateful to The Conservation Foundation, and in particular to its Director of Research, Mr R. G. Snider, for being selected to work on this study. The work was done under the direction of Professor Kingsley Davis and the writer remains deeply indebted to him for his guidance and encouragement at every stage. Special thanks are also expressed to Professor D. V. Glass for his critical reading of the manuscript; as a result of his criticism both the form and the content of the whole study have been improved.

Discussions with colleagues on the other Jamaican studies, Miss Judith Blake and Mr Peter M. Stern, have helped greatly in developing and clarifying many of the topics dealt with here, though, of course, errors of fact or interpretation must in no way be ascribed to them. In addition special thanks are due to Mr Stern for his assistance in preparing the manuscript for publication. Throughout Mr Ira Shere has served as assistant on the project and it is in no small way due to his industry and ability that the study has been completed within the scheduled time. Mr Shere also prepared the Figures used, with the exception of Figs. 8 and 9, which have been reproduced with the permission of the editors of *Population Studies*. Acknowledgments are also made to the staff of The Conservation Foundation for assistance so willingly given, and in particular to the Librarian, Miss Joan Carvajal, who so readily secured much of the historical and statistical records required for the study.

Special debts of gratitude are here expressed to two authorities on West Indian populations, whose help in the past has proved most valuable in work on the present study. The first is to Mr L. G. Hopkins, who directed the 1946 census of the West Indies, and with whom the writer was associated on this work. Many of the problems of fertility and mating in the West Indies treated here were first raised and discussed in Part A of the 1946 census report. The second is to Dr Brigette Long, who willingly gave advice on locating and using historical material on West Indian populations.

Institutions in Jamaica have also helped in various ways. In particular, thanks are due to the Registrar General and his staff for advance information and for special tabulations supplied from time to time, and to the statisticians of the Central Bureau of Statistics, who supplied data and gave advice on many problems.

GEORGE W. ROBERTS

NEW YORK, *January 1956.*

INTRODUCTION
BY KINGSLEY DAVIS

Most tropical areas today suffer from a malady so prevalent and so well known that it strikes no one as peculiar. Yet it is peculiar because it is uniquely modern, and because it exists despite the fact that it would seem easily remediable by techniques far simpler than those used to solve many other modern problems, including physical disease.

The nature of the malady is this: the combination of rapid population growth with widespread poverty. In the past the world has seen plenty of poverty, but the poverty was not associated with a fast increase in numbers. Instead, such demographic expansion as did occur tended to take place among the more successful peoples. Now, however, it is primarily the under-developed parts of the earth, the parts where additional population can be least provided for, that are exhibiting the fastest rates of increase—rates, indeed, which have never before been equalled by any country in human history. This relatively new disorder, this strange social imbalance, complicates the other ills of these regions, makes their future economic progress uncertain, and exacts a greater toll in human frustration than any mere organic illness. It cries aloud for both perceptive understanding and scientific research, but unfortunately it gets little of either.

The island of Jamaica exhibits the malady in its classic, though not in its most extreme, form. In that beautiful place one and a half million people occupy a territory of 4400 square miles. Their average density, 340 persons per square mile, is roughly six times that found in the United States and three times that found in Ireland. Yet, despite this already high density, the population has been growing at a rate which, if continued, would double the number of people on the island every 41 years. In fact, it seems likely that during the next 20 to 40 years the rate of human multiplication may even exceed that of recent decades. In any case, the prospect of continued population increase poses a serious problem. The Jamaican economy has long provided only a low level of living and is now characterized by chronic unemployment, with between

15% and 20% of the available labour force out of work. The International Bank for Reconstruction and Development says in its report on the island that 'a further reduction in unemployment can in the end be achieved only through emigration, the possibility of which is limited, and a limitation of population growth'. Certainly the rapid multiplication of human numbers puts a major obstacle in the way of alleviating Jamaica's poverty.

In view of the prevalence of the population malady in much of the world and its presence in Jamaica, the present volume by Mr George W. Roberts is most timely. His is one of the few thorough case studies of population dynamics in an under-developed area, and aside from the material in Volume III of Kuczynski's *Demographic Survey of the British Colonial Empire*, it is the first such major study that deals with Jamaica. Among other things, it delineates the historic growth of population on the island, describes the changing internal distribution and the fluctuating external migration, presents the main features of the population structure, and analyses the patterns of fertility and mortality which have contributed to past growth and will affect potential future growth. The demographic side of the island's economy and social organization is thus brought forth with a precision and thoroughness that only an expert can provide.

Before discussing this excellent volume, I should perhaps mention that it is the first of three projected studies dealing with Jamaica's population problems. These interrelated studies, which The Conservation Foundation undertook to sponsor in 1952, are, first, an analysis of the island's demographic history; second, a field study of attitudes and family relationships in Jamaica as they relate to reproduction; and third, a discussion of the resources and land-use patterns of the island as they relate to past and future population trends. In seeking a person qualified to accomplish the first task, the Foundation naturally turned to Mr Roberts, who was already a recognized authority on the demography of the British Caribbean region. He kindly consented to take a leave from his regular post to make this intensive study, for which he already had unexcelled preparation.

The three studies just mentioned were designed to supplement one another. The present volume approaches the subject from the standpoint of systematic demography, utilizing the official census returns and vital statistics to throw as much light as possible on population trends and structure. The field study, based on intensive interviews, will provide information of a different kind—on family relations, motives, and attitudes with respect to sex and children. Finally, the analysis of resources and land-use will fill in another part of the picture. The three studies combined should throw more light on the population problems of Jamaica than is available for nine-tenths of the other under-developed areas. Yet these studies represent only a beginning. It is hoped that they will provide a basis and a stimulus for still more research, because a many-sided investigation of one area, rather than a more superficial look at many areas, has much to recommend it from both a scientific and a practical standpoint.

Altogether it can be said that the amount of scientific research devoted to population problems in heavily peopled but impoverished regions is negligible in comparison to the gravity and magnitude of these problems; Jamaica is only one case among a great number which, all told, involve hundreds of millions of people. The value of Jamaica as an object of intensive study arises partly from the fact that its demographic pattern is in general typical of that of many under-developed countries. Its death-rate has declined sharply, particularly in recent years. Its birth-rate, while lower than that of some regions, is still high enough to provide, with the lessened mortality, a rapid natural increase which is added to an already crowded population. But Jamaica also has certain peculiarities (as a single case always has) which force the student of its population problems to blaze new trails in scientific analysis. Why, for example, has its birth-rate long been lower than that of Puerto Rico, which has had a greater degree of urbanization and industrialization and more encouragement of birth control? Again, since Jamaica's family structure differs in important respects from that found in Latin and Asian countries, what effects has this had on the island's birth-rate? How, in turn, has the institution of slavery, which endured through more than half the entire period of British rule, influenced the population trends? From the standpoint of the science of population, these particular features of Jamaica are as important as the parallels with other under-developed areas.

Another reason for giving attention to Jamaica is that none of the British territories in the Caribbean has previously been intensively studied from a demographic point of view. In contrast, Puerto Rico, with its Latin-American cultural background, has been thoroughly studied; and Japan, Malaya, India, Taiwan and Egypt have received at least some serious demographic attention. Jamaica, though the most populous of the British colonies in America, is yet small enough to be amenable to investigation as a unit. It has, furthermore, far better sources of demographic information than most under-developed areas. These sources are by no means perfect, but Mr Roberts, who has used internal checks wherever he could, has reason to vouch for the proximate accuracy of the censuses and vital statistics (though not for the external migration figures).

In the task of analysing Jamaica's demographic history, Mr Roberts has had several unique advantages. Reared in the British West Indies, he knows the people and their culture as only a native can. At the same time, his training in demography and his experience with West Indian censuses and vital statistics have given him a rare technical competence. For several years his regular position has been that of Vital Statistics Officer of the Colonial Development and Welfare Organization in the Caribbean headquarters at Barbados. In this position he has advised the colonial governments of this area on matters pertaining to registration and census statistics and population estimates. At the same time he has been active in population research, publishing a number of distinguished articles dealing with his region in the British Journal, *Population Studies*.

Mr Roberts has used to good advantage his knowledge of the comparative statistics and history of Britain's possessions in America. Frequently, he draws instructive comparisons between Jamaica and other territories. He is able to show, for example, that until 1919–23 the birth-rate in Jamaica was higher than that of British Guiana and Trinidad. After this period the Jamaican rate declined somewhat while that in British Guiana and Trinidad rose, so that the past relationship was reversed. The explanation he finds in the long-range effects of a different amount and kind of immigration into these two territories. From 1834 to 1914 British Guiana and Trinidad received a much heavier immigration than Jamaica, composed of a high proportion of East Indians. Once the East Indians were established in their new homes in Trinidad and British Guiana and their sex ratio had become normal, they exhibited a higher rate of procreation than the natives. Furthermore, they came to compose an ever larger proportion of the total population of the two territories, thus pushing up the overall birth-rate.

Occasionally the author is able to draw on the statistics of other territories to help fill in gaps due to missing information for Jamaica. In his very skilful study of mating and marriage as they bear on fertility in Jamaica, for instance, he has at least two technical problems. On the one hand, as he makes clear, the traditional apparatus of demographic analysis with respect to fertility has been worked out in societies where 'mating and marriage are largely synonymous terms'. In the West Indies, however, most children are born outside of legal wedlock, and a sizeable proportion of the mothers in the census report their marital status as 'single'. In this situation it becomes obviously difficult, on the basis of census and vital statistics, to include mating habits in the analysis of repro-duction in the way that would be done, say, in Britain or the United States. Such terms as 'marital fertility', 'births by duration of marriage', 'male net reproduction rates' and 'age at marriage', become almost meaningless from the standpoint of a statistical understanding of reproductive behaviour. Given the difficulties of analysing reproduction in relation to the various types of sexual unions he has to deal with, the author's second problem is that certain crucial data are not available, or not available in the best form, for Jamaica. In one such case, by calling upon superior data from Barbados where the mating patterns are generally similar to Jamaica, he is able to deduce something concerning the ages at which women enter the different types of relationships. He finds that women who enter the 'keeper relationships' tend to do so approximately at ages 19 or 20 (which is perhaps an overestimate). Those women who eventually settle down in a 'common law marriage' usually do so at a late age, around 30 years of age. Though many common law marriages eventually are legalized by formal wedlock, this does not occur until about 34 in the modal case, at which time the woman has already borne most of her children.

These illustrations suggest that the character of Jamaican society and the peculiarities of its statistics do not permit an easy demo-graphic analysis along traditional lines. Mr Roberts has brought a great deal of skill and ingenuity to the task. He has also brought sociological and historical knowledge to bear on his subject. The reader will soon find that the volume deals, for example, with the influence of slavery on nearly all aspects of the demography of the island. The author is not able to say as much or to be as certain about this influence as we would like, because the sources are not adequate. The sources, however, are not entirely fruitless, so that the connexions between slavery on the one hand and mortality, mating patterns, birth-rates and population structure on the other, can be treated with some hope of approximating the truth.

The result of the skills, experience, and interest which Mr Roberts has brought to his task is an unusually competent and original case study of a highly interesting population. He is to be congratulated by professional demographers, by specialists in under-developed areas and above all by the Jamaicans themselves for his skill in catching their lives in his demographic net.

KINGSLEY DAVIS

UNIVERSITY OF CALIFORNIA
BERKELEY, *January* 1956.

CHAPTER I

DEMOGRAPHIC MATERIAL AVAILABLE

THE DEVELOPMENT OF THE DEMOGRAPHIC RECORDS

'The Blue Books of Jamaica are the worst returns in the Colonial Office; there is a slovenliness . . . manifest in every document. . . . It is to be hoped that the authorities of Jamaica will in future pay more attention to the important subject of statistics.' In these words the eminent statistician R. M. Martin summed up the data on Jamaica at his disposal just before the first census of the island was taken.[1] These strictures are no longer true. During the century since Martin wrote, a body of fairly reliable demographic data has been built up, and its steady development and its reliability are surveyed in this chapter.

The inauguration of census-taking in Jamaica and the establishment of an efficient system of civil registration throughout the island were not the result of administrative decisions taken in Jamaica as distinct from the other territories of the West Indies. In fact, both census-taking and civil registration had their origin largely in the policy of the British Government in respect of the West Indies as a whole. There was no comparable policy aimed at developing a uniform system of recording migration in general, but the central direction exercised over indenture migration still assured, through the numerous controlling laws it introduced, the development of a uniform and reliable system of recording indenture migration. Consequently the origin and expansion of the demographic records of the island are here treated in the broader context of the initiation of such records throughout the British Caribbean.

Censuses. The records of the West Indies are rich in population estimates and these have been widely quoted by historians of the region. Such is the wealth of these that E. B. Burley has been able to collect upwards of fifty estimates of the population of Barbados between the date of the first settlement and 1844 when the first

[1] R. M. Martin, *History of the Colonies of the British Empire in the West Indies . . .*, London, 1843, p. 17.

censuses of the British West Indies were organized.[1] Numerous
estimates of the population of Jamaica in the eighteenth century
are also available. F. W. Pitman has collected eighteen estimates for
the period 1658–1787.[2] But most of these early estimates are of no
more than historical interest; they cannot be safely used to chart
population movements in the colonies. Some cover only the free
(or white) population, others cover only the slave population. It is,
however, the methods employed in securing these estimates that
lead one to suspect that most of them were no more than informed
guesses. Indeed, if the methods adopted by Barbados in 1812 are at
all representative of the procedures generally in use, then most of
these estimates can be dismissed as almost worthless. Faced with
the problem of presenting population estimates to the Home
Government for the years 1809–11, the Governor of Barbados
entrusted the work to the rectors of the several parishes. Most of
these embarked on a species of 'calculation' based on baptisms and
burials.[3] One rector complained that the subjects about which
information was sought 'are so far beyond my reach that even the
shadow of accuracy must be precluded from my report'. Though
one priest did in fact attempt a rough enumeration of his parish,
the data for the island as a whole remain useless. The Governor of
Jamaica, who was also called upon to present population estimates
for the same period, confessed that he knew 'of no mode by which
it [the number of white and coloured people] can be ascertained
with any tolerable accuracy'.[4]

Most of the early population estimates of Jamaica are probably
seriously defective. Discussing 'the first recorded census of Jamaica,
taken seven years after its capture', Bridges, after listing the
numbers of men, women and children, adds, 'There were likewise
a polinco of negroes, consisting of about 150 under one Boulo, as
lancers and archers, and many private men-of-war men, besides
many more comers and goers, Frenchmen and others.'[5] This
leaves little doubt as to the incompleteness of the count. The

[1] E. B. Burley, *Memorandum summarizing the Returns of a Census taken in the Island of Barbados in the year 1715*, Colonial Office Library, no date.
[2] F. W. Pitman, *The Development of the British West Indies, 1700–1763*, New Haven, 1917, p. 373.
[3] Letter from Governor Sir George Beckwith to Earl of Liverpool, 13 January 1812, *Parliamentary Papers (P.P.)*, 1813.
[4] Ibid. letter from Lieutenant Governor Morrison to Earl Bathurst, 28 January 1813.
[5] G. W. Bridges, *The Annals of Jamaica*, London, 1828, vol. I, note LXII.

reliability of the enumeration taken in 1673 is also questionable. Bridges described this as relating to 'the number of Christian men, women, children and negro slaves in the several parishes', which suggests that not all Christians are included. This description may equally well be taken to mean that the record refers only to Christians. Moreover, many of the estimates given by Long are derived by methods much less reliable than those relied on by the rectors of Barbados in 1812.[1] Even though many of the early estimates of slave populations were based on poll-tax records, they remain probably defective. There is, however, evidence that some of the early nineteenth-century estimates of population in the West Indies were the product of reasonably careful enumerations, but it is not always easy to determine which of these early records were so derived.[2]

Slave registration constituted the first systematic attempt at population enumeration in the West Indies. Though neither a system of civil registration nor a properly ordered series of census enumerations, it is of such historical importance and has yielded such interesting material that it cannot be ignored. Established partly to prevent the clandestine movements of slaves between the colonies and partly in the interest of securing better treatment for them, slave registration was of importance in virtue of the demographic material, admittedly very limited in extent, which it provided for the period 1816–32. Avowedly framed to ascertain 'all deductions from and additions to the former stock of slaves', the system appeared, in design at least, adequate to afford an accurate assessment of the movements of slave populations.[3] The methods of obtaining numbers of slaves may in a sense be

[1] Edward Long, *The History of Jamaica*, London, 1774.

[2] This is borne out by a manuscript in the Colonial Office Library summarizing the enumerations made for a census of Port-of-Spain and its suburbs in 1834. This evidently was taken with some care, and if representative of the earlier annual estimates published for Trinidad suggests that these are at least worthy of some study.

[3] Among records on slave populations of the West Indies the following are of particular interest: James Robertson, *General Summary of the Slave Population of the District of Demerara and Essequibo Colony of British Guiana, agreeably to the Registers of the Returns for the years 1817, 1820 . . .*, no. 700, 1833; *Return of the Number of Slaves in each of the West Indian Colonies*, 1833. In addition to such publications appearing in the *P.P.*, there are the publications of R. M. Martin, *Statistics of the Colonies of the British Empire*, London, 1839; and of G. R. Porter, *Tables of the Revenue, Population, Commerce, etc., of the United Kingdom and its Dependencies*, Supplement to Part III, London, 1835.

considered rough enumerations at triennial intervals. (Two colonies, Grenada and Tobago, gave annual movements of slave populations.) The first registration for Jamaica was in 1817, and the last year for which records are available is 1829. In effect, owners of slaves were supposed to return to the registrar of slaves 'a just and true return' under oath of the slaves in their possession at the official registration date.

As has been argued elsewhere, slave registration did not provide for the continuous registration of vital events in the modern sense.[1] Such records of deaths and births as appear in the slave registers were not actually true accounts of vital processes but emerged from rough attempts to break down the differences between the numbers of slaves at the beginning of an inter-registration period and the numbers at the end of such a period into components of growth. On these terms deaths in the registers do not represent all deaths in the slave populations between two registration dates. To see why this is so it is necessary to analyse more closely the underlying aim of slave registration: to ascertain 'all deductions from or additions to the former stock of slaves'.

Now the deaths occurring in any inter-registration period, say between 1817 and 1820, may be considered in two parts. There is first the group of deaths occurring among persons alive at the time of the first registration (1817). Secondly, there are the deaths among those children born between the two registration dates. The weakness of the mortality records of slave registration is that the only account of deaths taken is of deaths of the first group. The mortality experience of those children who died in the same inter-registration period in which they were born was completely lost to the record; indeed, under the system, the existence of such children could not even be acknowledged. The result is a gross understatement of mortality; no information on infant mortality, or indeed of mortality under 3 years in general, is available.

In the same way the fertility records of slave registration are defective. Under the system births merely indicated 'additions to the former stock of slaves'. Thus the births within a given inter-registration period, say between 1817 and 1820, represent only the children under 3 years of age at the date of the 1820 registration. There was in fact no registration of births during the inter-registra-

[1] G. W. Roberts, 'A life table for a West Indian slave population', *Population Studies*, vol. v, no. 3.

tion period. The number of 'births' were the population under 3 years of age at the end of a given registration period. The weaknesses of the slave registration records were fully realized by James Robertson.

Despite these weaknesses registration provided more reliable estimates of slave populations of the colonies than those previously in use. This improvement in the case of Jamaica is well summarized in the following words of Gardner: ' . . . there was a far greater number of slaves in the island than had been supposed. Hitherto returns had only been given in of those for whom the poll tax was paid; but of slaves in possession of small proprietors who paid no tax, returns were never made. This simple fact disposes of the assertion once made that the discrepancy of numbers between the old poll tax and the registration lists indicated clandestine importations. Had such importations taken place the discrepancy would of course have been more marked in the great sugar parishes; the very reverse was the case.'[1]

The passing of slavery brought to an end the only attempts so far witnessed to provide estimates of the populations at regular intervals. And it was not until 1844 that another general enumeration throughout the Caribbean was undertaken. This was the time when labour shortage was being acutely felt by the planters and when keen interest was shown in the possibility of inaugurating large-scale immigration of indentured workers, which, in the opinion of many contemporaries, could alone correct the situation brought about by the decline in the labour force available for work on the plantations. It was under these circumstances that the Secretary of State sent a circular letter, dated 7 February 1844, ordering that a census be taken in all colonies on 3 June 1844.[2] Evidently the advice of the Registrar General of England was sought on this matter. Graham states: 'On the 5th August, 1843, I transmitted for the use of the Secretary of State for the Colonies some suggestions respecting the mode of taking a census in each of our colonial possessions as requested by Lord Stanley.'[3] No copy of this

---

[1] W. J. Gardner, *A History of Jamaica*, New York, 1909, p. 253.

[2] See 'Copies of the last census of the population taken in each of the British West India Islands and in British Guiana, specifying their respective dates', *P.P.*, 1846, no. 426.

[3] Letter from Major Graham to Secretary of State Gray, dated 7 December 1848, *Minutes of the Barbados House of Assembly* (hereafter referred to as *M.B.H.A.*), 1849.

communication has been located. Indeed, it is not clear what specific instructions were sent to the colonies on the method of census-taking to be adopted. All except four of the West Indian colonies complied with the instructions in the Secretary of State's circular letter. The colonies that did not take censuses on the appointed date were British Guiana, British Honduras, Virgin Islands and St Lucia. The Virgin Islands had taken a census in 1841, and instead of taking another within the space of 3 years submitted the returns of the 1841 count. Because of constitutional difficulties and the scattered nature of the population of British Honduras, the Governor of that colony contented himself with presenting a rough estimate of the population. St Lucia had taken a census in August 1843, and therefore did not take another in 1844. The Governor issued instructions 'to have births and deaths ascertained to the 3rd June next...'. British Guiana had also taken a census in 1841, and instead of taking another presented the 1841 returns and some rough indications of the probable movements between 1841 and 1844.

Census-taking, essentially a new feature in the Caribbean, was in many colonies greeted with strong suspicion by the inhabitants. In Nevis the Act for taking the census 'caused considerable sensation amongst the Negro population, and many absurd conjectures were hazarded and discussed by them as to the intention and objects of the Bill', but the Governor appeared confident that his efforts to allay suspicion 'through the medium of the clergy' were successful.[1] In St Vincent the census was 'misrepresented' and 'has been considered by a portion of the laboring population as a preliminary step to imposing heavy taxation on labor'.[2] More disturbing conditions developed in Dominica. Here many people were under the impression that the census proclaimed the intention of reintroducing slavery and the island fell into a state bordering on riot.[3]

Under such circumstances inadequacies in these early censuses were to be expected; there is in fact a strong possibility that under-numeration was a characteristic of most of them. Many Governors expressed doubt as to their reliability. The Governor of Jamaica reported as follows: 'As this is the first enumeration of the inhabitants of the colony which has been made during freedom, it

[1] Despatch from President Graeme to Governor Fitz Roy, 29 July 1844, *P.P.* 1845.
[2] *Trinidad Standard and West India Journal*, 13 June 1844.
[3] Ibid.

may be presumed that the returns are not altogether to be depended upon. Nevertheless, it may be reasonably supposed that they approximate the truth and I regret to observe that they do not furnish any very satisfactory proof of progressive increase in the population of the island.'[1] But evidently grossly exaggerated estimates of the population of Jamaica were current just before the first census was taken, and probably it was the vast difference between such estimates and the enumerated population that led to criticisms of the accuracy of the 1844 enumeration. Even so careful a statistician as Martin, though admitting in 1843 that it was 'impossible to state with accuracy the actual population of Jamaica', based a calculation of density on an assumed population of half a million and described the resulting figure of 78 as 'a remarkably small proportion, particularly in comparison with Barbados, where there are 600 to the square mile'.[2] The less cautious Phillippo, writing in the same year, asserted that it was 'generally supposed that the aggregate population, including 30,000 whites, is now half a million, which is about 70 persons to the square mile'.[3]

Likewise the Governor of Barbados declared that 'in consequence of the mode of taking the census not being sufficiently searching and rigid the returns . . . fall short of the real number by some thousands'.[4] There is reason to believe that under-registration was a feature of many of the first censuses, but the comments of the Governors do not help in assessing their reliability, as in no instance are their criticisms soundly developed.

In favour of the censuses of 1841–4, it should be stated that unlike most of the earlier efforts they were apparently carried out in a manner similar to that used in the English census of 1841. They were evidently based on detailed enumerations, and the results are in most cases presented by enumeration districts. Apart from considerations of accuracy, their chief weakness is the lack of uniformity of presentation adopted in the several colonies.

When in 1848 the Registrar General of England was about to publish 'the population of England and Wales as lately arranged with respect to the Districts into which the country is now divided for the purpose of registering Births and Deaths . . .', he addressed

---

[1] Despatch from Earl of Elgin to Lord Stanley, 7 November 1844, *P.P.* 1845.
[2] R. M. Martin, op. cit. p. 8.
[3] J. M. Phillippo, *Jamaica, its Past and Present State*, London, 1843, p. 84.
[4] Despatch from Governor Grey to Lord Stanley, 4 October 1844, *P.P.* 1845.

a letter to the Secretary of State for the Colonies suggesting the desirability of including returns for the colonies as well.[1] He also urged that the colonies should take a census on the same date as that proposed for England in 1851. Accompanying this letter was a memorandum on census procedure in the colonies which deserves some attention as it probably influenced the whole course of census-taking in the region. Graham admitted that the method of census-taking could not be the same throughout all colonies 'as the conditions of the respective populations and the means of ascertaining the facts differ in different parts', but urged that as much uniformity as possible should be preserved. Among his main recommendations were:

(1) The enumeration should be carried out in 'a convenient number of enumeration districts, comprehended in the established divisions of the colonies', and the areas of these districts should be given. The population of towns should be enumerated within boundaries strictly defined.

(2) Where the 'habitations' were not fixed the population might be 'enumerated in Tribes and Families'.

(3) Enumerations should be made at equal intervals of time and the process should take no more than one day. If possible it should comprise 'the persons in each district on the previous night, at a season of the year when the facts can be recorded with the most facility and when there is no great displacement of the population by festivals or by other causes'. Persons away from their home on the census night were to be described as 'visitors' or 'travellers', while those out of the colony 'should also be enumerated at home, with the word "absent" after their names', but such persons were to be omitted from the abstract 'otherwise the same persons would be counted twice'. Graham emphasized that the treatment of these groups (which he termed 'floating population') required great care.

(4) In cases where it was difficult to obtain information on aborigines the inquiry should be confined to males aged 20 and upwards, 'the fighting men'. Here, he urged, partial and imperfect information about women and children would lead to 'confusion and error'.

[1] Letter from Registrar General George Graham to Secretary of State Grey, dated 7 December 1848; enclosure in despatch from Grey to Governor of Barbados, 20 January 1849, *M.B.H.A.* 1849.

(5) It was stressed that every individual should be enumerated by name, a procedure which, it was claimed, might prove useful for many purposes connected with police and defence. Here it seems the confidential nature of the returns was not admitted.

(6) The importance of obtaining age data was emphasized; if accurate returns of age were not available approximations should be used.

Graham also laid down rules for the preparation of census abstracts. Enumerators should not be entrusted with the preparation of abstracts. The schedules should, after collection by the enumerators, be sent to the seat of Government where abstracts were to be made 'on a uniform plan under proper supervision'. Among the tabulations suggested were those of age, sex and race, and the number of persons who entered the colony during the year immediately preceding the census.

Presumably copies of Graham's memorandum were sent to all the colonies. The fact that the series of censuses taken in 1851 had a larger measure of comparability than those taken in 1841–4 suggests that these censuses were taken in conformity with the rules outlined by him. Jamaica was one of the three colonies which did not take a census in 1851. The failure to do so was doubtless due to the severe cholera epidemic which made its appearance in October 1850, and continued throughout 1851, completely unsettling conditions in the island. It was not until 1861 that the second census was taken.

Further action to attain greater comparability of colonial censuses was taken in respect of those of 1891. 'The Secretary of State for the Colonies, by Despatch of 28 February 1889, called attention to the discussion at the Colonial Conference of 1887 on the subject of the Census, and with a view to apply uniformity of treatment in certain leading features, and so far as possible secure a homogeneous Census of the whole Empire, suggested, after consultation with the Registrar General of England, some heads of enquiry, and the extent of detail, for general application.'[1] The Registrar General of Jamaica, who from 1881 was entrusted with the supervision of censuses in the island, took these recommendations into account in preparing the 1891 Census Report.

The main features of the eight censuses of Jamaica will now be

---

[1] *Census of Jamaica and its Dependencies, 1891.*

noted.[1] The first census was taken on 3 June 1844, 'in compliance with the provisions of an Act, 7 Vict., c. 30, entitled "An Act for taking the Census of the Inhabitants of this Island"'. The published results consist only of three tables accompanied by a short letter of transmittal from Governor Elgin to Lord Stanley, dated 7 November 1844. The first two tables give cross-tabulations by sex and colour, sex and country of birth, sex and 'trade or avocation', and sex and age. The following age groups are used: under 5, 5–9, 10–19, 20–39, 40–59 and over 60. The tabulations are presented in terms of five parishes. A useful class of data presented for the whole island consists of the numbers of houses on the several types of plantations. The final table compares the population of 1844 with the last slave returns of the island (1834). The document is signed by the Island Secretary.

The *Summary of Census Returns*, the results of the second census taken on 6 May 1861, is a longer document, consisting of 8 pages. It distinguishes twenty parishes and covers in more detail the same categories dealt with in 1844. The breakdown by country of birth is in great detail, giving eighty-five countries. The attempt to make an exhaustive tabulation of 'rank, profession or occupation' yields a virtually useless list of hundreds of occupations. A finer age breakdown is introduced though no cross-tabulation by sex appears. Three new types of material are included: (1) data on marital status in terms of married, widowers, widows and unmarried; (2) those able to read, those able to read and write and the numbers attending school; (3) the numbers of deaf and dumb, blind, cripples, insane, affected with yaws and affected with leprosy. The document is signed by the secretary of the executive committee.

The *Summary of Census Returns, 1871*, giving the results of the enumeration of 4 June 1871, is a more ambitious document, extending to 51 pages, but carries no text. Its chief feature is the adoption of the fourteen parishes in terms of which all subsequent census reports have been framed, thus assuring full areal comparability from 1871 to 1943. In the main, the document, which carries no signature, represents a more elaborate treatment of the categories used in 1861. One improvement is that instead of the exhaustive list of occupations given in 1861 there appears a more useful tabulation in which certain broad categories, such as agriculturists, are

---

[1] Further details on the various censuses of Jamaica are given in *Eighth Census of Jamaica and its Dependencies*, 1943.

recognized; still the list of occupations remains too unordered to be of much use. A reduction in the number of countries of birth to fifty-six is also made and an entirely new tabulation 'Religions of the People' is included.

The *Census of Jamaica and its Dependencies, taken on 4 April 1881*, is the work of S. P. Smeeton, the Registrar General, who organized the enumeration. It includes a 7-page text and a description of the main administrative procedures taken, but is much shorter than the previous report, extending to 31 pages. For the first time also the census of the dependencies, the Turks and Cayman Islands, is reported on in the same document. A new age-grouping appears: by single years of age under 5, by 5-year age-groups up to 65, and thereafter by decennial groups. The breakdown called 'Colour' becomes more explicitly a racial classification in terms of White, Coloured, Black, Coolie and Chinese. The occupations present a more orderly picture, being grouped in the six major classes adopted in the English census of 1861, while the tabulation of houses is done in six categories.

Also the work of S. P. Smeeton is the *Census of Jamaica and its Dependencies* embodying the results of the census of 6 April 1891. This follows the general pattern of the previous report, but includes some improvements based on the recommendations of the Secretary of State for the Colonies, the most important of which is a breakdown of occupations by age-groups. A special return on the East Indian population 'prepared at the request of the Protector of Immigrants at Calcutta' is introduced. The tabulation of religious persuasions is dispensed with: 'In Jamaica enquiry [on religion] might fittingly be abandoned . . . on account of the too ready admission of such particulars.' The entries of the population of various electoral districts introduced in this report have formed a regular feature of all later censuses. For the first time the cost of the census is given, the average cost per head of the population amounting to 2·37*d*.

Longer than any previous report is *Census of Jamaica and its Dependencies taken on 3 April 1911*, which extends to 93 pages. In general, it follows the pattern of the previous report, the only substantial addition being the tabulation of educational status by age. Information on religion is reintroduced, but though 'the heading of the column in the Schedule was worded so as to show that it was not a return of church members that was required', the Registrar

General again expressed dissatisfaction with the returns under this head. It is interesting to note that the cost of the census is reduced to 1·97*d.* per head of the population. This census was organized by Registrar General David Balfour.

Census of Jamaica and its Dependencies, taken on 25 April 1921, also the work of David Balfour, is in general similar to the previous report. Its chief weakness is that it dispenses with the tabulation by age of persons employed which was given in the reports of 1891 and 1911. The cost is considerably higher than in the past, amounting to 3·38*d.* per head of the population.

The organization and results of the most recent census of the island (1943) are fully described in *Eighth Census of Jamaica and its Dependencies, 1943.* The information secured represents a considerable advance over past censuses, and the tabulations, which were done mechanically, are much more extensive than any hitherto attempted in the British West Indies. The work of enumeration and analysis was entrusted to a special department, known subsequently as the Central Bureau of Statistics. As it was much more exhaustive than previous censuses and financed on an entirely different basis, it proved more costly. In fact, the total costs work out at 15·12*d.* per head of the population.

Vital statistics. In the sphere of vital statistics there was no concerted effort to establish a system throughout the West Indies at the same time as in the case of census-taking. Of historical note is the fact that in many colonies there were early laws providing for registration of baptisms, burials and marriages. In Jamaica, for instance, efforts were made as early as 1661 to secure records of marriages, deaths and burials, while a form of registration was introduced in 1683 by 'An Act for the Maintenance of Ministers and the Poor and Erection and Repairing of Churches'.[1] But the early registration records of the West Indies are wholly ecclesiastical and in no sense the precursors of modern civil registration. In effect they applied mainly to the European populations. For not only was the concept of registration of vital events entirely alien to the slave codes, in terms of which the lives of the majority of the population were ordered, but the exorbitant registration fees made general registration among the slaves impossible. In Jamaica the following recorder's fees for registration were fixed towards the end of the

[1] R. R. Kuczynski, *A Demographic Survey of the British Colonial Empire,* Oxford, 1953, vol. III, p. 240.

eighteenth century: burials from £1 6s. 8d. to £2 13s. 4d.; marriages from £1 6s. 8d. to £4; baptisms from 5s. to £1. And as these were specifically charges on the owners (they could not be collected from the slaves) the cost of registration alone precluded the establishment of the institution of marriage among slaves. Registration was primarily a means of remunerating the rectors, not a means of providing vital records of the population.[1]

In Jamaica an early attempt was made to record births and deaths among slaves, but it was in no sense a genuine measure of vital registration. It was basically part of the pro-natalist policy designed to stimulate population growth among the slaves, which was initiated in the slave code of 1792. Apparently no systematic collection of records of births and deaths on the plantations was made in accordance with this law.[2]

The first comprehensive attempt to assess the level of mortality in the West Indies, though not based on vital registration, deserves passing notice, as it constitutes one of the two classes of mortality data available for the West Indies during the early nineteenth century, and some use will be made of the data in the study of mortality in Jamaica. It was the concern of the British Government over the conditions of the early nineteenth century that led to the execution of the detailed analysis of sickness and mortality among troops in the West Indies for the period 1817–36; that is, for a period slightly longer than that covered by slave registration. This is A. M. Tulloch's and H. Marshall's *Statistical Report on Sickness, Mortality and Invaliding among Troops in the West Indies* (London, 1838). In the preface to this work it is stated: 'In October 1835, the the Secretary at War deemed it requisite that an enquiry should be instituted into the extent and causes of sickness and mortality among the troops in the West Indies, with a view of founding thereon such measures as might appear likely to diminish the great loss of life annually experienced in these colonies.' It is possible to disagree with some of the techniques of analysis used, but it cannot be

---

[1] Details of registration in Jamaica in the slave days are given in G. W. Bridges, *The Annals of Jamaica*, London, 1828, vol. I, Appendix. Similar high fees of registration were charged in other colonies. In Dominica, where marriage could cost as much as £8, 'An Act to settle a stipend on the Rev. George Clarke . . .' (11 Feb. 1830), after laying down registration fees, states 'Provided always that nothing herein contained shall prevent the rector or clerk from receiving further gratuities for Baptism, Marriages and Burials if the parties concerned think fit to offer such'.

[2] This law is discussed in Chapter 7.

denied that their report constitutes a landmark in the medical history of the region. Indeed, it is as far as is known the only attempt ever made in the nineteenth century to survey mortality throughout the West Indies.

The earliest effort to introduce a system of registration was in the case of marriage. The fact that marriage among slaves was either prohibited by law or not countenanced by the clergy in general constituted an important feature of the slave regime. A clause providing for the solemnization of marriage among the slaves was incorporated into the Jamaica slave code of 1826, but evidently this device, aimed at the moral improvement of the slaves, remained largely inoperative. And with the abolition of slavery this law ceased to be effective, so that the whole issue of solemnization of marriage among the ex-slaves and the status of the small number of marriages among them during slavery were shrouded in legal obscurity. And when the progress made under apprenticeship came under careful and often critical review, it became imperative that the problem of marriage among the ex-slaves be satisfactorily settled. Consequently Lord Glenelg's despatch of 15 September 1838 sought to establish the legal basis of marriages already in being and to provide for a system of formal registration of marriages among all the populations.[1] He declared that 'it is expedient and necessary to amend the said marriage laws, and to adapt the same to the altered state and conditions of society . . .'. And the first general marriage law was passed in Jamaica in 1840; it was aimed at establishing a system of marriage registration for the population in general.

No such dramatic pronouncement heralded the introduction of registration of births and deaths. In the memorandum of Registrar General Graham, already referred to, the introduction of a system of vital registration throughout the West Indies was urged in the following terms: 'I would suggest that the marriages, births and deaths should be registered and abstracted annually. In some colonies the age at death and the cause of death could probably be obtained; which would render the information complete.' Before the appearance of Graham's memorandum, in fact, at least two colonies had made tentative attempts to establish registration systems.

---

[1] Despatch from Lord Glenelg to Governors of British Guiana, Trinidad, St Lucia and Mauritius, 15 September 1838, given in *Extracts from Papers Printed by Order of the House of Commons, 1839, relative to the West Indies*, London, 1840. This despatch is discussed further in Chapter 7.

In Jamaica two laws were passed in 1844, the true precursors of modern registration in the island. The first, 'An Act for registering births and deaths in this island', provided for the voluntary registration of births and deaths by civil authorities. The second, 'An Act to register births and deaths in this island . . .' (8 Vict. c. 47), aimed at increasing the efficiency of registration by having the clergy forward quarterly returns to the parochial Registrars.[1] Under the latter law, 'General Abstracts of Births and Deaths Registered in the Island' were published in the *Votes* from 1844 to 1848, but these were acknowledged to be incomplete and unreliable. 'The Abstract cannot be looked upon as correct in numerical detail, inasmuch as it appears by the return upon which it is founded that very few of the Assistant Registrars have performed the duty assigned to them by the 8th Vict. c. 47, and a general belief also prevails that the Registering of a birth or a death has been omitted in many instances by individuals.'[2] Consequently the publication of these records was soon discontinued and an attempt was made to repeal the registration laws in 1846, but this was disallowed by the Queen in Council.[3] In the words of Milroy, the registration of births and deaths remained 'a nullity' and the laws were finally repealed in 1855.

Secondly, a registration law was passed in Trinidad in 1847. Though based on a more satisfactory basis than those of Jamaica, it also failed to assure adequate registration throughout the island. Not only did the law fail to provide for all registrations, but it was several years before Registrars were appointed throughout the island, and the first published returns (1867) were incomplete.

In the 1850's many of the smaller islands established systems of vital registration; in fact most of the colonies of the eastern Caribbean did so before Jamaica. In the meantime it appears that

[1] Kuczynski, op. cit., gives details of these and earlier registration laws.

[2] 'General Abstract of Births and Deaths registered in the Island Secretary's Office between 4 June 1844 and 4 June 1845', *Votes of the House of Assembly of Jamaica*, 1845 (hereafter referred to as *V.H.A.J.*). See also *Report of the Central Board of Health*, 1852.

[3] See *The Report on the Cholera in Jamaica*, by Dr Gavin Milroy, which also shows conclusively that the records of these registrations were useless. Dr Milroy's exhaustive survey of the cholera epidemic of 1850–52 and of kindred problems of health and mortality in Jamaica appears as an enclosure in a Circular Despatch from the Duke of Newcastle to Governors of the West India Colonies, 31 March 1853, *P.P.* 1854, vol. XLIII.

the Secretary of State was corresponding with Governors on the subject of introducing civil registration in all colonies. Mention is made in the *M.B.H.A.* of a circular letter of 27 April 1867, which apparently summarized the existing registration practices in the colonies then maintaining registration systems. It was evidently a perusal of this document that led the Registrar General of England to draw up a memorandum dated 31 March 1868, which did for registration in the colonies what his memorandum of 1847 did for census-taking in the region.[1]

In this memorandum Graham observed that the same considerations of public and private advantage which led to the establishment of registration in the United Kingdom and elsewhere had equal weight in the case of the colonies. But there were many inadequacies in the colonial systems. The following desiderata for adequate registration were laid down:

(1) Civil registration of marriages, births and deaths should apply to all classes and religious persuasions.

(2) All registration forms should avoid 'needless multiplication of statement and clearly establish the identity of persons recorded and assist medical and statistical inquiries in useful research'.

(3) Registration districts should be formed and registrars appointed to each district.

(4) All registers should be kept in duplicate or certified copies of them made, and one set should be retained by the registrar while the other should be sent to the central office where 'the superintending authority' should be located and where alphabetical indices and abstracts should be prepared.

(5) The registration of births, deaths and marriages should be enforced by suitable penalties. Registration, it was stressed, was for 'all ranks of society, without interference with the institutions and rites of religionists of any class'. In conclusion, it was observed that Ceylon had an excellent system in force which was 'worthy of imitation'.

All the registration laws adopted in the West Indies were in fact closely modelled on the English Acts and therefore incorporated all the principles outlined by Graham.

Effective civil registration in Jamaica, with full provisions for

[1] Enclosure in letter from Duke of Buckingham and Chandos to Governor of Barbados, 14 April 1868, *M.B.H.A.* 1868.

adequate enforcement, was introduced by a law of 8 October 1877, which has remained substantially unaltered until 1950. The legal basis of registration in the island has been discussed in detail else-where and need not be considered at length here.[1] Its main weak-ness has been the failure to provide for the registration of still-births. And in contrast to the situation that developed in other colonies where, despite the absence of explicit provisions for the registration of such events, registration of still-births grew up, apparently as an administrative necessity, it was not until 1950 that steps were taken to register these events.

Certain general features of registration will now be noted. The first Registrar General appointed to the island was S. P. Smeeton, who continued to serve until 1908 and under whom a sound system of vital registration was built up. At the commencement of registra-tion in 1878, 102 registration districts were set up and a Registrar appointed to each. The securing of suitable Registrars was, as Smeeton points out, not always easy. It was necessary 'to select a large number of persons to act as Registrars of Births and Deaths each of whom should be residing at a convenient centre and be qualified by good character, intelligence and education, for the duty of collecting and recording the necessary information. For many districts this was ... a difficult task, and in many cases the remuneration by fees is too small to induce anyone to seek the work except as supplementary to some other occupation. The duties of the Registrar consequently have in many instances to be entrusted to persons not really fitted for them and in spite of the most detailed instructions and examples given, faulty registers are sent up, further explanation is given, correction or re-registration is directed and then just as all this educational work is apparently successful, the Registrar resigns and the same tedious round of selection, instruction and correction is again entered upon in this office.'[2]

The Registrar General was also entrusted with the duty of registering marriages. Up to 1879 a very loose system of marriage registration was in force. The Ministers of Religion as marriage officers in general failed to comply with the regulations under which they were supposed to transmit records of marriages performed to

<hr/>

[1] Complete summaries of the registration laws and practices are given in *Abstract of Arrangements respecting Regulations of Births, Marriages and Deaths.* ... H.M.S.O. 1952. See also Kuczynski, op. cit.
[2] *Annual Report of the Registrar General's Department*, 1891.

the Island Secretary's Office. In effect there was no authority specially deputed to maintain records of marriages performed. This weakness was corrected by Law 15 of 1879 which made the Registrar General the central agent for the registration of marriages. Under this law also the appointment of marriage officers was vested in the Governor, and persons so appointed became registrars responsible for sending records of marriage to the General Register Office. The Registrar General maintained 'constant watchfulness' over the marriage register and had to devise a special system to ensure that copies of all marriage registers reached his office.

Reviewing the progress made in the first 10 years of registration in the island, Smeeton could with some assurance remark, 'Doubtless [there are] cases of neglect or avoidance of the Law ... as in England and elsewhere, but the number of registrations secured ... gives evidence of a satisfactory thoroughness in the application of the measures and of the law-abiding disposition of the people in a matter which must cause inconvenience and appear inconsequential to not a few.'[1] Completeness of registration improved with the passage of time. But one persistent weakness of the system has been the low proportion of deaths medically certified as to cause. In his first Report, Smeeton deplored 'this lack of correct information [which] so seriously threatens to render unavailable the most valuable field of future labour for the Department, viz. the tabulation of the causes of death'. Both the Registrar General's *Reports* and the *Reports of the Island Medical Department* discussed this weakness, frequently offering explanations and suggestions for improvements, but the proportion of deaths not medically certified has remained, by comparison with the colonies of the eastern Caribbean, very high (41 % in 1951). This, however, does not mean that the mortality data for Jamaica are less reliable than those for the eastern group. It means rather that a more rigorous system of medical certification is maintained in the island, a system which does not make medical certification a necessary prerequisite for the interment of a corpse.

Migration statistics. The first migration statistics of the West Indies were the records of the slave trade. As we shall see in Chapter 2, these were, so far as Jamaica was concerned, probably reliable, and in fact constituted the only data on population

---

[1] *Annual Report of the Registrar General's Department*, 1887–8.

movements in the island with claims to accuracy before the establishment of slave registration. But apart from the unique records of the slave trade there were no migration statistics available before the nineteenth century.

During the nineteenth century interest in migration in the West Indies stemmed not so much from the desire to control population growth in general as from the concern over the changing numbers of the population available for labour on the sugar plantations. This was especially true of those colonies which continued to depend on sugar as their staple product, but even Jamaica, which early diversified its agriculture at the expense of sugar, evinced a keen interest in the movements of the population of working age Consequently records of migration were in most cases records of workers (and usually of their families) introduced in order to supplement the colonies' population engaged in agriculture, and of those leaving the colonies to seek work elsewhere. The value of instituting a system of recording all migration affecting the colonies was only slowly recognized, and in most cases no attempt was made to collect such data until the present century. Except for indenture immigration, the British Government never took steps, as in the case of census-taking and vital registration, to have a satisfactory system of recording migration adopted in the colonies.

Migration statistics of the West Indies can be considered in terms of three categories of migration affecting the region in the post-emancipation period. The first type—the immigration of indentured workers and their families from India, China, Africa and elsewhere —gave rise to the only reliable data of this class in the nineteenth century. Closely controlled by law at all stages, the course of indentured migration has been carefully covered in published records. Though the *V.H.A.J.* carry accounts of immigrants introduced into Jamaica under indenture from 1840, it was apparently Lord Stanley's circular letter of 28 February 1843 which resulted in the systematic presentation of records of indenture migration in the West Indies. This circular letter requested from the colonies returns of immigrants introduced at public expense since 1 August 1834, and the money expended on their introduction.[1] It was the replies to this circular letter which comprise the first comprehensive

[1] 'Returns of the number of immigrants into the British West Indies from 1 August 1834 and of all votes of money for purposes of immigration and annual expenditure of the same . . .', *P.P.* 1844.

information on indenture migration. From 1844 the Colonial Land and Emigration Commission began publishing detailed returns of indenture migration into the West Indies, and for the period up to 1872 these records provide the main source of data on immigration into the region. After 1872 the Colonial Land and Emigration Commission ceased to supervise West Indian migration, but local reports—in the case of Jamaica mainly the *Reports of the Indian Agent General*—continued to record the numbers of indenture immigrants. In addition, very comprehensive accounts of the numbers who emigrated from India are available in the *Annual Reports on Emigration from the Port of Calcutta to British and Foreign Territories* from 1875 onwards.

The second type of migration affecting the British Caribbean during the past century was intercolonial migration. Though most of the historical background of these movements can be studied from available records, there has never been any general attempt to record or even to prepare rough estimates of such movements. Actually the only colony which tried to gather such information in the nineteenth century was Trinidad, and here the patent un-reliability of the published returns was admitted.[1] So far as Jamaica is concerned, the lack of intercolonial migration statistics is of no importance, because there never was any substantial movement between Jamaica and the eastern colonies.

The third type of migration affecting the British West Indies was that to foreign territories such as Panama, Cuba and the United States. Here again records are very scanty. Apparently the earliest attempts to trace such migration were initiated in Jamaica. It is significant to note that the recording of these data was entrusted not to the Protector of Immigrants but to the Collector General. According to the Registrar General, 'The information of arrivals and departures, which relates to Deck passengers only, is derived from Returns furnished to the Collector General by the various Steamship Companies, and as it is certain that many persons went to and fro as ordinary passengers and by Sailing vessels the net migration loss might possibly be somewhat heavier than as shown by the Official Returns.'[2] The early figures appearing in the *Annual Reports of the Collector General* are specifically stated to cover only 'labourers'.

[1] See *The Annual Reports of the Protector of Immigrants*.
[2] *Census of Jamaica*, 1891.

It was largely the efforts of Smeeton that resulted in the establishment of a general system of migration recording. In 1901 he commented, 'It is desirable ... that a close watch and record should be maintained as to the numbers of persons leaving and coming to the Island. The matter has been submitted to Government and it is hoped that ... complete returns of emigration and immigration will be secured relating to all the Island ports. ...' In the following year the numbers of labourers listed in the *Collector General's Report* were published in the *Registrar General's Report*, and it was announced that arrangements were in hand to have shipping companies furnish returns of all incoming and outgoing passengers. However, it was not until 1909 that the system of collecting migration returns was devised, and for the first time the Registrar General's Report carried returns purporting to cover the whole island. Though probably indicating correctly the direction of migration, these returns continued for a long time to be grossly defective.

From the foregoing account of available demographic data four broad periods can be distinguished in terms of the extent and quality of demographic material.

(1) *The early slave period.* Here data were in general defective or wholly non-existent. In view of the tenuous methods used to derive population estimates at this period, their reliability is very questionable. There were no collections of vital statistics during this period, and the records of the slave trade (essentially a category of immigration) were the only reliable measures of population movements.

(2) *The period covering the last years of slavery and apprenticeship.* During these years some suggestive records became available. The first, the slave registration, gives some idea of population movements during the period. The second was the detailed mortality analysis of Tulloch and Marshall. As yet, however, there were no organized recordings of migration.

(3) *The period of census records and indenture migration.* From the date of the first census it became possible to determine population movements on a limited scale. The records of indenture kept since the early 1840's enhance the value of the demographic material available, especially as this was virtually the only migration then in progress. Though the absence of any system of civil registration

severely limits the quality of any analysis that can be made for this period, it is still possible to glean from the data available rough estimates of vital rates for the island.

(4) *The period of civil registration.* This period opens with 1878, the date of the introduction of effective civil registration. The general collection of migration returns was started in 1910, but this remains the least satisfactory of the demographic material available.

RELIABILITY OF THE RECORDS

There is no reason to believe that gross deficiencies in enumeration mark any of the eight censuses of Jamaica. Its regularity of growth, in such marked contrast to the experience of the territories of the eastern Caribbean, attests to the basic reliability of these censuses. For unlike the eastern colonies, where population growth throughout the century of census-taking was dominated by changing patterns of migration, Jamaica has, at least with the exception of the period 1891–1921, been only to a limited degree influenced by sizeable migration. Its relatively even rates of population growth are in fact fully consistent with all available information on population movements since 1844.

A test which is at once an indication of the accuracy of the age data and of the general reliability of enumeration is provided by the ratios between the decennial age-groups at one census and the decennial age-groups 10 years younger at the census taken 10 years earlier. These in effect are survival ratios based on the census populations and should all be less than unity, if the populations are not affected by large-scale migration.[1] In the absence of such migration, ratios in excess of unity suggest either under-enumeration or misstatement of age. The age structure of Jamaica has been influenced by migration in varying degrees throughout the period covered by demographic records, but it was not until the last decade of the nineteenth century that migration attained very large dimensions. And it is probably safe to assume that the small net immigration during 1861–81 and the net emigration during 1881–91 were not large enough to invalidate tests of the accuracy of the census data based on survival ratios. Series of such ratios calculated from the censuses of 1861–91 appear in Table 1. These indicate

[1] Cf. J. D. Durand, 'Adequacy of existing census statistics for basis demographic research', *Population Studies*, vol. IV, no. 2.

that the census data, both in terms of age breakdowns and completeness of enumeration, are reasonably reliable. For not only are the ratios, with two exceptions, below unity, but they also decline with advancing age. The ratio of females aged 20–29 in 1881 to those aged 10–19 in 1871 amounts to 1·054. Taken in conjunction with the adjoining ratios, this suggests that it is misstatement of age at 1881 rather than under-enumeration at 1871 that gives rise to this high ratio. This is supported by the comparatively large proportion of the population returned as of unknown age in 1881, 1·44% in the case of females and 1·30% in the case of males. It is also of interest that more than half of the population of unknown age occur in two contiguous parishes.

Table 1. *Survival ratios from census populations*

Ratio	Both sexes			Male		Female	
	1861–71	1871–81	1881–91	1871–81	1881–91	1871–81	1881–91
P'' 10–19/P' 0–9	0·949	0·963	0·996	0·953	1·000	0·973	0·992
P'' 20–29/P' 10–19	0·956	1·009	0·903	0·964	0·816	1·054	0·988
P'' 30–39/P' 20–29	0·787	0·741	0·702	0·758	0·671	0·725	0·730
P'' 40–49/P' 30–39	0·825	0·822	0·822	0·829	0·834	0·816	0·811
P'' 50–59/P' 40–49	0·675	0·652	0·677	0·633	0·657	0·670	0·697

Note. In these ratios P'' indicates the decennial age-group 10 years older than the decennial age group P' and referring to a census date 10 years later than the latter.

A comparison of age data for the 1943 census of Jamaica with data for other West Indian populations indicates that the accuracy of age reporting at recent censuses was much less satisfactory in Jamaica than elsewhere in the West Indies. Thus the index of preference at age 60 for Jamaica mounted to 1·81, which was higher than in most of the other territories.[1] The question of the completeness of vital registration in Jamaica early engaged the attention of Registrar General Smeeton, who concluded that it was probably nearer realization in the case of deaths. 'It is easy for some births to escape registration. With many persons the only inducement to register is the knowledge of the law's requirements and its penalties ... not an appreciation of the

[1] *West Indian Census*, 1946, part A.

importance or the advantages of the public record . . . and hence
in cases where the birth occurs away from the possible cognizance
of a Registrar, it is not strange that the duty should be neglected.
In the case of deaths this escape from registration is not easy . . .
the surrounding circumstances of the case, the funeral, and in all
but a few instances the services of the minister, give a greater
publicity than in the case of births. Further than this, the minister
of religion performing the burial service is bound by law to report
the burial to the nearest Registrar, unless he receives at the same
time a certificate showing that registration has been attended to.
With these safeguards the losses to the death registers are probably
few. . . .'[1]

Tests of the reliability of birth registration in Jamaica have
already been published.[2] One series of tests compares births
registered in intercensal periods with corresponding estimates of
births derived by applying appropriate life-table values to census
population under 10. Births estimated in this way are very close to
the numbers of births actually registered and, indeed, suggest a high
degree of reliability of the basic records. A second set of tests,
intended to be more crucial, however, proves inconclusive. These
tests, based on the assumption that the census counts of children
under 10 are complete and that mortality records are reliable,
consist of comparisons of census populations under 10 with corresponding populations estimated from births and deaths. It is evident
that the inconclusive nature of these tests must be ascribed to the
very wide age range in terms of which they are framed.

In an analysis of the completeness of birth registration in
England, D. V. Glass has shown that tests of this nature limited to
populations aged 2–4 have certain advantages over tests based on
wider age intervals.[3] Series of tests for Jamaica limited to this age
interval are therefore given in Table 2. The differences between the
census populations aged 2–4 and the corresponding estimated
populations are much more regular than in the case of the more
extensive tests already presented. On the assumption that the census
counts of children are accurate it appears that for male births
under-registration around 1891 and 1921 was about 2·3% and in

---

[1] *Annual Report of the Registrar General's Department*, 1880–1.
[2] G. W. Roberts, 'A note on mortality in Jamaica', *Population Studies*, vol. IV,
no. I.
[3] D. V. Glass, 'A note on the under-registration of births in Britain . . .',
Population Studies, vol. V, no. I.

Table 2. *Tests of reliability of birth registration, 1891–1943*

Age	1891 populations			1911 populations			1921 populations			1943 populations		
	Census (a)	Estimated (b)	(b)/(a)	Census (a)	Estimated (b)	(b)/(a)	Census (a)	Estimated (b)	(b)/(a)	Census (a)	Estimated (b)	(b)/(a)
Male												
2	8,310	8,590	1·034	11,880	12,250	1·031	11,040	11,590	1·050	15,630	15,650	1·001
3	8,720	9,260	1·062	11,530	10,480	0·909	11,490	11,160	0·971	16,610	15,810	0·952
4	9,120	7,690	0·843	11,720	11,040	0·942	11,610	10,590	0·912	16,110	16,010	0·994
2–4	26,150	25,540	0·977	35,130	33,770	0·961	34,140	33,340	0·977	48,350	47,470	0·982
Female												
2	8,080	8,570	1·061	11,680	12,120	1·038	10,830	11,540	1·066	15,600	15,650	1·003
3	8,940	9,190	1·028	12,230	10,450	0·854	11,980	11,290	0·942	17,070	15,890	0·931
4	8,950	7,750	0·866	11,400	11,350	0·996	11,460	10,730	0·936	15,560	15,700	1·009
2–4	25,970	25,510	0·982	35,310	33,920	0·961	34,270	33,560	0·979	48,230	47,240	0·979

1911 about 3·9%. Slightly lower percentage deficiencies are shown for female births during the same period; the estimated under-registration being 1·8% in 1891, 3·9% in 1911 and 2·1% in 1921. At the most recent census (1943) the estimated under-registration is about 1·8% for the males and 2·1% for the females. The years since 1891 therefore show no evidence of severe under-registration of births.[1] There is in fact a strong suggestion that within a short time after its inception civil registration attained a reasonable degree of completeness.

An indication of the general consistency of the demographic material available since 1881 is afforded by Table 3. This is in no sense an evaluation of the reliability of the basic data. It merely seeks to establish the consistency of the data on which most of the analyses presented in this study are based: registration data (births and deaths), census enumerations and the series of life tables from 1891 to 1946. The discrepancies take the form of differences between two sets of population estimates at the end of each intercensal interval.

The first set of population estimates (column 2) are prepared on

Table 3. *General consistency of available demographic data, 1881–1943*

Intercensal interval	Estimated population over 10 at end of interval	Census population over 10 at end of interval	Estimated migration balance	Total	Discrepancy	
					Annual	
					No.	% average intercensal population
1881–91	494,400	477,700	− 24,800	8,100	810	0·13
1891–1911	621,800	598,400	− 43,900	20,500	1,000	0·14
1911–21	711,100	625,500	− 77,100	8,500	850	0·10
1921–43	852,000*	868,800*	+ 25,800	9,000	400	0·04

* Population over 12.

[1] As vital registration was introduced in 1878 it is not possible to present similar tests in respect of the census year 1881. However, it may be noted that on the basis of the births during the first year of vital registration the male population aged 2 at September 1881 is estimated at 7050, whereas the census population aged 2 according to the 1881 census is 7170. Similarly, the estimated female population aged 2 is 7000, which is close to the corresponding census figure (7170).

the assumption that the population over 10 at the end of a given 10-year intercensal interval (or over some higher age when the intercensal interval exceeds 10 years) is the total population at the beginning of the interval, less the deaths occurring in the period; they are in effect populations that would result if there were no migration and if the mortality conditions estimated for that period on the basis of life-table experience actually obtained. (Throughout it is assumed that migration affects only the population over 10 years.) Thus for 1881–91 and 1911–21 the populations in column 2 represent the total census populations of 1881 and 1911, each aged 10 years. In the case of 1891–1911 the initial (1891) population is aged 20 years to give estimates of the 1911 population over 20, while the age-group 10–19 at that year constitutes the survivors of children born between 1891 and 1900. Similarly, the estimated 1943 population in column 2 is derived by carrying the total 1921 population forward 22 years to give the population over 22 in 1943, while the survivors of the children born between 1921 and 1930 are taken as the population aged 12–21 in 1943. (The comparison for the interval 1921–43 is therefore in respect of the total populations over 12.) For all years prior to 1921 the survival ratios used to age the populations involved are derived from Jamaica life tables for 1889–92 and 1910–12. The use of the averages of these stationary populations for the computation of all survival ratios employed between 1881 and 1921 is justified in view of the stability of mortality rates over this whole period. In the case of the interval 1921–43 the survival ratios are averages obtained from the Jamaica life tables for 1920–2 and 1945–7.

The second set of population estimates comprises merely the census population within the relevant age-group to which has been added estimated migration balances. These balances are the differences between the intercensal population increases and the total natural increases over the corresponding period.

Resulting differences entered in column 5 can therefore be taken as measures of the total discrepancy over each intercensal period. As the intercensal intervals vary the differences can best be considered in terms of annual averages. These are largest (1000) in 1891–1911 and smallest in 1921–43 (400). Though furnishing no conclusive proof of the reliability of the census and registration data, these differences suggest a sufficient measure of consistency among the types of data employed in this study.

As use will be made of slave registration data at several stages of this study it is of interest to examine the reliability of these also. Complete success in the scheme would have assured in effect a full and accurate breakdown of the movements of slave populations into three components: mortality, fertility and manumission. (Migration of slaves from one colony to another need not be considered here, as up to 1825 the numbers involved were too small to affect appreciably the numbers of slaves in any colony, while after 1825 such migrations were prohibited.) With complete records of births, deaths and manumissions it would be possible to arrive at populations at the last registration date by adding to the initial slave population the total births during the period of registration and subtracting the numbers of deaths and of manumissions during the same period (here the terms births and deaths are used in the special sense already indicated). From this it follows that the discrepancy between the population at the last date of slave registration as estimated on this basis and the numbers given in the registers affords a measure of the reliability of slave registration as a record of population movements. Judged by these discrepancies slave registration cannot be said to be grossly defective. In fact only three colonies show discrepancies large enough to discredit the system; in the case of the Virgin Islands, Trinidad and Dominica the estimated population at the last registration date differs from the registered population by 35, 11 and 10% respectively. Other colonies show much smaller discrepancies. In the case of three of them the discrepancies are 1% or less, while four show discrepancies between 4 and 6%. In the case of Jamaica the discrepancy is 3·2%.

CHAPTER 2

GROWTH OF THE POPULATION

Fundamentally, the peopling of Jamaica, as indeed of the whole British West Indies, has been effected by means of the introduction of immigrants of one kind or another. The indigenous inhabitants have contributed nothing to its present population. Three such waves of immigration can be conveniently distinguished. Firstly, there was the introduction of Europeans, a small though highly important influx of planters, soldiers, indentured workers of various kinds, administrators, traders, professional men and others. These, coming singly or with their families, constituted the dominant class of the population. Immigration of this sort attained numerical significance only in the days of slavery when numerous efforts were made to induce their entry into the island. However, an influx of Europeans, mostly administrators and professional men, has persisted up to modern times, though the numbers involved have remained small. Secondly, there was the slave trade, which surpassed in scale all other movements into the island and which has provided the majority of its present population. Thirdly, there has been the more recent introduction of indentured workers and their families from foreign territories which, though small, represents an interesting phase of West Indies migration as a whole and which on these grounds is reserved for separate consideration in Chapter 4.

Immigration of Europeans and the slave trade dominated population movements in the regime of slavery. Under the conditions of mortality then prevailing, population increase among both classes, and indeed even the maintenance of existing numbers, could be assured only by constant large-scale recruitment from outside. After the termination of the slave trade population growth in Jamaica was almost exclusively determined by the balance of births and deaths. European immigration was of negligible importance, and though the introduction of indentured labourers did help to sustain the numbers of plantation workers, its effect on population growth as a whole was small. However, a new phase of migration developed in the last quarter of the nineteenth century— emigration to foreign territories—and this, as we shall show, was sufficient to affect sensibly the rates of growth between 1891 and 1921.

POPULATION GROWTH IN THE PERIOD OF EARLY SLAVERY

In the absence of any exact knowledge of the numbers of the indigenous inhabitants of the island, it is futile to attempt to trace population movements before the conquest of the island by the Spaniards in 1494.[1] The occupation of the island by the Spaniards, far from inducing an increase in population, actually resulted in its decline. For lacking the gold that was the main attraction of the mainland territories, Jamaica proved much less favoured and settlers were few. Moreover, the harsh policy pursued by those who did settle in the island had the effect of completely wiping out the Indians within a period of 50 years.[2]

The capture of the island by the British in 1655 did not lead to any immediate large-scale settlement either. Conditions remained disturbed for some time after 1655. Guerilla warfare against the remnants of the Spaniards continued for some time. And the first settlers were not the kind who would attract others to the island. Among the earliest to settle after the conquest were 'raw soldiers, vagabonds, robbers and renegade servants'. The island remained for a long time the centre of buccaneering, and in the words of Governor Lynch, 'Privateering was the sickness of Jamaica, for that and planting a country are absolutely inconsistent'. It is true that during these years there was some trading with the Spanish mainland, but general conditions were too unsettled to attract immigrants in any numbers, especially the type who would develop the agriculture of the colony.

During slavery the population consisted mainly of two distinct classes: the dominant white group and the slaves. And measures to increase the numbers of these two demand special attention, as it was mainly these policies that determined population growth in the regime of slavery. Sandwiched between the much more numerous slave population and the socially superior whites was the small group of free people, coloured as well as black, who, though physically free, laboured under severe social disabilities.

[1] Most of the discussion of population growth in the period of slavery is based on F. W. Pitman, *The Development of the British West Indies*, 1700–1763, New Haven, 1917, and unless otherwise stated the quotations are from this source.
[2] R M Martin, op. cit. p. 7.

White population. By the end of the seventeenth century conditions in Jamaica had improved to such an extent that settlers were entering in greater numbers and agricultural settlement was proceeding. With the establishment of civil government, the decline in privateering and the dispersion of the remaining Spaniards, the island enjoyed a measure of prosperity. After the restoration the Government undertook to bring settlers from Barbados, and by 1671 there were in the island 41 cocoa walks, 19 indigo estates, 57 sugar plantations, 3 cotton plantations and several smaller estates.

Though never in great demand as sugar-plantation workers, the white labourers were assured an important position in the agriculture of the island as producers of foodstuffs for the population as a whole. Moreover, they were required to form a bulwark against the slaves in times of insurrection. With the growing number of slaves it was essential to maintain a corresponding increase in the white population, upon whom the task of suppressing disturbances might devolve.

By the opening of the eighteenth century the maintenance of adequate numbers of the island's white population became a settled policy. The earliest measures to assure this took the form of deficiency laws, the first of which was passed in 1703. This required every planter to keep 1 white servant for his first 10 Negroes, 2 for the first 20 and 1 for every 20 thereafter. It also decreed a definite proportion between the numbers of white servants kept and the number of livestock on the plantation. Amendments to this original act, varying the prescribed ratio of whites to slaves, were passed from time to time.

These deficiency laws, which were renewed annually, failed to provide the colony with any substantial population increments. Moreover, it proved wholly uneconomic, as the cost of transporting one man from England amounted to £15, while his maintenance called for an annual expenditure of £50. As these indenture servants were subject to extremely high rates of mortality, population recruitment of the white element proved evidently more costly than the maintenance of the slave population, the maintenance of which depended on the slave trade. The success of the scheme was further impaired by the unsatisfactory character of the whites secured, 'a lazy useless sort of people who come cheap and serve for deficiencies and their hearts are not with us', as Governor Hunter characterized them. By 1736, indeed, the deficiency laws came to be regarded more as sources of revenue than as measures for recruiting the white population.

Quite different attempts to increase the white population of the island centred around the introduction of convicts from England. An Act of Parliament of 1717 provided that English courts might contract for the transportation of convicts whose labour should be sold in the colonies for a period of years. However, convicts brought under these terms were even less satisfactory immigrants than the indentured servants. In the words of the Governor, 'These people have been so farr from altering their Evil Course and way of living and becoming an Advantage to Us, that the greatest part of them are gone and have Induced others to go with them a Pyrating and have Inveigled and Encouraged several Negroes to desert from their Masters . . . so that I could heartily wish this Country might be troubled with no more of them.'

Much alarm was expressed over the deficiency of the white population following the disturbances of the 1730's, and efforts along other lines were made to induce agricultural settlement in the island. The planters petitioned Parliament for an act to encourage the cultivation of coffee, essentially a small farmer's crop not calling for extensive cultivation, in the hope that this might attract white settlers. But this also failed to induce any considerable immigration.

A more realistic approach to land settlement was made by Governor Hunter, who ascribed the small white population to the scarcity of good agricultural land for new settlers. For by this time much of the good agricultural land had already been absorbed by the sugar plantations, even though it proved possible for them to cultivate only small portions of their holdings. To increase the land available for settlement planters with large holdings of uncultivated land had to dispose of this to persons able to bring it under cultivation, or in default the land would revert to the crown. Grants were henceforth limited to 1000 acres and the Government acquired 30,000 acres of unoccupied land for allotments to families settling in the island. Under the acts setting up these schemes each family entering the island was provided with free transportation and received up to 100 acres of land and one year's supply of provisions.

Up to 1752 it was schemes of these types that were mainly relied on to stimulate increases in the white population. Between 1739 and 1752 £17,300 was spent on land-settlement schemes of this nature.

And between 1749 and 1754 £16,000 was spent on introducing 347 white immigrants. Thus it appears that immigrants secured under these plans were very few in number. On the other hand, Governor Thomas of the Leewards wrote in 1733, 'From the Temptations thrown out of Jamaica we are daily losing numbers of our inhabitants.' However, the overall failure of these schemes was widely acknowledged in Jamaica. This failure was summed up by Governor Knowles: 'The several laws passed since 1734 relating to grants and for enforcing cultivation of land and encouragement of newcomers, have been so far from answering the purposes intended by them that they have done more hurt than good to the country, by encouraging a number of idle dissolute wretches to resort here . . . and few who have shown some inclination towards industry.'

Records of the movements of white population during the days of slavery are scanty and defective. There were no documents similar to the poll-tax rolls that could be utilized for estimating their numbers. Nor were there any migration records comparable to the slave-trade records. Many of the estimates often quoted, due mostly to Long, are no more than informed guesses, the validity of which can, under the circumstances, hardly be assessed. The estimates collected by Pitman are given in Table 4. These suggest that

Table 4. *Growth of the white population*

Year	Population
1658	4,500
1664	6,000
1670	8,000
1673	8,564
1677	9,000
1694	7,000
1722	7,100
1730	7,658
1734	7,644
1739	10,080
1746	10,000
1754	12,000
1762	15,000
1768	17,949
1778	18,420
1787	25,000

Note. These estimates, given in F. W. Pitman, op. cit. p. 373, are described as 'gathered mainly by Long'.

throughout the period 1664–1734 the numbers of Europeans changed very little. The increases after this date may indeed indicate some success of the policy aimed specifically at attracting new settlers. These estimates indicate that between 1734 when the policy was first actively pursued and 1762, by which time it was discontinued, the European population doubled, increasing from 7600 to 15,000. However, the tenuous nature of all these estimates must be emphasized. A complete lack of migration records makes it hazardous to put forward any detailed estimates of rates of growth, especially in view of the probable enormous mortality prevailing at the time.

Slave population. Slaves provided the labour force for the cultivation of the major plantation crop, sugar. The profitable cultivation of this crop required a larger and more highly organized labour force than any other crop in the eighteenth century. The early failure of the Indians as plantation labourers and their eventual extinction led the Spaniards to employ slaves, a policy also urged by Las Casas in the belief that this might mitigate the lot of the Indians. The English at first relied on white indentured labourers as well as on slaves. But it soon became apparent that it was cheaper to employ slave labour on the plantations. Early in the eighteenth century it seemed generally agreed that so far as agriculture in the island was concerned white and slave labour complemented each other. The slave proved most economical as a producer of sugar, whereas the white worker could most profitably engage in the production of foodstuffs necessary for the island as a whole. Thus slaves were in constant demand. This demand stemmed not only from the necessity for extending the sugar plantations; it was essential to maintain existing numbers. For under prevailing conditions of mortality the cessation of the slave trade would have meant a drastic reduction in their numbers (see Chapter 7).

Jamaica enjoyed a unique position so far as the importation of slaves was concerned. The Asiento with Spain in 1713 provided that the South Sea Company should furnish the Spanish American colonies with slaves, to the number of 4800 annually. Jamaica, favourably situated to form the base of this trade, was chosen as the depot for the whole movement. Slavers brought their cargo to the island, where the slaves were sorted out and refreshed before being despatched to their destination in the Spanish territories.

It was claimed by the planters that, though the island was the centre of this important part of the slave trade, its annual supplies of slaves were only meagre. It proved more profitable to sell the slaves in the Spanish colonies as the Spaniards paid in ready cash, and usually in gold. The less opulent planters of Jamaica often asked for long credit or offered payment in sugar, terms which were far from acceptable to the slave dealers. As a consequence only those not suited to the Spanish market were sold to Jamaica planters. 'This Asiento . . . deprives the planters of the best and only sells them the worst of the Negroes', claimed Governor Lawes.

In order to increase their share of the slaves imported into the island the planters resorted to a duty of £1 per head on all slaves re-exported. (There was, in addition, an import duty of 10s. per head on all slaves.) It remains doubtful whether this had any effect on the proportion of slaves retained in the island. In any event, contrary to the claims of the planters, the proportions of slaves they secured were considerable. Certainly the numbers that were retained in the island far exceeded those that were re-exported.

Two classes of data are available for the study of the movements of slave populations in Jamaica before the period of slave registration: estimates of the slave populations and returns of exports and imports of slaves. The latter are certainly the more accurate. For the taxes on imports and exports of slaves made careful listing of such migrations essential. Compilations from the official lists have been made and these are given for 1702–75 by Pitman and Bridges.[1] As regards estimates of the slave populations, these are generally based on the rolls from which the poll taxes were levied. But, as Bryan Edwards cautions, 'It appears . . . from the report of a committee of the assembly . . . that in most of the parishes it is customary to exempt persons not having more than 6 negroes from the payment of taxes on slaves, whereby many of the negroes (especially in the towns) are not given in.'[2] He illustrates the limitations of the records by the case of Kingston. Whereas the number of slaves in Kingston was 16,659, the estimates based on the poll-tax rolls showed only 6162.

As returns of slave imports and re-exports are available only for the period 1702–75, we shall consider separately the position of the

---

[1] F. W. Pitman, op. cit., and G. W. Bridges, op. cit. It should be noted that these returns contain several discrepancies.

[2] Bryan Edwards, *The History, Civil and Commerical, of the British Colonies in the West Indies*, 1793, vol. I, p. 922.

slave population at dates after 1787. Estimates of slave populations and net importations are given in Table 5. The population in 1703 was estimated at 45,000, but the dubious nature of this estimate must be emphasized.[1] The estimate for 1722 is also probably

Table 5. *Growth of the slave population, net importation of slaves and estimated rates of increase*

Year	Slave population	Net importation of slaves	Annual rate of increase in slave population (%)	Estimated rate of natural decrease (%)
1658	1,400	—	—	—
1673	9,504	—	—	—
1703	45,000	—	—	—
1722	80,000	55,536		
1730	74,525	32,379	0·7	3·7
1734	86,546	19,754		
1739	99,239	20,341	2·8	1·5
1746	112,428	36,510	1·8	2·8
1754	130,000	43,295	1·8	2·3
1762	146,464	54,908	1·5	3·0
1768	166,914	41,472	2·2	2·0
1778	205,261	54,951	2·1	2·1
1787	210,894	—	0·3	—

Note. The estimates of slave populations and of net importations of slaves are taken from F. W. Pitman, op. cit. p. 373, and Appendix 2. The net importations entered in the table are the totals between successive dates of the population estimates. The records of slave imports and re-exports given by Pitman begin with 1702 and end with 1775. Martin (op. cit. p. 7) gives a series of seven estimates of slave populations which differ slightly from those quoted by Pitman.

defective, for it exceeds the more carefully compiled estimate prepared by Governor Hunter for 1730, which put the total population at 74,500. From 1734, when the population was estimated at 86,500, there was, according to these estimates, a steady increase in the slave population, which by 1787 stood at 211,000. Thus within little over half a century (1730–87) the slave population increased nearly threefold. During most of the period covered by these estimates annual rates of increase exceeded 1%. Rates were lowest in 1722–34 (0·7%) and in 1778–87 (0·3%) and highest in 1734–9 (2·8%) and in 1762–8 (2·2%).

[1] In the words of the Lieutenant Governor, 'At my first coming to the Government I did compute by the Rolls that the Island had 45,000 men but cannot find now above half.' Quoted in Pitman, op. cit. Appendix 1.

It was manifestly the slave trade which dominated population growth during this period. Between 1702 and 1775 the total number of slaves imported into the island was 494,800, while re-exports totalled 135,600. The fact that 73% of the total number of slaves imported (359,100) were retained in the island assured the appreciable increases indicated from the available data. Annual importations of slaves increased from 4500 in 1702–21 to 10,300 in 1730–3. Over the remainder of the period it amounted to between 6000 and 9000. Up to the 1730's there was some truth in the planters' claim that the Asiento deprived the island of too many of the slaves imported. In fact during 1702–21, 39% of the slaves imported were re-exported, and this proportion increased in subsequent years, reaching 52% during 1730–3. Following the war with Spain, however, re-exports declined appreciably, and of those imported after 1739, 84% were retained in the island. As a result of the decline in re-exports the net increments to the slave population rose. Thus during the years 1702–21 average net importations stood at 2800. In 1722–9 average net increments from the slave trade amounted to 4000 and during the succeeding eight years did not change much. However, after 1739 net importations rose steadily and by the end of the period amounted to 7900. It is clear that, if the estimates of the slave populations are reliable, the slave trade sufficed not only to maintain the slave population but even to ensure a substantial rate of increase in the face of extremely high rates of natural decrease.

If the available estimates of the slave population are accepted, approximate rates of natural decrease can be derived on the basis of the population increases and the net importations between successive dates for which population estimates are given. These rates of natural decrease, entered in Table 5, are extremely high. During the period 1722–34, when slave importation was highest, there was, according to our estimates, an annual rate of natural decrease of nearly 4%. In only one period (1734–9) did the rate fall below 2%. Rates of this magnitude attest to the severe mortality suffered by the slaves and presumably also to their low fertility. These two inevitable concomitants of slavery during this era of the slave regime resulted in a severe wastage of human life but were, as will be shown in Chapter 7, compatible with the attitudes towards reproduction and the easy acquisition of slaves afforded by the slave trade.

No returns of slaves imported after 1775 have been located and so no detailed treatment of the growth of slave population between this date and 1817 can be attempted. Martin gives a series of slave population estimates for each year from 1800 up to 1816.[1] These show a steady increase from 300,900 in 1800 to 312,300 in 1806; and thereafter irregular movements appear, due doubtless to the cessation of the supplies of new slaves. The estimate for 1816 (314,000) is considerably lower than the more accurate estimate obtained from the first year of slave registration, thus raising strong doubts as to the reliability of all slave population data before 1816.

Free population. These were of two types. The first, the free coloured, the offspring of white males and female slaves, owed their freedom to their masters; indeed, it became general for a white father to secure the freedom of his coloured offspring and even to educate them. The second, the free black population, purchased their freedom or became free through the wills of their masters, or by manumission during the lifetime of their masters, or by special enactments as rewards for some public service they performed.[2] Clearly the growth of the free coloured population was dependent to some degree on the rate of miscegenation, but it is impossible to state definitely the extent of miscegenation during slavery or its implications for population growth. It should be noted that some observers held that mulattoes were infertile. 'This elegant cross-breed seldom are (*sic*) productive with each other; so that if they are to propagate effectually it must be with a black man or a black woman . . .'.[3] The increase in the numbers of the free black population would, presumably, depend mainly on the rate of manumission. But in view of the obstacles in the way of manu-mission in the British colonies, this probably contributed very modestly to the increase of the free population.[4] In any case the growth of the free black population would be mainly at the expense of the slave population.

In view of the scanty and contradictory nature of the data available on the numbers of these people, no clear picture of their

---

[1] R. M. Martin, op. cit. p. 7.

[2] The position of the free population is discussed in great detail by Charles H. Wesley, 'The Emancipation of the Free Colored Population in the British Empire', *The Journal of Negro History*, vol. XIX, 1934, pp. 137–70.

[3] H. M'Neill, *Observation on the Treatment of the Negroes in the Island of Jamaica*, London, 1788, p. 42. See also Long, op. cit. vol. II, p. 335.

[4] An account of the obstacles to manumission in the British colonies is given in F. Tannenbaum, *Slave and Citizen*, New York, 1947, pp. 65 et seq.

growth during the days of slavery can be obtained. According to Gardner, an act passed in 1761 required all free persons to take out certificates of freedom, and from these it appears that the number of black and coloured free people amounted to 3408, of whom 1093 were in Kingston.[1] Long, after trying 'various modes of calculation' of the population of the island, arrived at 'a general estimate' which put the 'free black and mulatto' population at 37,000 (this estimate presumably refers to the late eighteenth century).[2] Martin gives 'the free coloured population' of the island in 1775 and 1788 as 4100.[3] Wesley quotes estimates from the Parliamentary Debates to the effect that in 1825 there were 30,000 free coloured and 10,000 free blacks in the island. From such contradictory figures it is difficult to obtain any reliable indication of the numbers of free persons in the days of slavery. However, in view of the fact that at the first census (1844) the number of coloured persons was 68,500 and the average rate of increase of this group between 1844 and 1861 was 1%, it is clear that the estimates given by Martin and Wesley are too low.

SLAVE POPULATION GROWTH DURING THE PERIOD OF REGISTRATION

With the advent of registration, more reliable indications of the movements of slave populations became available. Table 6 gives the numbers of slaves registered at triennial intervals in Jamaica

Table 6. *Slave population movements in Jamaica during slave registration*

Year of registration	Slave population	Increase due to births	Decrease due to	
			Deaths	Manumissions
1817	346,150	—	—	—
1820	342,382	24,346	25,104	1,016
1823	336,253	23,249	26,351	921
1826	331,119	23,026	25,170	957
1829	322,421	21,728	25,137	1,117

Note. These are given in a return of the Office for Colonial Registry in the *Parliamentary Papers* of 1833, and in R. M. Martin, op. cit. p. 7.

[1] W. J. Gardner, op. cit. p. 173. See also Long, op. cit. vol. II, p. 337.
[2] Edward Long, op. cit. vol. I, p. 377.
[3] R. M. Martin, op. cit. p. 7.

between 1817 and 1829, together with the breakdown of the movements into three components, increase by births, and decreases due to deaths and manumissions. (The terms 'births' and 'deaths' are used in the sense previously indicated.)

The slave population declined steadily throughout the period of registration, from 346,150 in 1817 to 322,400 in 1829, the total loss over the 12 years being 23,700. Evidently this decline continued after 1829, as the compensation returns of 1834 showed a total slave population of 311,100. There was therefore a 10% reduction in the slave population during the 17 years following 1817. Clearly manumission accounted for only a small portion of the decline, the total number of slaves manumitted during the 12 years 1817–29 being only 4000. It was the excess of deaths over births that resulted in this decline. For each triennial period the total number of deaths registered exceeded 25,000, but in view of the limitations of the system of slave registration these considerably understate the true number of deaths. Nevertheless, because of the comparatively small measure of inconsistency in the slave data of Jamaica (3·2%) (see Chapter 1), attempts to determine, from the three components of decline available, rough estimates of the rate of decline due to vital processes can be made with some assurance. Had there been no manumissions, the slave population in 1829 would have been 4000 more than the number actually registered. This suggests an annual rate of natural decrease of about 5 per 1000. With the exception of Barbados, indeed, the excess of deaths over births was general throughout the slave populations of the West Indies. In the fever-ridden coastlands of British Guiana, the indicated rate of natural decrease was as high as 11 per 1000. Though somewhat lower in other colonies, it was in most cases in excess of 4.

The vast difference between the indicated rate of natural decrease during the period of registration and the figures during the period 1702–75 suggests strongly the unreliability of early estimates of the slave population. At the same time, it is not improbable that the efforts made since the late eighteenth century to increase fertility and reduce mortality did in fact effect a more economical type of human reproduction than the essentially wasteful one that characterized all slave populations in the heyday of the slave trade.

A special demographic factor, the immediate consequence of the cessation of the slave trade in 1807, might have contributed to the loss of slaves through deaths in the period 1817–29. For the populations that formed the subject of registration differed in one fundamental respect from those existing prior to 1807. The latter were being constantly recruited by importations of slaves of working age, as a consequence of which they were, despite the heavy wastage experienced during the period of 'seasoning', probably heavily weighted by persons of working age. The only comprehensive data available for a slave population, those for British Guiana, fully bear this out. The cessation of the slave trade meant that the groups of African-born slaves surviving into the period of registration tended to exert special effects on population growth and mortality among slaves between 1817 and 1829. For as these groups aged, they expanded the numbers of advanced ages without contributing to the reproductive performance of the population as a whole. Such a development tends at once to push up the crude death-rate and to reduce the population. This process, clearly illustrated in the slave population of British Guiana, probably obtained in most colonies. The increase (with the implied lower death-rate) in the case of Barbados may be associated with the fact that the large-scale importation of slaves into this colony ceased long before it did in the case of the other colonies.[1] Under such conditions, most of the slave population of Barbados during the years of registration would be native-born and therefore not unduly weighted by those of advanced age.

It is impossible to arrive at any reliable estimate of the overall population of Jamaica during the period covered by the slave registration, because uncertainty shrouds the numbers of the white and free coloured population at this time. Wesley quotes an estimate for 1825 showing 25,000 white, 30,000 free coloured, 10,000 free black and 340,000 slaves.[2] But that is evidently an overestimate. It was most unlikely that the population of the island would decline from over 400,000 in 1825 to 377,000 in 1844. Martin presents what he calls 'an imperfect view of the population . . . prepared from various documents laid before the Finance

---

[1] This at least is suggested by Hugh Hyndman in his evidence before the Select Committee of 1832. 'It [the slave trade] ceased by law at Barbados at the same time that it ceased elsewhere; but in point of fact it ceased . . . before the abolition of the trade, for when I resided there in 1806 and 1807 several slave vessels came to that colony but the purchasers were so inconsiderable as not to be worth mentioning.' Answer to Question 777, *Report of the Select Committee on the Commercial State of the West India Colonies*, 1832.

[2] C. H. Wesley, op. cit.

Committee of 1828', according to which the island population was about 342,000.[1] But in view of the reserve with which this is put forward and its obvious incompleteness, not much credence can be attached to it.

POPULATION GROWTH IN THE CENSUS PERIOD

Interest in population movements in the West Indies heightened in the early 1840's when the planters were experiencing difficulty in securing labour. Committees both in the West Indies and in England turned their attention to this question, but in the absence of any reliable population estimates all that could be elicited were the dubious opinions of observers. Thus the Rev. W. Knibb declared before the Select Committee of the House of Commons in 1842 that the population of Jamaica was growing by natural increase, 'especially where the influence of Christianity is felt; and free villages are rising up even on the top of Mount Diable; it is covered with free cottages'.[2] In his detailed analysis of the health of the island following the cholera epidemic of 1850–2, Milroy could do no more than record 'the general impression among the best informed . . . that for many years past the population of the island has not been on the increase, if it has not actually been diminishing'.[3] Though in Jamaica, as in all the other West Indian territories, population movements between the period of slave registration and the date of the first census cannot be confidently assessed, it is probable that the excess of deaths over births that prevailed throughout the slave regime was by 1844 completely arrested. One factor that might have contributed to such an improvement in mortality was the reduction in the numbers of the large body of African-born persons (former slaves) introduced into the island before 1807.

It was not until 189 years after the capture of the island by the British that the first comprehensive census of the island was taken. Though the eight censuses taken since 1844 give a picture of a century of uninterrupted population growth, the long lag between the date of the first census and the time of the establishment of effective civil registration leaves a gap of over 35 years for which no records of natural increase are available. This hampers the

---

[1] R. M. Martin, op. cit. p. 7.
[2] Evidence of Rev. W. Knibb, Question 6077, *Report of the Select Committee appointed to inquire into the State of the Different West India Colonies . . .*, 1842.
[3] Gavin Milroy, op. cit.

Table 7. Summary of population movements in Jamaica, 1844–1943

Year of census	Census population	Intercensal increase		Births, deaths and natural increase during intercensal interval			Migration balance	Rates per 1000 population		
		No.	Annual rate, %	Births	Deaths	Natural increase		Birth	Death	Natural increase
1844	377,433	—	—	—	—	—	—	—	—	—
1861	441,264	63,800	0·92	275,400	224,400	51,000	+ 12,800	40	32	8
1871	506,154	64,900	1·38	184,800	127,900	56,900	+ 8,000	39	27	12
1881	580,804	74,600	1·38	208,200	139,200	69,000	+ 5,600	38	26	12
1891	639,491	58,700	0·97	224,200	140,700	83,500	− 24,800	36·7	23·1	13·6
1911	831,383	191,900	1·32	581,100	345,300	235,700	− 43,900	39·5	23·5	16·0
1921	858,118	26,700	0·32	320,200	216,400	103,800	− 77,100	37·9	25·6	12·3
1943	1,237,063	378,900	1·67	765,300	412,200	353,200	+ 25,800	33·2	17·9	15·3

Note. Births for all intercensal periods prior to 1881 are estimated by applying appropriate survival factors from the 1891 life-tables to the census populations of 1861, 1871 and 1881. For 1844–61 the population used is that aged 0–16; for the other two the populations used are aged 0–9. Births given for all later intercensal periods are the total numbers registered. Migration balances for 1844–81 are the net indenture immigration revealed by published records (see Appendix II). For later periods migration balances are derived from intercensal population increases and registered births and deaths. Deaths for periods prior to 1881 are estimated from intercensal increases, estimated births and recorded migration balances. For later periods deaths entered are the total numbers registered. Discrepancies for some periods are due to rounding.

43

detailed study of population movements during the pre-registration years 1844–81. However, it is possible to obtain rough estimates of such movements on the basis of the published migration records and estimates of the numbers of births during the three intercensal intervals before 1881. Estimates constructed on this basis are presented in Table 7, which summarizes the various population movements during the century following 1844.

Three broad periods of population growth during the century following 1844 can be distinguished: the first extends from 1844 to 1881, the second one extends from 1881 to 1921, while the third period covers the years following 1921. These three divisions will now be discussed in detail.

1844–81. What in the main characterizes this period is that it is the only one in the post-emancipation years (apart from the years following 1921 when very special conditions obtained) during which any appreciable net immigration took place. The total net inward movement amounted to 26,400 and declined steadily as indenture immigration was reduced. Clearly this net immigration contributed only modestly to population growth in the island, but increase to the population from migration during these years is a feature Jamaica has in common with British Guiana and Trinidad and it warrants the separation of this period from later ones.

According to the first census, the population in 1844 stood at 377,400. By the time of the second census (1861) the island population had increased by 63,800, the annual rate of growth being 0·9%. Even though, due to the introduction of Africans and small numbers of East Indians, there was a recorded net immigration of 12,800, the population growth was due mainly to the excess of births over deaths, which amounted to over 51,000, equivalent to an annual rate of natural increase of about 8 per 1000. This period witnessed the severe cholera epidemic of 1850–1, which undoubtedly resulted in heavy loss of life. As there was no system of effective registration in force at the time, the magnitude of this disaster can never be accurately known. But the fact that the estimated death-rate (32) was much higher than that of the succeeding intercensal interval emphasizes the degree to which population increase was retarded as a result of this epidemic.

A much higher rate of population growth was witnessed from 1861 to 1871, during which time the population increased by 64,900 to reach 506,200, the average rate of intercensal growth being 1·4%. Increments from net immigration amounted to 8000, and again it was natural increase that in the main determined the extent of population growth. According to the present estimates, births exceeded deaths by 56,900, a figure higher than that for the longer intercensal interval of 1844–61. In the absence of any epidemics there was a notable decline in mortality; the death-rate stood at 27, while the rate of natural increase rose to 12 per 1000.

Though the years 1871–81 were marked by a pronounced fall in the rate of net immigration (the total recorded net immigration was only 5600), the population increased by 74,600 to reach 580,800, while the rate of growth (1·4%) was as high as in the preceding period. African immigration had now ceased, and although small introductions of East Indians were still in progress, the significance of this was greatly reduced by the numbers returning to India during these years. According to the present estimates, the death-rate at this time was 26 per 1000 and the rate of natural increase 12. The fact that the estimated rate of natural increase is so close to the intercensal annual rate of growth emphasizes that migration as a factor in population growth in the island was of negligible significance.

1881–1921. The dominant feature of this period is a reversal of the direction of net migration. In place of the small net inward movement that characterized the previous 40 years, there appeared a net outward movement, which increased with the passage of time, attaining considerable dimensions during the decade 1911–21. In fact, this is the period which has been most strongly influenced by migration, the total net outward movement being nearly 146,000 or about 3600 a year. As a consequence of this outward movement, rates of population growth declined appreciably.

Between 1881 and 1891 the population of Jamaica increased from 580,800 to 639,500, which represents a rate of growth of just under 1% per year. This is much lower than the rate experienced over the previous 20 years and reflects the effects of emigration on population growth. This outward movement, coupled with the fact that East Indian immigration was now reduced to a mere trickle, resulted in an annual net emigration of about 2500, so that even though the rate of natural increase was the highest so far observed (14), the total addition to the population (58,700) was nearly 16,000 less than in the preceding intercensal period.

The absence of a census in 1901 reduces considerably our detailed knowledge of population movements during 1891–1911, which must in fact be treated as one intercensal interval. The evidence is that throughout this period there was a net emigration of about 44,000, but it is possible that most of this took place between 1901 and 1911. It is doubtful whether any heavy emigration occurred between 1891 and 1901; indeed, the cessation of work on the Panama Canal might have induced many who emigrated before 1891 to return to Jamaica. Despite the appreciable emigration, the population increased by 191,900, equivalent to an annual rate of growth of 1·3%, which is appreciably higher than the rate prevailing in the preceding intercensal interval. The rate of natural increase (16 per 1000) for this period was the highest so far recorded.

It was within the last intercensal interval, 1911–21, that emigration from Jamaica reached its zenith. Here, as in Barbados, it was the opening of large-scale emigration to the United States and other foreign territories that added so greatly to the numbers leaving the island. Another very marked feature of these years was the rise in mortality. The years 1918–22 were exceptionally unfavourable, and the highest death-rates ever recorded occurred at this time. As a result of large net emigration (77,100) and the increased death-rate (26), the addition to the population between 1911 and 1921 was only 26,700, the smallest intercensal increase ever recorded, and equivalent to an annual rate of growth of 0·3%.

After 1921. Much of this period is covered by the longest intercensal interval since the establishment of census-taking in the island. The fact that 22 years elapsed between 1921 and the date of the next census means that many aspects of population change after 1921 are obscured. Nevertheless, it is clear that three factors confer a degree of unity on the whole period subsequent to 1921. These are declining fertility, declining mortality and a change in the direction of net migration. In addition, there was a pronounced increase in internal migration. These changes will be dealt with in detail in later chapters. Here, however, it is necessary to indicate briefly that the altered pattern of external migration and the declining mortality were by far the chief factors making for higher population growth after 1921.

Due to the immigration restrictions introduced in the United States and other areas, emigration ceased, and in its place there was a small net immigration, representing probably the return of emigrants who had left the island before 1921. The fact that in place of the appreciable and at times very substantial losses due to emigration, witnessed over the preceding 40 years, there appeared a net immigration of 1100 annually, added appreciably to population growth.

But undoubtedly the most important element making for the acceleration of population growth was the decline in mortality. In this connexion very special significance attaches to the year 1921. In Jamaica, as in many other Caribbean territories, this marks the end of a long period of high and stationary mortality and the opening of an era of declining mortality. A demographic phase much less wasteful of human life than that prevailing in the past was emerging. The transition from what W. S. Thompson calls the 'pre-industrial' stage of demographic evolution into the 'expanding' phase was being effected.[1]

Between 1921 and 1943 the island population increased from 858,100 to 1,237,100, the annual rate of growth being 1·7%, by far the highest so far recorded. Indeed, the addition to the population during these 22 years (379,000) was greater than the increase recorded during the preceding 50 years. The reduction in the death-rate to 17·9 was accompanied by a decline in the birth-rate from 37·9 in 1911–21 to 33·2, so that the rate of natural increase (15), though higher than that of the preceding intercensal interval, was lower than that recorded during 1891–1911. The net immigration, probably composed of returning emigrants, amounted to 25,800. Though this constitutes only a small addition to the population, it is significant in that it is nearly as much as the total increment to the population during 1844–81 from indenture immigration (26,400).

To summarize the record of population growth during the century

[1] W. S. Thompson, *Population and Peace in the Pacific*, Chicago, 1946, p. 31.

covered by census records, it can be said that the population increased from 377,400 in 1844 to 1,237,100 in 1943, the average rate of growth being 1·2%. Clearly most of this increase resulted from the excess of births over deaths, for though rates of natural increase varied slightly from one intercensal interval to the next, the excess of births over deaths was throughout sufficient to ensure a continuous increase in the population even in the face of appreciable emigration. According to these estimates, the total increment to the population resulting from natural increase during the century amounted to 953,100, as compared with an overall net outward movement of about 93,600.

A comparison of population growth in Jamaica with population growth in other West Indian territories brings out interesting contrasts. The 125 censuses taken in the region between 1841 and 1946 reveal diverse patterns of growth. The populations of these censuses are given in Appendix I, while Fig. 1 depicts in summary form the growth of the various territories. Two features strongly differentiate Jamaica from the others, its much larger population and the comparatively stable rates of growth it has enjoyed. All the other populations reveal, at one time or another during the past century, marked irregularities in rates of growth, most of which have been associated with migration. This relative stability of growth in the case of Jamaica is emphasized by the summary of intercensal rates of growth for the territories given in Table 8. Clearly the populations of the eastern Caribbean can be divided into two categories. The first, Trinidad and British Guiana, constitutes areas heavily influenced by immigration, both of indentured workers and their families from foreign countries as well as of workers from neighbouring islands. And during the periods when such immigration was in full force these territories showed rates of growth far above those for Jamaica. For instance, throughout the period 1844–1921 Trinidad shows annual rates of growth ranging from 3·1% to 0·9%, which are generally more than twice as high as the corresponding rates for Jamaica, which range from 1·4% to 0·3%. British Guiana also showed rates of growth higher than those for Jamaica for the period when sizeable increments to its population from indenture immigration were being received. With the cessation of immigration into these two territories their rates of growth declined appreciably. The second category of the eastern populations, comprising Barbados, the Windwards and the Leewards, represents areas of heavy

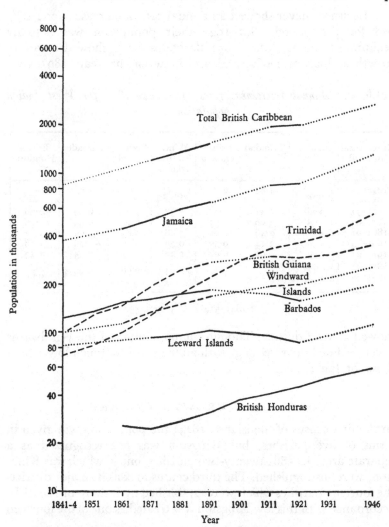

Fig. 1. Population growth in West Indian territories, 1841–1946

Note. Dotted lines indicate the absence of one or more censuses on appointed dates.

emigration. Much of this outward movement was to British Guiana and Trinidad, but emigration to foreign territories became increasingly important after 1891. It is clear that rates of growth for these smaller territories are generally much lower than those for Jamaica, the contrast being most marked between 1881 and 1921.

The Leewards never showed an annual rate of increase over 1·0%, and for the period 1891–1921 their population was actually declining. Similarly, after 1851 Barbados never showed a rate of growth as high as 1·0%, and also between the years 1891–1921

Table 8. *Annual intercensal rates of growth* (%) *for West Indian populations*

Intercensal interval	Jamaica	Trinidad	British Guiana	Wind- ward Islands	Leeward Islands	Barbados	British Honduras
1844–51	} 0·92 {	1·84	2·50*	} 0·82 {	—	1·53	—
1851–61		1·87	1·66		—	1·17	—
1861–71	1·38	2·41	2·71	1·37	—	0·59	−0·36
1871–81	1·38	3·06	2·37	1·25	0·16	0·59	1·06
1881–91	0·97	2·46	1·03	0·88	0·65	0·62	1·38
1891–1911	1·32	2·14	0·33	0·79	−0·39	−0·29	1·26
1911–21	0·32	0·92	−0·02	0·42	−0·91	−0·95	1·13
1921–31	} 1·67† {	1·21	0·48	} 0·94 {	0·98	} 0·84 {	1·26
1931–46		2·03	1·15				0·96

* 1841–51.   † 1921–43.

showed rates of decline. Though the population of the Windwards never declined the rates of growth after 1871 were much lower than those for Jamaica.

POPULATION GROWTH BY PARISH

In the first census of the island (1844) the population was given in terms of five parishes, but Kingston was not recognized as a separate area. In 1861 twenty-two parishes, one of which was Kingston, were distinguished. The third census took as its major districts the fourteen parishes which were established by Law 20 of 1867.[1] In accordance with this law Kingston was treated as one of the fourteen

[1] Law 20 of 1867, 'A Law to Reduce the number of Parishes', stated 'Whereas some of the parishes . . . are much smaller both in extent and population than other parishes, and the equalization of the several parishes in extent and population and their reduction in number will tend to a better and more economical administration . . .'. This law carefully defined the boundaries of all parishes. The boundaries of Kingston, considered in the law as a parish, were also defined. But the boundaries of Kingston as a city were defined by Law 27 of 1870, 'A Law to Define the Limits of the City of Kingston'. It is not clear which definition has been followed by census authorities or indeed whether the same boundaries have been maintained from 1871 to 1943.

parishes. All subsequent censuses of the island have maintained this fourteen-fold division. It is possible that slight changes in the boundaries of Kingston or Port Royal have been made from one census to another, but probably no alterations sufficiently far-reaching to invalidate comparisons of population growth since 1871 have been made. The growth of the island population in terms of these fourteen parishes is shown in Table 9. Of outstanding interest

Table 9. *Parish populations, 1871–1943*

Parish	1871	1881	1891	1911	1921	1943
Kingston	34,314	38,566	48,504	59,674	63,711	110,083
St Andrew	31,683	34,982	37,855	52,773	54,598	128,146
St Thomas	32,673	33,945	32,176	39,330	42,501	60,693
Portland	25,313	28,901	31,998	49,360	48,970	60,712
St Mary	36,495	39,696	42,915	72,956	71,404	90,902
St Ann	39,547	46,584	54,127	70,651	70,922	96,193
Trelawny	28,812	32,115	30,996	35,463	34,602	47,535
St James	29,340	33,625	35,050	41,376	41,946	63,542
Hanover	26,310	29,567	32,088	37,432	38,240	51,684
Westmoreland	40,823	49,035	53,450	66,456	68,853	90,109
St Elizabeth	45,200	54,375	62,256	78,700	79,281	100,182
Manchester	38,925	48,458	55,462	65,194	63,945	92,745
Clarendon	42,747	49,845	57,105	73,914	82,555	123,505
St Catherine	53,972	61,110	65,509	88,104	96,590	121,032
Total	506,154	580,804	639,491	831,383	858,118	1,237,063

has been the growth of Kingston. This town, originally laid out in 1695 by Colonel Christian Lilly, soon became the chief port of the island, increasing in size and importance after the destruction of Port Royal in 1703.[1] For a brief period (1755–8) it became the seat of the island Government, but this was soon returned to Spanish Town, which continued until 1872 to be the capital of the island. In 1872 Kingston became the permanent seat of the island Government, though several administrative departments remained at Spanish Town. An early estimate (1788) of the population of Kingston given by Bryan Edwards was 26,500. Martin's 'imperfect view' of the population of 1828 put the population of Kingston at 35,000; but the first enumeration of the town (1861) showed a total population of 27,400. Port Royal, which in all subsequent censuses appears as distinct from Kingston though the two are throughout

[1] F. W. Pitman, op. cit. p. 18.

this study treated as one unit, had a population of 7900 in 1861, almost identical with Martin's estimate for 1828. From a total of 34,300 in 1871 the population of Kingston increased to 38,600 by 1881 and to 48,500 by 1891. Even as early as 1881 there was evidence of appreciable in-migration into the town, for though the intercensal increase during 1881–91 amounted to 9900 there was a small excess of deaths over births (100). Again during 1891–1911 there was a natural increase of 6900, whereas the population of the town increased by 11,200. In-migration became of much greater importance after 1911, and this will be considered fully in Chapter 5. It is evident that the greatest population growth experienced was between 1921 and 1943, from 63,700 to 110,100. By the latter date 9% of the island's population was located in the town, as compared with 7% in 1871. But recent evidence is that the growth of the town has passed its zenith. Within the past 10 years the rate of growth of Kingston has declined steeply, and as it assumes the status of the centre of a rapidly expanding urban area it is losing population to the suburban area of St Andrew, and may ultimately decline in size.

The expansion of St Andrew as an urban area constitutes the most significant factor of population growth in the island since 1921. Between 1871 and 1891 population increase in this parish was small, from 31,600 to 37,900. Unlike Kingston, St Andrew showed no evidence of in-migration during this period, for whereas the addition to its population between 1881 and 1891 amounted to 2900, the recorded natural increase was 4200. Between 1891 and 1911 there was an appreciable increase of 14,900, and here for the first time there was evidence of in-migration, for the recorded natural increase fell far short of this, being 7900. As a consequence of the heavy emigration from the island the increase in population was very small between 1911 and 1921 (only 1800) and this was due wholly to natural increase. The important transformation of St Andrew into an urban area took place between 1921 and 1943. This was indeed part of the complex of social and demographic changes that began in the island during this period. From 1921 to 1943 the population of the parish more than doubled, increasing from 54,600 to 128,100. Its annual rate of growth was nearly 4%, whereas the rate of growth for the island as a whole was 1·7%. By 1943 there were still certain areas of the parish defined in the census as rural; in fact only 92,900 or 72% of its total population was classified as urban.[1] Whereas in 1871 only 6·3% of the island's population was located in St Andrew, the percentage had risen to 10·4 by 1943. Thus it may be said that the two parishes comprising the urban centre of the island supported 19% of the total population. As is to be expected, population growth of the parishes has been generally determined by natural increase. But important differentials in rates of growth are to be noted, and these result from migration, both external and internal, which has affected the island since the late nineteenth century. Estimates of internal and external migration balances for the several parishes, which will be derived in Chapters 4 and 5, make it possible to trace the elements of increase in parish populations since 1911. These are shown in Table 10. These emphasize the importance of external migration as a determinant of population growth during 1911–21. All parishes experienced losses due to external migration. St Mary recorded the greatest loss (13,000), while St Ann, Kingston, Portland, Manchester and St Elizabeth all experienced losses in excess of 6000. Moreover, in Kingston, St Andrew, Portland, St Mary and Trelawny net emigration was greater than natural increase. Though natural increase and external migration in the main determined population growth in the several parishes during these years, there was also some internal migration. Three parishes were strongly affected by out-migration, St Elizabeth, Manchester and St Ann, while in-migration into Kingston and St Andrew exceeded 14,000. As a consequence of these factors, only two parishes, Clarendon and St Catherine, showed any appreciable growth, 8600 and 8500 respectively, while four parishes, Portland, Manchester, St Mary and Trelawny, actually experienced a decline in population.

After 1921 declining mortality resulted in much greater population growth than in the past. As before, St Elizabeth, Manchester, Clarendon and St Ann experienced the largest natural increase, but everywhere births exceeded deaths to an extent sufficient to promote population increments on an impressive scale. Changes in the patterns of migration, however, aided materially in producing the differential rates of parish growth, which, despite the general natural increase, are to be noted during the years 1921–43. St Andrew gained considerably from both currents of migration. Most of the

---

[1] Although the whole of St Andrew was not an urban area in 1943 the whole parish will be so treated in this study, chiefly because in the analysis of vital rates it is impossible to distinguish between rural and urban areas.

Table 10. Elements of increase in parish populations, 1911–43

Parish	1911–21				1921–43			
	Natural increase	Net internal migration	Net external migration	Total increase or decrease	Natural increase	Net internal migration	Net external migration	Total increase
Kingston	1,300	+ 10,300	− 7,500	+ 4,100	28,100	+ 21,500	− 3,200	+ 46,400
St Andrew	1,800	+ 4,100	− 4,100	+ 1,800	10,700	+ 47,500	+ 15,500	+ 73,700
St Thomas	3,600	+ 800	− 1,200	+ 3,200	12,000	+ 5,500	+ 700	+ 18,200
Portland	5,100	+ 1,300	− 6,800	− 400	16,500	− 3,000	− 1,700	+ 11,800
St Mary	10,500	+ 900	− 13,000	− 1,600	25,000	− 8,700	+ 3,100	+ 19,400
St Ann	13,700	− 4,700	− 8,800	+ 200	35,800	− 14,300	+ 3,700	+ 25,200
Trelawny	3,800	− 400	− 4,300	− 900	15,600	− 2,900	+ 200	+ 12,900
St James	4,700	− 300	− 3,800	+ 600	20,200		+ 1,400	+ 21,600
Hanover	4,900	− 800	− 3,300	+ 800	17,300	− 4,200	+ 300	+ 13,400
Westmoreland	8,600	− 1,200	− 5,000	+ 2,400	29,300	− 8,000		+ 21,300
St Elizabeth	13,300	− 6,200	− 6,500	+ 600	37,800	− 19,600	+ 2,700	+ 20,900
Manchester	10,500	− 5,100	− 6,600	− 1,200	31,900	− 10,200	+ 7,000	+ 28,700
Clarendon	12,800	+ 700	− 4,900	+ 8,600	41,300	− 2,700	+ 2,400	+ 41,000
St Catherine	9,200	+ 600	− 1,300	+ 8,500	31,700	− 900	− 6,300	+ 24,500
Total	103,800		− 77,100	+ 26,700	353,200		+ 25,800	+ 378,900

Note. The estimates of internal and external migration balances for the parishes are derived in Chapters 4 and 5. The figures of births and deaths on which the estimated natural increases are based are not corrected for usual place of residence. Both birth and death returns show that the errors involved here are small and are limited mainly to the Kingston and St Andrew areas. Moreover, these errors largely offset one another and the estimates of natural increase are believed to be fairly reliable. The slight discrepancies in the totals for 1921–43 are due to rounding.

internal movement represented a convergent migration into St Andrew, while of the net immigration of 25,800, a very large proportion (15,500) settled in that parish. On the other hand, outmigration greatly retarded population growth in all the other parishes.

POPULATION DENSITIES

One of the most important implications of population growth has been the rising densities it has produced for an island of only 4410 square miles, a large proportion of which consists of mountainous terrain. At the first census (1844) the density was 86 per square mile, and this has risen steadily during the ensuing century, amounting to 280 by 1943. According to the population of 1951 (1,443,700) the density was 327 persons per square mile.

It is instructive to compare the crude densities of Jamaica over the century covered by census records with densities of other West Indian islands. Such densities are shown for selected dates over this period in Table 11. In 1844 Jamaica was one of the least densely settled islands of the Caribbean; only Dominica and Trinidad showed lower densities, 74 and 37 respectively. A steady rate of population growth assured the island a density higher than that of Dominica by 1881, but Trinidad still remained a much less densely settled island. By 1921 population density in Jamaica (195 per square mile) was higher than Dominica, Antigua, and Trinidad, but still much lower than the high levels shown by Barbados, Grenada, Montserrat and St Vincent. The substantial population increases witnessed in most of the islands since 1921 resulted in further rises in densities, and by 1943–6 Jamaica ranked ninth in the scale of densities of the ten units. The most important change since 1921 has been the increase in the density of Trinidad to 282, that is, to a higher level than that of Jamaica. On the showing of crude densities, therefore, mounting population pressure on resources appears to be more acute in the eastern Caribbean than in Jamaica, and the importance of this is reinforced by the fact that in general, fertility in the eastern group is higher than in Jamaica.

There have been appreciable variations in density among the fourteen parishes of the island. The position of St Andrew demands special attention. Density here has since 1871 been higher than in any other parish except Kingston, but it has increased strikingly after 1911, that is, with the development of the parish as an urban centre. In 1943 the density of St Andrew (707) was four times what it was in 1871. At the time of the census of 1871 there were five parishes with densities under 100. Increasing population growth throughout the island reduced these to three in 1881 and two in 1891. By 1911 the density exceeded 100 in all parishes. Excluding

Table 11. *Densities (persons per square mile) in West Indian Islands at selected dates*

Island	1841–4	1881	1911	1921	1943–6
Jamaica	86	132	188	195	280
Trinidad	37	86	168	185	282
Barbados	735	1033	1036	943	1159
Grenada	217	319	502	499	544
St Lucia	90	165	209	221	301
St Vincent	182	270	279	296	411
Dominica	74	92	111	121	156
Antigua	215	205	189	175	245
St Kitts–Nevis	214	289	283	250	302
Montserrat	230	315	381	379	448

St Andrew, we see that in 1911 three parishes had densities in excess of 200 and by 1943 densities exceeded 200 in all parishes except Portland and Trelawny. No striking patterns of densities over the island emerge, but clearly the parishes with consistently the highest densities have been Hanover, Westmoreland and St Mary, while those with the lowest densities have been Portland, St Ann and Trelawny.

A measure useful in amplifying the statistics on population densities is the number of persons engaged in agriculture per square mile of cultivated land. The measures used here are calculated on the basis of males alone, though the considerable importance assumed by females in agricultural enterprise in the past might seem to suggest that both sexes should be considered. But, as will be argued in Chapter 3, this heavy involvement of females in agriculture in the past derives from slavery, and their subsequent dramatic withdrawal from it signifies probably no more than a break with slave tradition; it thus appears that the position of women in agriculture in the past might be easily exaggerated, and their inclusion in the calculation of agricultural densities would yield an unduly favourable rise in the efficiency of agriculture. Consequently it seems more satisfactory to base the measures on males alone.

From Table 12 it appears that with the substantial rise in the acreages of cultivated land there has been a marked reduction in the number of males employed per square mile of the cultivated area. Thus between 1891 and 1943 the total acreage more than doubled, whereas the number of males per square mile declined from 550 to 290. This emphasizes the appreciable rise in the efficiency of agriculture over the period. Most of this has occurred in the sugar industry. According to G. E. Cumper, the yield (tons cane per acre) increased from 0·9 in 1897 to 2·9 in 1943, while over the same period the man-days required to produce 1 ton of sugar declined from 155 to 40.[1] Even though nothing approaching this took place in the cultivation of other crops, the significance of such progress for the agriculture of the island as a whole should not be minimized.

Table 12. *Acreages of cultivated land and agricultural densities, 1881–1943*

Year	Acres of cultivated land	Males engaged in agriculture per square mile of cultivated land	Cultivated acres per head of the population
1881	123,300	—	0·21
1891	158,700	550	0·25
1911	265,400	380	0·32
1921	291,600	350	0·34
1943	382,900	290	0·31

Note. The acreages for 1881–1921 are from the records of tilled land on which taxes were paid in accordance with the Property Tax Law of 1868, and are published in the *Reports on the Excise, Internal Revenue and Customs Department*. These acreages refer only to land actually under crops and do not include pasturage or woodlands. Each entry is the average for 5 years centred on the appropriate census year. The acreage for 1943, which also covers only land under crops, is taken from Table 186 of the census report.

The number of males in agriculture per square mile of cultivated land cannot be calculated for years prior to 1891, as breakdowns of the gainfully occupied population by sex are not available for these years.

Another way of assessing the pressure on agricultural land of mounting population is in terms of cultivated acres per head of the population. This, in effect, reflects the capacity of the island to sustain its population from its own agricultural resources. This

[1] G. E. Cumper, 'Labour demand and supply in the Jamaican sugar industry 1830–1950', *Social and Economic Studies*, vol. 2, no. 4.

measure increased markedly between 1881 and 1911 from 0·21 to 0·32 acre per person. After 1911 the measure remained almost unchanged at about one-third of an acre. It should be added that as livestock has since 1921 come to play a more important part in the agriculture of the island the figure of 0·31 acre for 1943 understates the capacity of the island to support its population.

CHAPTER 3

MAJOR CHARACTERISTICS OF
THE POPULATION

Inevitably the demographic characteristics of a country must be powerfully influenced by the pattern of its historical development. This is very clearly demonstrated in the case of Jamaica. Above all, its long history of slavery, extending for 178 years of the three centuries of British rule, has had important and long-lasting effects on its population. Population growth in the days of slavery was, as we have already seen, largely determined by two historical factors, the slave trade and the policy of increasing the white population for reasons of security. But, as will appear in the course of this chapter, the influence of slavery was evident in many phases of population structure long after its abolition. Even demographic characteristics such as sex and age, which ultimately are to a large degree determined by strictly biological factors, did not escape the pervasive influence of the slave trade. Its effects on mating habits, perhaps the most indelible marks of slavery on the island's population, will be considered in Chapter 7.

But slavery is not the only historical element to be reckoned with in analysing the demographic position of the island. For instance, even though indenture immigration into Jamaica was of only small importance compared with the part it played in determining population movements in the eastern Caribbean, it still produced the small but important influx of East Indian and Chinese immigrants, which has affected not only the racial composition of the population, but also the economic and social structure of the island as well. Similarly, the decline of sugar as the staple of the island in the nineteenth century and the rise of the fruit trade were of some consequence for the demography of the island, in particular stimulating external migration, which in turn altered appreciably the sex composition of the population and retarded its rate of growth.

It is not intended here to treat explicitly the interrelation between historical development such as those outlined above and demographic movements in the island. But for the purpose of the analysis of its major demographic features offered in this chapter—

age structure, racial and sex composition, education and occupational status—it is well to recall the historical roots of many of the distinctive patterns that emerge.

AGE STRUCTURE

As Jamaica was never influenced by indenture immigration to the same extent as British Guiana and Trinidad, its population has never shown that heavy concentration of males of working age which has characterized the population of many eastern colonies of the West Indies. Nor has emigration to foreign territories been on a scale comparable to that which Barbados experienced. Uninfluenced therefore by exceptionally heavy migration movements, the population of Jamaica has throughout the century following 1844 shown an age structure relatively stable in composition. The age pyramids of Fig. 2, depicting the age structure of the island as a whole and of the urban centre Kingston for 1881 and 1943, emphasize this stability of age structure. On the other hand, the concentration of females between the ages of 20 and 30 in Kingston is well brought out here. In fact, the population of Kingston shows the typical age-sex structure of a modern city; the numbers are being constantly recruited from in-migration from rural areas while natural increase adds only comparatively small increments.

Nevertheless, it is instructive to examine the changing numbers in the various age-groups over the period covered by censuses. This can be conveniently done in terms of five broad age-groups: 0–4, 5–14, 15–29, 30–64 and 65 and over, though the age distributions given in 1844, 1861 and 1871 censuses are not sufficiently fine to permit the consideration of the movements of all these five age-groups over the whole period. The only group for which a whole century's records are available is that of children under 5. The groups 5–14 and 15–29 can be traced from 1861, while the 30–64 and the 65 and over groups can be traced only after 1881. The census populations in terms of these broad age divisions are shown in Table 13.

The first group, children under 5, has, in the absence of any marked variations in levels of fertility over the century, been influenced mostly by conspicuous reduction in death-rates among children and by the rising numbers of births consequent on the expansion of the numbers of persons of reproductive age. In 1844

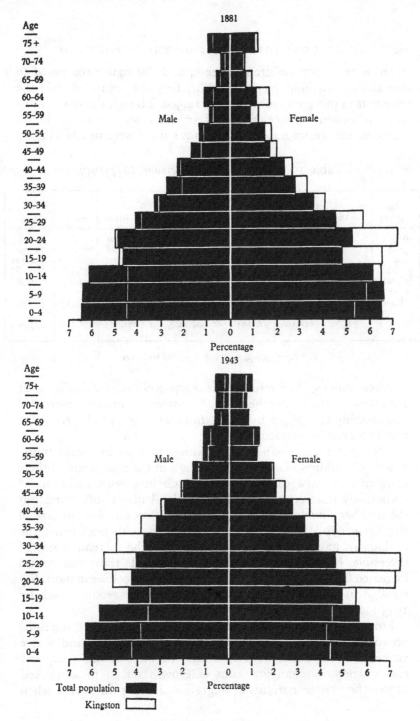

Fig. 2. Age pyramids of the populations of Kingston and of the total island, 1881 and 1943

there were 51,700 children under 5, and this figure continued to rise slowly, reaching 75,700 by 1881. Between 1861 and 1881 the numbers in this group increased by 14,500. Though the years 1881–1921 witnessed no acceleration of growth rates, the steady increments to this group meant that by 1921 there were nearly 112,000

Table 13. *Population by age-groups, 1844–1943*

Age-group	1844	1861	1871	1881	1891	1911	1921	1943
0–4	51,707	61,137	70,984	75,653	83,081	114,810	111,653	156,365
5–14	⎫	110,518	123,824	146,934	164,452	216,356	226,010	295,921
15–29	⎪	123,122	149,271	165,357	182,834	233,471	238,670	338,492
30–64	⎬ 274,018	115,873*	127,722*	160,193	183,916	237,109	249,336	394,183
65 and over	⎪	30,546†	33,029†	24,700	24,104	29,153	32,030	51,761
Unknown	⎭ —	—	1,324	7,967	1,104	484	419	341
All ages	325,725	441,196	506,154	580,804	639,491	831,383	858,118	1,237,063

* Aged 30–59.   † Aged 60 and over.

children under 5. As a result of the sharp declines in mortality after 1921, particularly in the case of infants, numbers increased considerably to 156,400 by 1943, thus reaching a level three times that of a century previously.

The second age-group (5–14) consists, at least in recent times, mainly of children of school age, though in the past many of these children were engaged in agricultural activities. Some children did accompany immigrants entering the island under indenture, but the numbers involved were too small to affect sensibly the size of this age-group, the controlling factors of growth here being the declines in mortality over the century and the increasing size of successive birth cohorts. Between 1861 and 1881 this group increased from 110,500 to 146,900. Subsequent declines in mortality resulted in much greater increments to this age-group, which by 1943 totalled 295,900.

From the standpoint of child-bearing and employment, the third group (15–29) comprises by far the most important and active elements of the population. It is also the most mobile part, and consequently its changing rates of growth reflect to a marked degree the various currents of migration. During the period when indenture immigration was in progress, this division of the population showed a high rate of growth, increasing from 123,100 in 1861 to 165,400 in 1881. The appreciable outward movement that developed after 1881 brought about a pronounced fall in the rate of growth. But with the cessation of emigration after 1921, considerable increments to this age-group have been recorded, and by 1943 there were 338,500 as compared with 123,100 in 1861.

Consisting mostly of mature members of the working population and those past the prime of their reproductive life, the fourth group (30–64) has shown a large expansion since 1851. As some of the immigrants introduced prior to 1881 were over 30 years of age, immigration probably meant appreciable increments to this group, but nothing definite can be said about movements here until after 1881. Doubtless there was some loss due to emigration between 1881 and 1921, but still the numbers increased in this period from 160,200 to 249,300. The reversal of the direction of emigration after 1921 and the declining mortality resulted in marked increments to this age-group. Particularly striking was the increase in the male population, which from a total of 114,200 in 1921 rose to 191,300 in 1943, an increase of 68% in 22 years. The female, much less subject to the effects of migration, experienced a smaller increase: from 135,100 in 1921 to 202,800 in 1943, an increase of 50%.

Increases in the dependent population over 65, those past working age and child-bearing age, can again be traced only after 1881. Between that date and 1921, the group increased slowly from 24,700 to 32,000. In common with the population of younger ages, the rate of growth of the dependent group accelerated considerably after 1921. In fact between 1921 and 1943, the population past working age constituted the most rapidly expanding element of the island's population. By 1943, the total over 65 stood at nearly 52,000 or about twice what it was in 1881.

Though no notable changes in age structure have been witnessed in the island, the slight declines in fertility in 1921–43 have resulted in a small reduction in the ratio of the dependent to the productive elements of the population. Thus the ratio of the dependent group (taken here as those under 15 and over 65) to the productive group (those aged 15–64) ranged from 78% to 74% during the period 1844 to 1921, but fell to 69% in 1943.

RACIAL COMPOSITION

Affected only to a limited extent by migration after the cessation of the slave trade, the population of Jamaica has since 1844 retained a greater degree of homogeneity than most of the other populations of the West Indies. But such significant contrasts emerge from comparisons between the basic Negro elements established by slavery and other smaller racial groups that a brief analysis of the racial composition of the island population is essential. Breakdowns by race in the censuses of 1844, 1861 and 1871 were limited to a threefold division into black, coloured and white, though from the data on place of birth some idea of the numerical importance of indenture immigrants at these dates can still be secured.

From 1881, when full racial breakdowns were first made available, up to 1943 the dominant position of the black and the coloured groups has been evident. As can be seen from Table 14, these two groups account for about 96% of the total population at all censuses. The proportion returned as black (Negro) has remained very steady since 1844, being highest (78·5%) in 1861 and lowest (75·8%) in 1911. The coloured element, composed mainly of the product of Negro and white intermixture, has likewise constituted a uniform proportion of the total. This proportion was greatest (19·8%) in 1871 and smallest (17·5%) in 1943. However, this low figure in 1943 is due to the inclusion in the East Indian and Chinese divisions of certain elements of the population which in the past might have been classified as coloured. In fact if these, termed respectively East Indian coloured and Chinese coloured, are included in the coloured element, the latter accounts for 18·3% of the total population, that is, the same as in 1921. Both the black and the coloured elements experienced considerable increases between 1844 and 1943. The Negro population of 1943 (966,000) was 3·3 times its size a century earlier (293,100). Similar in magnitude has been the increase of the coloured group, which increased from 68,500 in 1844 to 216,200 in 1943, a rise of more than threefold. It is, therefore, increases in the Negro and coloured elements that have determined population growth in the island.

In contrast to the two major racial groups of the population, the whites have shown no rise in numbers since 1844. In that year there were 15,800 whites in the island, the highest number ever recorded at a census. An irregular reduction in their numbers has continued

Table 14. *Racial composition of the population*

Year	Black No.	Black % Total	Coloured No.	Coloured % Total	White No.	White % Total	Chinese No.	Chinese % Total	East Indian No.	East Indian % Total	Others No.	Others % Total	Total No.
1844	293,128	77·7	68,529	18·1	15,776	4·2	—	—	—	—	—	—	377,433
1861	346,374	78·5	81,065	18·4	13,816	3·1	—	—	—	—	9	0·0	441,264
1871	392,707	77·6	100,346	19·8	13,101	2·6	—	—	—	—	—	—	506,154
1881	444,186	76·5	109,946	18·9	14,432	2·4	99	0·0	11,016	1·9	1,125	0·2	580,804
1891	488,624	76·4	121,955	19·1	14,692	2·3	481	0·1	10,116	1·6	3,623	0·6	639,491
1911	630,181	75·8	163,201	19·6	15,605	1·9	2,111	0·3	17,380	2·1	2,905	0·3	831,383
1921	660,420	77·0	157,223	18·3	14,476	1·7	3,696	0·4	18,610	2·2	3,693	0·4	858,118
1943	965,960	78·1	216,348	17·5	13,809	1·1	12,394	1·0	26,507	2·1	2,045	0·2	1,237,063

Note. The populations of the racial groups for 1943 are taken from Table 46 of the census report. Slightly different figures of racial composition are given in Table 48, which shows the following: black 965,944; coloured 216,250; white 216,250; Chinese 12,394; East Indian 26,507; others 2573.

ever since, the only notable departure from this downward trend being the small increase recorded between 1891 and 1911. By 1943 the white population (13,400) was 15% lower than it was a century earlier. As a consequence of this decline in numbers a reduction in the proportion of the population classified as white has been noted; this has fallen from 4·2% in 1844 to 1·1% in 1943.

Though East Indians were first introduced into the island in 1845, it was not until 1881 that these were recognized as a separate racial group; at this time there were 11,000 in the island. This number declined slightly to 10,100 in 1891 but has increased appreciably since then, being 26,500 in 1943. However, this increase has been partly the result of including 5100 coloured East Indians in the group. Clearly this group has been involved in appreciable miscegenation. As a result of the expansion of the numbers the proportion of the population classified as East Indian has risen slightly from 1·9% in 1881 to 2·1% in 1943.

The growth of the small but socially very important Chinese element has taken place mostly in the present century. Unlike the East Indian immigrants, who entered the island as indentured immigrants, the Chinese came as free immigrants. (There were, in fact, only two occasions on which indentured Chinese were introduced, once in 1854 when 472 were brought in and in 1884 when 680 were obtained; see Chapter 4.) The biggest increase in the Chinese population took place between 1891, when there were only 480 in the island, and 1911, when the number had risen to 2100. Here again much of the increase is to be explained in terms of increasing miscegenation, an inevitable consequence of the strong imbalance between the sexes among the early immigrants.

Several small groups of miscellaneous racial types appear for the first time in 1943. For instance, the Syrians, an important small group, number 1000; but it is not improbable that many others of this group may be confounded with the larger group termed Jews (1300).

Of special interest in an analysis of racial elements of the population is their geographical distribution. This can be conveniently treated in terms of three areas, Kingston, St Andrew and the rest of the island. The proportions found in these areas appear in Table 15. The predominantly rural character of the Negro population is evident. From 1861 to 1921 the proportion of this group living in Kingston remained between 6% and 4%, and in

Table 15. *Geographical distribution of the racial groups*

Year	% black in			% coloured in			% white in			% Chinese in			% East Indian in		
	Kingston	St Andrew	Other parishes	Kingston	St Andrew	Other parishes	Kingston	St Andrew	Other parishes	Kingston	St Andrew	Other parishes	Kingston	St Andrew	Other parishes
1861	5·7	5·4	88·9	13·9	4·7	81·4	32·6	6·0	61·4	—	—	—	—	—	—
1871	4·2	6·7	89·1	13·2	4·1	82·6	33·6	9·6	56·8	—	—	—	—	—	—
1881	4·3	6·3	89·4	12·7	4·6	82·7	34·6	10·3	55·1	84·9	4·0	11·1	1·9	4·3	93·8
1891	4·6	6·4	89·0	14·5	3·8	81·7	40·6	11·1	48·3	61·3	1·9	36·8	8·6	3·5	87·9
1911	4·6	6·3	89·1	14·5	6·1	79·4	37·8	15·1	47·1	35·7	9·4	54·9	2·8	2·8	94·4
1921	5·3	5·9	88·8	14·0	6·9	79·1	32·2	22·9	44·9	31·9	10·0	58·1	4·7	3·4	91·9
1943	6·9	8·8	84·3	16·5	13·7	69·8	12·5	47·9	39·6	33·5	16·8	49·7	7·2	13·3	79·5

contrast with the development among the rest of the population there has been only a slight increase to 7% by 1943. As already seen, the important feature of St Andrew, especially since 1921, has been its expansion as an urban centre. The strongly rural character of the black population is therefore emphasized by the fact that only 9% of its total had by 1943 established themselves in this parish.

The coloured population has always shown a much higher concentration in urban areas than the black. Between 1861 and 1921 the proportion located in Kingston remained fairly steady, between 13% and 14%, increasing slightly to 17% in 1943. But the most striking shift in the location of the coloured people has been their growing concentration in St Andrew; the proportion established here rose from 7% in 1921 to 14% in 1943. So that about 30% of the coloured population can now be classified as urban dwellers, as compared with only 16% for the black.

As is to be expected, the socially dominant white group has been much more strongly urbanized than any other. Prior to 1943 the proportions located in Kingston were very high by comparison with those for other groups, being greatest (41%) in 1891 and lowest (32%) in 1921. It is interesting that the proportion living in St Andrew showed a rise much earlier than in the case of other racial groups, increasing from 11% in 1891 to 15% in 1911 and to 23% in 1921. By 1943 nearly half were located in this parish. Accompanying this rising concentration in St Andrew there has been a withdrawal from Kingston. In 1891, 41% of the white population lived in Kingston; but the decline since then has been marked, particularly between 1921 and 1943, when it declined from 32% to 12%. As will be shown in Chapter 5, the extensive development of St Andrew as a suburban area did not begin until after 1921, but clearly the higher social European groups were beginning to settle in this parish long before 1921.

Even the small numbers of Chinese in the island before 1911 suggest that from the first these people have sought to establish themselves in urban areas. Coming as free immigrants they were in a much better position to do so than the East Indians. Later developments, however, indicate a spread of their commercial and trading activities to other parts of the island; there has, in fact, been a rising proportion of Chinese located in rural areas up to 1921. But their essentially urban location remains. In 1943 half of the population were living in the Kingston–St Andrew area, a higher proportion than for any other group except the whites.

Introduced as indentured labourers for the sugar estates, the East Indians have up to 1921 remained largely rural, between 88% and 94% of their numbers being found outside Kingston and St Andrew during these years. Between 1921 and 1943, however, there has been a marked influx into the urban areas. Whereas only 3% lived in St Andrew in 1921, the proportion so located increased to 13% by 1943.

To sum up the present geographical distribution of the racial elements, it can be said that the white remains the most urban, with 60% of their numbers in Kingston or St Andrew. The Chinese group shows the second highest urban concentration with 50% located in the urban area. Third in terms of urban concentration comes the coloured group, 30% of whom live in urban areas. The comparatively rapid urbanization of the East Indians is illustrated by the fact that though brought into the island exclusively as agricultural labourers, they are now more highly urbanized than the black group. In 1943, 20% of the East Indians were urban dwellers as compared with 16% for the blacks; the latter thus remain the predominantly rural element of the population with 84% living outside the main urban areas.

The homogeneous racial composition of the population of Jamaica is again reflected in the extremely small numbers returned as foreign-born. Indeed, with the exception of 1844 when there was still an appreciable number of African-born ex-slaves in the island, the proportion of the native-born population has been very high. From 88% in 1844, it increased to 96% in 1861 and has increased slightly at later censuses, being just under 98% between 1891 and 1943. As can be seen from Table 16, the African-born population has declined drastically in number since 1844, the number in 1861 being less than one-third of those returned at 1844, despite the appreciable introduction of liberated Africans in the intercensal period. In 1844 there were nearly 8000 persons born in the British Isles, but this number also declined steeply to less than 3000 in 1861. The relatively stable numbers at later censuses, however, emphasize the fact that there has been a small but steady influx of Europeans throughout the past century. Between 1871 and 1921 the numbers of foreign-born Indians ranged from 5000 to 9000, but with the cessation of the indenture movement the numbers declined rapidly to 2200 in 1943. The only racial group among those considered here which showed an increase in the size of its foreign-born section is the Chinese. Whereas Chinese of foreign birth amounted to only 300 in 1891, the number increased to 2400 in 1921. Between 1921 and 1943, however, the increase was small. Still the Chinese constitute the second largest racial group of foreign-born population in the island at 1943. The negligible numbers of the population born in other British Caribbean colonies show that at no time was there any significant interchange of population between the eastern and western colonies of the Caribbean; in 1943 only 1600 of the island's population were returned as natives of other British Caribbean territories. The fact that by far the largest contingent of foreign-born population in 1943 were Cubans shows the growing links between Jamaica and Cuba in recent years. Natives of Cuba amounted to 6700, or more than one-quarter of the total foreign-born population.

Table 16.   *Important elements of the foreign-born population*

Year	Africa	British Isles	B.W.I.	India	China	Cuba
1844	33,519	7,960	—	—	—	—
1861	10,515	2,955	247	2,262	239	40
1871	6,859	2,285	604	7,793	141	750
1881	4,402	2,715	750	8,695	140	1,005
1891	1,839	2,452	985	5,193	347	459
1911	591	2,508	1,374	7,797	1,646	245
1921	237	2,410	1,004	7,138	2,413	591
1943	191	2,063	1,563	2,245	2,552	6,713

It is instructive to note that the pattern of foreign-born population of 1943 differs markedly from that of past censuses. Though the numbers remain small (25,800), the increased numbers from Central America and Cuba bear witness to increasing contacts with Latin American countries.

SEX COMPOSITION

It is known that the demand for male slaves in the heyday of slavery resulted in a preponderance of male slaves, but the limitations of the data make it impossible to estimate the extent of this imbalance between the sexes. Though this imbalance was the direct result of the policy of relying on the slave trade rather than on reproduction as a means of recruiting the slave population, many contemporary writers naïvely ascribed the low rates of natural increase at the time to the shortage of females. Some also saw in this imbalance a cause of the 'immoral' aspects of Negro conduct under slavery, promiscuous mating and infrequency of marriage.

With the cessation of the slave trade and the ensuing policy of stimulating reproduction as the only means of maintaining the plantation labour force, a reduction in the preponderance of males was inevitable. This is borne out by the first available data on sex composition of the Jamaican population, the slave registers. At the time of the first registration (1817) there was still a small excess of males, the sex ratio being 1003. The decline in subsequent years has been steady. The sex ratio was 992 in 1820, 982 in 1823, 966 in 1826 and 964 in 1829. It is important to note that the declining sex ratio at this time cannot be attributed to differential rates of manumission between the sexes. Indeed, more females than males were manumitted. Between 1817 and 1829 the number of females manumitted amounted to 2566, as compared with 1445 for the males.

The excess of females that emerged in the last years of slavery has distinguished the Jamaican population ever since. The decline in the sex ratios noted between 1817 and 1829 continued, and at the time of the first census there were only 928 males to every 1000 females. A slight rise in the sex ratios followed the introduction of indentured immigrants, the ratios increasing to 938 in 1861 and to 950 at the two succeeding censuses. With the reversal of the direction of net migration after 1881, which involved a much greater loss of males than females to the island, the sex ratio declined and by 1921 was down to 881, the lowest ever recorded. But with the cessation of migration after 1921 the imbalance between the sexes was redressed somewhat, and by 1943 the sex ratio had risen to 937, nearly the same as it was in 1861.

Sex ratios for several broad age-groups are shown in Table 17. Clearly ratios are lowest for the ages over 55, that is, in age-groups where the death-rates are more favourable to the females. The greatest changes occur in the ratios for the age-groups 15–29 and 30–54, those most strongly influenced by migration. Thus within the period of indenture immigration sex ratios for these ages tended to be high. But the strong outward movement witnessed between 1881 and 1921 resulted in sharp falls in these ratios. Following the halt in emigration after 1921, the deficiency in males between the ages of 15 and 55 was drastically reduced. In fact, the sex ratio for the age-group 30–54 in 1943 (960) was the highest ever recorded for this group. Clearly these sex ratios underline the extent to which migration affected the composition of the population of working and child-bearing age.

Table 17. *Sex ratios (males per 1000 females) by age-groups*

Age	1881	1891	1911	1921	1943
		Total island			
0–4	991	998	994	1003	1003
5–14	994	1016	1011	999	1008
15–29	929	849	851	791	878
30–54	950	901	890	846	960
55 +	848	787	811	781	787
Total	950	917	916	881	937
		Kingston			
0–4	857	839	998	962	977
5–14	744	762	782	808	845
15–29	700	707	671	671	695
30–54	706	688	756	758	835
55 +	508	427	466	470	560
Total	704	708	723	721	775

An excess of females appears in all parts of the island but is greatest in Kingston, thus reflecting the almost universal pattern of female concentration in urban areas. In 1871 the sex ratio for Kingston stood at 682, whereas that for the rural areas (excluding St Andrew) was 972. That the sex-selective emigration occurring between 1881 and 1921 affected the rural areas more than the urban is clear from the fact that between 1881 and 1921 there was a steady decline in the sex ratio of the rural areas (excluding St Andrew) from 970 to 901. On the other hand, the sex ratio of Kingston during these years increased from 704 to 721, and by 1943 amounted to 775. This decline in female concentration in Kingston, when viewed in conjunction with the growing concentration of females in St Andrew, does not signify any fundamental change in sex composition of the urban population, but merely the spread of the urban centre beyond the boundaries of Kingston. Actually the sex ratio for the suburban population of St Andrew in 1943 (774) is almost the same as that for Kingston (775).

Another important aspect of sex composition is the sex ratios among the several racial groups recognized at the censuses. As Table 18 clearly indicates, it is the coloured group that shows the lowest ratios; these range from 799 in 1921 to 919 in 1871. Somewhat higher are the ratios for the black group, which range from 885 in 1921 to 957 in 1943. As is to be expected, both of these racial groups show an increased proportion of females at the censuses of 1891–1921, that is, during the period of heavy sex-selective emigration. Though the strong preponderance of females is confined to the population of Negro descent, it is interesting to note that in the case of the white group the sex ratio has declined markedly within the past 50 years. In fact the 1943 census shows a sex ratio of 836, which is almost identical with that shown by the coloured group (832). This decline in the proportion of males among the white probably derives from more of the female coloured population electing to class themselves as white, presumably because of the enhanced social status such identification usually assures; it is unlikely that it is due to any fundamental alteration in the sex composition of the white element. The preponderance of males has been most marked in the case of the Chinese among whom female immigrants were rare until 1921. However, the sex ratio among this group had by 1943 been reduced to 1265. Most of the East Indians entering the island were also males, and the sex ratio among these during the period of indenture immigration was high, ranging from 1703 in 1881 to 1214 in 1921. By 1943 this imbalance had largely disappeared; the sex ratio was down to 1023.

In general the overall sex composition characteristic of each racial group tends to obtain throughout the island. But certain

Table 18. *Sex ratios (males per 1000 females) among racial groups*

Year	Black	Coloured	White	East Indian	Chinese
1844	923	858	1432	—	—
1861	934	892	1119	—	—
1871	952	919	1116	—	—
1881	950	875	1124	1703	4211
1891	936	800	1139	1330	3454
1911	919	838	1104	1332	5436
1921	885	799	1023	1214	3205
1943	957	832	836	1023	1265

differences in the concentration of females in the urban area are to be noted. The population of Negro descent has long shown a high proportion of females in the urban areas. Thus between 1871 and 1943 sex ratios for the black group in Kingston ranged from 676 in 1911 to 801 in 1943. Even more pronounced has been the preponderance of females among urban coloured dwellers; here sex ratios over the same period ranged from 567 in 1891 to 673 in 1943. Up to 1911 there was a considerable excess of males among white urban dwellers, but this disappeared after 1921. By 1943 the sex ratios were down to 892 (Kingston) and 744 (St Andrew); this, as has already been noted, may be merely the result of the increasing indefiniteness of the division between coloured and white. Both the East Indians and the Chinese show that entry into urban areas was up to 1921 largely confined to males. But by 1943 the East Indians showed an excess of females in Kingston (888) and St Andrew (938). Though the imbalance between the sexes among the Chinese has been considerably reduced in recent years the excess of males in the urban population persists, the sex ratios in 1943 being 1306 in Kingston and 1141 for St Andrew. In fact the Chinese now constitute the only exception to the rule that females tend to outnumber males in the urban areas.

Three factors can be identified as being associated with the imbalance between the sexes in the population of Jamaica: the low sex ratios at birth, the mortality differentials between the sexes and sex-selective emigration. Mortality differentials and the course of external migration will be taken up in later chapters. It is necessary to consider here, however, the important question of the sex ratio at birth. This is a characteristic which the Jamaican population, and indeed all the non-Indian populations of the West Indies, share with the United States non-white populations. It is also interesting that the so-called birth statistics of the slave registers revealed the same feature. The 'births' between 1817 and 1829 showed a sex ratio of 1013. These low sex ratios were general throughout the West Indies; only three of the smallest islands (Montserrat, Nevis and the Virgin Islands) showed sex ratios in keeping with what now obtains for European populations; here the ratios were 1065, 1065 and 1072 respectively. It must be stressed, however, that though highly suggestive of the persistence of a supposedly biological phenomenon of racial basis, the slave data might be to an unknown degree affected by differential mortality under 3.

According to the registration records male births in Jamaica have never exceeded female births by more than 4·7%. In 1880 and 1895 sex ratios of 1047 were returned, the highest ever recorded. On three occasions female births exceeded male births: in 1883, 1885 and 1928, when the sex ratios were 995, 991 and 997 respectively. The total births registered between 1878 and 1950 yield a sex ratio of 1023·46. This is very close to the ratio among United States Negro births during 1915–48 (1029), but much lower than the ratio among United States white births over the same period (1056).[1] Other West Indian populations show sex ratios at birth closely paralleling that for Jamaica:

Population	Period	Sex ratio
British Guiana	1883–1950	1030·52 (excluding East Indians)
Trinidad	1901–1950	1025·25 (excluding East Indians)
Grenada	1878–1950	1014·02
Barbados	1891–1950	1024·66
Dominica	1909–1950	1017·49

The combined effects of low sex ratios at birth and marked differentials in mortality between the two sexes can be conveniently demonstrated by life-table populations constructed on the basis of radices which are in the same ratio as the sex ratios at birth during the 5-year periods covering the various census dates and on the basis of appropriate life-table mortalities. Sex ratios for stationary populations of this nature are compared with sex ratios among the actual populations in Table 19. The former show that as a result of sex ratios at birth and mortality patterns a preponderance of females in the population is to be expected. It also appears that this expected preponderance is larger than would be secured for non-Negro populations in general. This is emphasized by the fact that the stationary population for Jamaica (1946) shows an excess of females from age 42, which is close to the corresponding figure for the United States Negroes for 1930 (36), but much lower than the corresponding age for the United States white population of the same year (50).[2] As will be seen in Chapter 6, the mortality differentials by sex do contribute to the maintenance of the essentially feminine population of the island, but it remains obvious

[1] C. A. McMahan, 'An empirical test of three hypotheses concerning the human sex ratio at birth in the United States, 1915–48', *Milbank Memorial Fund Quarterly*, July 1951.

[2] See J. Yerushalmy, 'The age-sex composition of the population resulting from natality and mortality conditions', *Milbank Memorial Fund Quarterly*, January 1943.

that, as in the case of the United States Negro population, it is the low sex ratio at birth that is the chief factor determining this important characteristic of the population.[1]

Table 19. *Sex ratios (males per 1000 females) in stationary and in actual population*

Year	Sex ratio in stationary population	Sex ratio in actual population
1881	956	950
1891	991	917
1911	965	916
1921	972	881
1946	956	937*
1951	967	946

* Ratio according to 1943 census population.

However, an analysis of Table 19 shows that the wide divergence between the sex ratio in the actual and in the stationary populations cannot be accounted for by sex ratios at birth and mortality patterns, but are due largely to sex-selective emigration from the island. Thus in 1881, before strong outward movements had commenced, there was very little difference between the two sex ratios; 956 in the case of the stationary population and 950 in the case of the actual population. But as emigration increased the divergence between the two ratios became marked; by 1921 the sex ratio for the actual population (881) was considerably lower than that for the stationary population (972). With the cessation of emigration the imbalance between the sexes has been progressively reduced, and by 1951 there was very little difference between the sex ratios of the actual and the stationary populations, the former being 946 as against 967 for the latter. With most avenues of emigration now closed it is highly unlikely that any severe imbalance between the sexes will reappear, though as a consequence of factors which are both biological and social in nature there will always tend to be an excess of females in the island's population.

EDUCATIONAL STATUS

As is the case with so many features of West Indian populations, the present educational status of Jamaica must be viewed against

[1] Cf. T. Lynn Smith, *Population Analysis*, New York, 1948, p. 125.

the historical background of slavery. The provision of education was alien to the slave codes; for the performance of plantation labour called for no more than a people sufficiently robust to withstand a long and arduous daily regime of work. And unlike ancient slavery it was unnecessary to cultivate a class of educated slaves, since such duties on the plantation as required an educated personnel were delegated to a lower order of European worker, the overseer and the bookkeeper. Moreover, as J. Stewart expressed it, education of the slaves, though desirable 'as a means of accelerating moral and religious instruction', was considered dangerous, as educated Negroes would 'feel their strength' and the safety of the white population would then be threatened.[1]

Isolated cases of slaves and free men of colour receiving education were known. Edward Long recounts the case of Francis Williams, a Negro freeman, 'a boy of unusual lively parts [who] was pitched up to be the subject of an experiment which... the Duke of Montagu was curious to make, in order to discover whether by proper cultivation and a regular course of tuition at school at the University a Negro might not be found as capable of literature as a white person'. The result of this experiment, as Long interpreted it, constituted 'an unfortunate example to show that every African head is not adapted by nature to such profound contemplation'.[2] White fathers frequently had their mulatto offspring educated in England, a practice which contributed materially to raising the social status of the coloured classes in the days of slavery. But even the more enlightened slave laws passed during the last 20 years of slavery did not comprehend an educated class of slaves.

In its Report, presented in 1836, the Select Committee of the House of Commons on Apprenticeship listed among its criticisms of that institution the lack of facilities for education. It was reported that though free children in the town were being sent to school estate managers were averse to this practice.[3] But many improvements under apprenticeship reported by the magistrates, to the effect that churches and schools were 'crowded', were probably exaggerations designed not to present apprenticeship in a too unfavourable light. It is of interest that provisions for education

[1] J. Stewart, *View of the Past and Present State of the Island of Jamaica*, Edinburgh, 1823, p. 343.
[2] Edward Long, *The History of Jamaica*, London, 1774, vol. II, pp. 375–8.
[3] *Extracts from Papers printed by Order of the House of Commons, 1839, relative to the West Indies*, London, 1840, p. 171.

were not among the categories of 'laws calculated to meet the new exigencies of society' outlined in Lord Glenelg's despatch of 29 October 1838.[1] The development of public policy on education came at a later date. Still considerable rises in the educational status of the island have been witnessed throughout the nineteenth century. A convenient way of tracing these improvements is in terms of the population able to read and write. In keeping with the approach adopted in the 1943 census this is here measured in terms of the proportions over 5 able to read and write. In 1861, the first year for which such returns were made available, the literacy proportion in these terms amounted to 13·3% for the island and this increased slightly to 16·3% in 1871. The subsequent improvement is traced in Table 20. By 1943 the proportion had risen to 67·9%.

Table 20.   *Literacy proportions (percentage of population over 5 able to read and write)*

Parish	1871	1881	1891	1911	1921	1943
Kingston	40·4	51·2	59·2	76·9	79·4	88·6
St Andrew	14·1	23·0	32·7	56·6	64·3	82·9
St Thomas	7·9	14·7	18·6	41·0	48·7	68·5
Portland	15·4	19·6	31·7	51·4	55·1	72·4
St Mary	15·7	14·3	29·1	45·3	51·8	66·3
St Ann	17·7	19·0	33·0	44·7	49·1	62·4
Trelawny	17·3	25·0	35·0	44·7	47·6	60·2
St James	18·4	27·3	32·5	47·7	54·0	67·0
Hanover	17·5	21·1	29·1	41·9	45·6	63·2
Westmoreland	16·9	22·2	30·3	42·3	43·9	61·3
St Elizabeth	13·6	20·3	27·4	40·2	42·2	58·4
Manchester	8·4	25·2	32·0	46·7	54·5	65·6
Clarendon	13·5	20·3	26·9	43·1	49·8	59·9
St Catherine	13·8	19·1	27·8	40·7	46·5	62·7
Total	16·3	22·8	32·0	47·2	52·2	67·9

Literacy proportions for the fourteen parishes show that throughout the period for which data are available Kingston has been in the most favourable position. Indeed, before 1911 the difference between Kingston and the other parishes was striking. For instance, in 1881 the proportion for the former stood at 51%, whereas that for the second highest parish (St James) was only 27%. Two important features appear over the period 1881–1943. In the first

[1] Ibid. p. 341.

place the gap between Kingston and the other parishes has narrowed considerably. Thus in 1891 the literacy proportion in St Elizabeth, consistently a parish of low educational attainment, was less than half that of Kingston. By 1943, however, though still the lowest, St Elizabeth showed a literacy proportion (58%) much nearer to that of the capital (89%). In the second place the parishes nearest to Kingston have in general shown greater improvements than the more distant ones. In particular, the relative positions of St Andrew (where literacy rates are now similar to those of Kingston), St Thomas and Portland have registered impressive gains in literacy. Similarly, the present relatively favourable position of St James probably derives from the growing urban centre (Montego Bay) located in this parish.

As the standard of literacy presents one of the most significant indications of the social advancement of a population it is fitting to compare the latest literacy rates of Jamaica with rates for other countries. This can be conveniently done in terms of the proportions of the population over 10 returned as illiterate (unable to read and write). At the last census the proportions throughout the West Indies ranged from 7·3% for Barbados to 29·1% for the Windward Islands.[1] With the exception of the last mentioned group all the territories of the British Caribbean show illiteracy rates lower than that for Jamaica (23·9%). It is, however, necessary to make a broader assessment of the literacy position of the island. Of course, by comparison with North American and European populations, where illiteracy is now of negligible dimensions, the illiteracy proportion shown by Jamaica stands high. But comparisons with less highly developed countries place Jamaica in a much more favourable position. Thus the proportion for Jamaica is very close to that of Cuba (22·1%), and much lower than the proportions for Venezuela (56·6%), Mexico (51·6%), Panama (35·3%) and Puerto Rico (31·5%). Comparisons with more distant countries also show Jamaica in a favourable position. For instance, the illiteracy proportion for Ceylon (42·2% for age 5 and over) is nearly twice that for Jamaica, while that for Egypt (85·2%) is again many times the level of Jamaica.[2]

Overall rates of literacy or illiteracy give only a partial picture of the educational level of a population. It is necessary in this context

[1] See *West Indian Census, 1946*, vol. I, p. 44.
[2] See United Nations, *Demographic Yearbook*, 1948.

to consider the quality of the education received. Data bearing on this subject, secured for the first time at the 1943 census, emphasize that the educational attainment of a very large section of the population of Jamaica has not gone further than the elementary level. Of the total population over 10 years, 218,000, or 24%, did not attend school (see Table 21). But it is important to note that of the population who did attend school 94% went no higher than the level of the elementary school. Actually 72% of the adult population of the island received elementary schooling. Though this proportion was somewhat higher in the case of Kingston (85%) and St Andrew (74%), the proportion in the rural areas was also considerable (70%). But it is in regard to the higher forms of education that the population appears in a less satisfactory position. Only 25,500 of the population over 10 received any secondary education, equivalent to a proportion of 28 per 1000. Moreover, the

Table 21. *Population over 10 who received various levels of education, 1943*

Parish	Elementary schooling	Secondary schooling	Practical training	Professional training	No schooling
Kingston	75,183	5,723	2,348	719	4,852
St Andrew	75,622	10,155	2,591	1,875	11,464
Other parishes	511,061	9,660	3,845	3,829	201,811
Total	661,866	25,538	8,784	6,423	218,127

disparity between the rural and urban areas is striking. In St Andrew, evidently the seat of the elite of the island's population, 10% of the population over 10 attended secondary schools. The proportion in Kingston, though lower (6%), is still considerably in excess of that shown by the rural areas (1·3%). In fact 62% of the population who attended secondary schools were located in Kingston and St Andrew. The census also lists persons who received professional and pre-professional training, and here also the proportions are low. It is further possible that many of these professionally trained persons were not natives of the island. Only 6400 persons were professionally trained, or less than 7 per 1000 of the adult population. These were much more evenly distributed throughout the island than those attaining secondary levels of education. Most of the professionals were located in St Andrew, where 18 out of every 1000 persons were so classified, but the corresponding proportion in Kingston (8 per 1000) was not markedly in excess of the proportion in the rural areas (5 per 1000). A fourth category of education listed in the census was 'practical training', an ill-defined category which might indicate those trained in some occupational skill. These numbered 8800 and were highly concentrated in Kingston and St Andrew, where 56% of the total were located.

Though the foregoing emphasizes that, particularly in the rural areas, much remains to be done in advancing the level of education in the island, there has evidently been a continuous rise in the proportion of children attending school. By 1943 66·4% of the children aged 5–14 were at school. The rise in this proportion, together with the general population growth, has resulted in marked increases in the numbers attending school, as can be seen from Table 22. Between 1891 and 1943 the school population more than doubled, increasing from 99,800 to 206,500. By far the greatest increase occurred in St Andrew, where in fact the school population of 1943 (20,900) was 3·6 times the level of 50 years previously. Such has been the advance in schooling in the parish of St Andrew that by 1943 more than one-tenth of the school population of the island was located here, whereas in 1891 less than 6% were to be found in this parish. In Kingston also there has been an appreciable rise in the numbers attending school, from 6500 in 1891 to 17,300 in 1943. The progress in rural areas has been much less; the school population of 1943 (168,400) was less than double that of 50 years earlier. An interesting feature of the school population is its sex composition. Throughout there has been an excess of females. The imbalance between the sexes, though less striking than in the actual population, still reflects its fundamentally 'feminine' character.

A fairly wide range of educational data for 1943 is given in terms of ages 5–24. But in view of the fact that education other than that at the primary level is centralized mostly in the Kingston–St Andrew area, it seems more appropriate to consider school attendance for the age-group 7–14, which encompasses most of the children attending elementary schools. Several aspects of educational status based on this age-group appear in Table 23. Again the proportion attending school is highest in Kingston and the surrounding parishes and comparatively high in St James. Again St Elizabeth appears as the most unfavourable. However, an important qualification

Table 22.   *Growth of population attending school*

Year	5–9		10 and over		All ages		
	Male	Female	Male	Female	Male	Female	Both sexes
			Kingston				
1891							6,452
1911	2,184	2,156	2,192	3,156	4,376	5,312	9,688
1921	2,134	2,426	2,305	3,162	4,439	5,588	10,027
1943	3,344	3,775	4,464	5,748	7,808	9,523	17,331
			St Andrew				
1891							5,863
1911	1,838	1,817	2,480	2,194	4,318	4,011	8,329
1921	2,021	2,104	2,550	2,744	4,571	4,848	9,419
1943	3,930	4,124	6,236	6,567	10,166	10,691	20,857
			Other parishes				
1891							87,454
1911	24,151	24,336	29,839	29,153	53,990	53,489	107,479
1921	22,286	22,960	31,576	33,355	53,862	56,315	110,177
1943	30,527	32,148	52,056	53,628	82,583	85,776	168,359
			Whole island				
1891							99,769
1911	28,173	28,309	34,511	34,503	62,684	62,812	125,496
1921	26,441	27,490	36,431	39,261	62,872	66,751	129,623
1943	37,801	40,047	62,756	65,943	100,557	105,990	206,547

Note. Age breakdowns of the population attending school were not available for 1891.

of the data on school attendance must be noted. It appears that not all the children classified as attending school were at school during the months of September to December 1942, a period taken to measure the average attendance throughout the island. As can be seen, though the proportion returned as not at school during these 4 months amounts to only 6% of the total school population in the case of Kingston, it attained much higher dimensions in other parishes. In St Ann, Trelawny, Clarendon and St Catherine, one-fifth of the school population did not attend school in the last session of 1942. When these children are excluded from the school population, the proportion attending school drops drastically, thus emphasizing the fact that in view of the infrequent attendance of many children current statistics of school enrolment are largely artificial. On this basis the proportion attending school throughout the island is reduced to 68%. Among the parishes the proportion is lowest in the case of Trelawny (61%) and highest for

Kingston (89%). When the educational position of the island is considered in terms of the average number of months of school attendance per head of the school population for September to December 1942, a striking difference between the urban and the

Table 23. *School attendance, 1942–3*

Parish	% population 7–14 attending school*		Average no. of months at school Sept.–Dec. 1942	% of school population not at school Sept.–Dec. 1942
	(a)	(b)		
Kingston	94·6	89·0	3·29	5·9
St Andrew	88·7	78·8	3·00	11·1
St Thomas	84·3	69·7	2·54	17·3
Portland	85·7	72·0	2·55	16·0
St Mary	80·8	65·9	2·56	18·5
St Ann	79·9	63·9	2·46	20·0
Trelawny	76·1	60·6	2·54	20·3
St James	82·3	69·4	2·71	15·7
Hanover	80·3	69·1	2·68	13·9
Westmoreland	75·2	64·1	2·71	14·9
St Elizabeth	72·9	62·1	2·56	14·8
Manchester	79·2	67·8	2·69	14·4
Clarendon	77·5	61·4	2·49	20·9
St Catherine	77·8	61·7	2·46	20·7
Island	80·6	67·6	2·66	16·1

* Proportions under (*a*) include all children returned as attending school. Proportions under (*b*) exclude those children who, though returned under the school population, did not actually attend school in the period September to December 1942.

rural districts emerges. The average school child in Kingston showed an attendance of 3·3 months, while in St Andrew the figure stood at 3·0 months. Elsewhere the figures range from 2·7 (in St James and Westmoreland) to 2·5 (for St Ann and St Catherine).

Racial breakdowns of educational status were first presented in 1943. A convenient way of comparing the racial groups is in terms of the proportion aged 5–24 at school. The levels are such as to be expected in view of the close association between race and social status in the island, though as a result of the broad age-group involved the white group is not the most favourably placed. The Chinese show the highest proportion at school (57%). Second in importance comes the white, with 54% at school, and third the coloured, with 43% at school. It is significant that the proportion of East Indians at school (36%) is only slightly lower than that shown by the black group (38%), thus attesting to the rapid progress made by the small East Indian group in improving their social position in the island.

An effective way of summarizing the schooling experience of the island is by means of a life-table application. By applying the proportions at school to the stationary populations of the 1945–7 life-tables two instructive measures can be computed, the average age at entry into school and the average number of years' schooling a new-born child can expect during the course of its life from age 5 to age 20. These are shown for the most favourably placed parish,

Table 24. *Average age at entering school and average number of years of schooling per new-born child, 1943*

Parish	Average age at entering school	Average no. of years of schooling between ages 5 and 20 per new-born child
Kingston	6·5	8·1
St Elizabeth	8·1	5·3
Whole island	7·8	6·0

Kingston, for St Elizabeth, consistently one of the most unfavourably placed, and for the island as a whole in Table 24. The average age at entering school is considerably lower in Kingston (6·5 years) than in St Elizabeth (8·1 years). A much more striking difference appears in the case of the years of schooling to be expected by a new-born child. The value for Kingston (8·1 years) is 2·8 years above that for St Elizabeth.

Manifestly the expansion of educational facilities will loom large in the near future. This, as we shall argue in Chapter 9, must be the case in virtue of the considerable increments to the population of school age to be expected as a result of population growth. But if in addition to keeping pace with population growth it is desired to make available in rural areas a greater range of educational facilities, especially in regard to secondary education, then even greater efforts are called for. For in this case due account will have to be taken of the position in the rural areas: prevailing low literacy rates, the low proportions of children attending school and the large numbers who attend school only irregularly. And special efforts will have to be made to provide the added facilities which will be required as these limitations are removed.

OCCUPATIONAL STATUS

In dealing with the occupational status of the population we are again forced to consider briefly the possible influence of slavery on the patterns revealed during the century following 1844. The concept of a labour force of course had no meaning in the context of slave society. As Jaffe and Stewart rightly point out, 'Where working force status derives from social authority, there is no phenomenon or mystery requiring explanation.'[1] Virtually every able-bodied adult was, under slavery, impressed into service, as the compensation returns of 1834 illustrate. These indicate that with the exception of children under 6, who numbered 39,000, run-away slaves numbering 1100 and a group termed 'aged, diseased or otherwise non-effective', all slaves were considered 'work-units'. Of a total slave population of 311,100 compensation was paid on 17,100 head people, 187,700 field labourers, 16,100 tradesmen, 2300 employed on the wharves and 32,000 domestic servants. The fact that 82% of the population were in one capacity or another considered to be at work underlines one of the basic elements of slavery, the exaction of the maximum amount of labour from the population. This feature of slavery Richard Hill characterized as one of its 'heavy incumbrances'. Moreover, 'To have them available when a multitude was wanted, they were subsisted and employed at times when their labour was little profitable.'[2] That the transition to a completely free economy was gradual is abundantly demonstrated in the slow decline in the proportion of the population (particularly the females) employed. It is worth noting that the introduction of indenture labour long militated strongly against the formation of a truly free and mobile labour force on the sugar plantations and indeed in the colonies as a whole.

Unfortunately, in analysing occupational data we enter a domain in which census material proves often treacherous and unrewarding. Changing concepts of the working population, changing definitions of its major classes, and the persistent attempts to fit the essentially simple occupational pattern of the island into elaborate classifications, more suitable to countries on the road to full industrialization, impose severe limitations on the available data. As will be seen, these

---

[1] A. J. Jaffe and C. D. Stewart, *Manpower Resources and Utilization*, 1951, p. 2.
[2] Report from Stipendiary Magistrate Richard Hill, enclosure in despatch from Earl of Elgin to Lord Stanley, dated 5 January 1846, *P.P.*, vol. xxviii, 1846.

limitations apply with particular force in the case of agricultural classes. It is probable that not all the occupational categories listed in the Jamaica censuses, particularly those of the nineteenth century, cover persons continuously engaged in the production of marketable goods and services, but, with certain exceptions to be noted presently, all the census categories are taken here as forming part of the island's working force.[1]

Changing dimensions of the working force. It appears that one of the early results of emancipation was a release from the widespread compulsion of all adults to work. According to the first census (1844) there were about 234,500 persons occupied at that time, a smaller number than the 255,300 predial and non-predial slaves listed in 1834. Despite the steadily falling proportion of the total population occupied after 1844, there has been a marked rise in the numbers occupied since then (see Table 25). With the exception of the period from 1871 to 1881 when the total declined from 303,100 to 282,800, each census has recorded a rise in the population occupied, and by 1943 the total population declared to be at work numbered 484,300, that is, more than twice what it was a century earlier. Breakdowns by sex, introduced in 1891, show that from this date there has been a declining proportion of females in the occupied population. The only departure from this decline was noted in 1921 when, probably as a consequence of the depletion of the male labour force by heavy sex-selective emigration, an increased demand for female labour probably developed. (A similar development is to be noted in the case of Barbados, which also experienced heavy sex-selective emigration during 1911–21.) The change in sex composition of the labour force between 1921 and 1943 was profound. Whereas in the former year the sex ratio among the labour force stood at 1022, by 1943 it had increased to 1954.

In 1881 the fivefold occupational classification of the English census was adopted and this continued to be used until 1921. New and more elaborate classifications based on occupational and on industrial classes were introduced in 1943. For the purpose of tracing the size of the various classes an attempt has been made to group the occupational classifications of 1844–71 into fivefold classifications similar to those of 1881–1921. The regrouping in the

---

[1] The general term 'working force' is used by Jaffe and Stewart, op. cit., to cover that portion of the population usually engaged in the production of goods and services for the market.

Table 25. *Growth of the working force*

Year	Professional	Domestic (personal)	Commercial	Agricultural	Industrial	Total
			Both sexes			
1844	2,000	20,800	5,800	188,300	17,500	234,500
1861	3,200	16,300	6,300	183,600	29,500	238,900
1871	4,300	16,400	5,700	232,100	44,600	303,100
1881	4,700	15,000	7,400	208,600	47,100	282,800
1891	7,000	26,700	10,900	271,300	57,600	373,500
1911	9,200	38,000	19,800	271,500	72,400	410,900
1921	11,400	52,600	20,600	285,700	73,600	443,900
1943	20,000	60,500	52,000	228,600	123,100	484,300
			Males			
1891	5,900	5,300	8,100	137,600	26,500	183,400
1911	6,800	5,500	13,700	158,400	36,300	220,600
1921	7,100	7,200	13,500	160,300	36,300	224,300
1943	11,700	9,300	28,400	183,000	87,900	320,300
			Females			
1891	1,100	21,400	2,800	133,700	31,100	190,200
1911	2,400	32,500	6,000	113,100	36,200	190,300
1921	4,300	45,400	7,200	125,400	37,300	219,600
1943	8,200	51,200	23,600	45,600	35,300	163,900

Note. The discrepancies in some of the totals are due to rounding.

case of the 1871 census seems the least satisfactory; here several defects in the basic census data are to be encountered. The 1943 census data have also been regrouped to conform to the 1881–1921 classification. For this purpose use is made of the industrial distribution rather than of the occupational distribution. One advantage in this procedure is that the large indefinite class termed 'unspecified (odd jobs)' amounts to only 28,500, whereas in the case of the occupational breakdown there is a much larger indefinite class termed 'labourers' which amounts to 55,900. It is instructive to consider the changing size of the five classes from 1844 to 1943, which is shown in Table 25 and Fig. 3.[1]

[1] The regrouping of the industrial divisions of 1943 in order to bring them into conformity with the occupational classes used at earlier censuses has been done as follows. The professional class is composed of those engaged in public and professional services. Following the procedure adopted in the 1946 census report the domestic (personal) class is taken as the personal service class exclusive of the unpaid domestic servants. The commercial class comprises those engaged

Steady increases in the numbers returned as professional are in evidence. From a total of 2000 in 1844 it had increased to 20,000 by 1943; in 1843 professionals accounted for less than 1% of the total labour force, whereas in 1943 they accounted for 4% of the total. Also of importance has been the growing proportion of females in this class. Whereas in 1891 the sex ratio of the professional class stood at 5374, it had by 1943 declined to 1427. Even the recorded increase, however, understates the true position, for in the past the inclusion of military and naval personnel in the professional class unduly inflated these numbers, particularly in the case of males.

The domestic (personal) class, composed mostly of domestic servants, has been predominantly female in composition, and changes in its size have been primarily reflexions of the changing numbers of females involved. Actually the proportion of males in domestic service has remained very stable, being highest (3·2%) in 1921 and lowest (2·5%) in 1911. On the other hand, the proportion of females so engaged has increased appreciably from 11% in 1891 to 31% in 1943. However, it is easy to exaggerate the significance of female participation in domestic occupations, in view of the general decline in numbers of females occupied; actually the number of female domestics increased from 45,400 in 1921 to 51,200 in 1943.

Obviously the delineation of a commercial class in an essentially agricultural population is not easy, especially as the major crop involved, sugar, calls for a varied range of activities in its cultivation and manufacture. Still the growing numbers in the commercial class is consistent with diversification of enterprises, growth of service occupations attendant on enhanced urbanization and improving standards of living. There has been a steady increase in the commercial class during the century from 5800 to 52,000. Its past predominantly male character has been gradually altered by the growing number of women engaged. This, indeed, is the only class where a substantial increase in female participation between the crucial years 1921 and 1943 has been witnessed.

in transportation and communications, trade and finance, together with those employed in recreational services. Included in the agricultural class are the small numbers employed in quarrying, fishing and forestry. The industrial class is perhaps the least satisfactorily delineated; it is here taken to include those in manufacturing and mechanical undertakings, in electricity, gas and water, as well as the indefinite class termed 'labourers'.

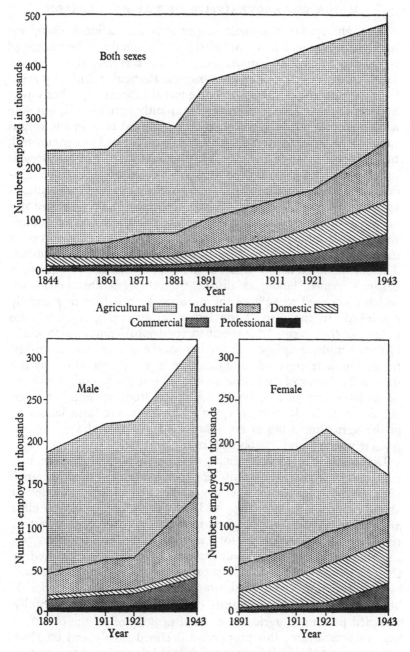

Fig. 3. Growth of the working force by occupational class, 1844–1943

Though by far the most important occupational class, the agricultural (in which are included here the small numbers engaged in fishing, mining and quarrying), is also certainly the most unsatisfactorily delineated in the censuses. Particular difficulty arises in the treatment of the so-called 'general labourers', a body who apparently have always been predominantly agricultural, but who are sometimes not included in the agricultural class in the census tabulations. Prior to 1891 these were not considered as agricultural, but censuses of 1891–1921 classed them as such. (In the present tabulation these general labourers are treated as part of the agricultural classes.) Again in 1911 a new subclass termed 'assistants in agriculture' was introduced. Although these accounted for nearly one-sixth of the total employed in this class there was no indication as to whether they were genuine agricultural labourers or unpaid assistants, or how they were treated at previous censuses. The occupational classification of the 1943 census also recognized a category termed 'labourers' distinct from the agricultural elements, which amounted to 12% of the number defined here as gainfully employed. In the case of the industrial grouping, however, the indefinite class termed 'unspecified (odd jobs)' amounts to 5·9% of the gainfully employed. A further source of confusion in regard to the numbers engaged in agriculture in 1943 is provided by the data in Table 205 of the census report. This shows a total number of 'farm labourers' of 107,000, which differs from the number given elsewhere in the Report (121,900). Moreover, the farm labourers in the agricultural tables are subdivided into two classes: 59,600 regular workers and 47,400 temporary workers.

Despite their many limitations, the data on agricultural employment suffice to show the declining dependence on agriculture since 1844. In that year the number engaged in agriculture amounted to about 188,300. Between 1844 and 1881 the numbers in this class moved irregularly, being highest in 1871 (232,100) and lowest in 1861 (183,600). In 1891 the total in agriculture was about 271,300 and this remained virtually unchanged in 1911, but increased to 285,700 in 1921. After 1921 a pronounced decline took place and by 1943 the number engaged in agriculture was down to 228,600. More important is the constantly declining proportion of gainfully occupied persons in agriculture. In 1844 80% of all the occupied were in agriculture; this proportion declined slowly and by 1891 was down to 73%. Still further reductions followed, the percentage in agriculture being 66 in 1911 and 64 in 1921. However, it was between 1921 and 1943 that the most considerable reduction in employment in agriculture took place; in fact by 1943 less than half (47%) of the gainfully employed were so classified.

Though the declining participation in agriculture is characteristic of both sexes, it is the large-scale withdrawal of females after 1921 that has in the main resulted in the great reduction in the numbers engaged in agriculture. Thus between 1921 and 1943 the number of females in agriculture declined from 125,400 to 45,600; a reduction of 64% within 22 years has thus taken place. Whereas in 1921 57% of all females gainfully employed were in some form of agriculture, the proportion declined to 28% by 1943. As we shall see, it is not improbable that some of this reduction may be explicable in terms of changing definitions of the gainfully employed or by changing definitions of employment in this class. At the same time this shrinking dependence on agriculture on the part of females appears in other West Indian populations as well.[1] The movements in the number of males employed have on the contrary been entirely different. In fact, the number of males in agriculture has continued to rise ever since 1844. And between 1921 and 1943 males so engaged increased from 160,300 to 183,000 or by 14%. But despite the continued increase in the number of males engaged in agriculture, the proportion so engaged has declined, especially between 1921 and 1943. Between 1891 and 1921 the proportion of males in this class ranged from 75 to 71%, but by 1943 the proportion declined to 57%. Low though this is compared with the proportions prevailing in the past, it still emphasizes that agriculture remains the predominant industry of the island.

In view of the considerable changes in the number in agriculture between 1921 and 1943 it is instructive to examine the numbers employed more closely. The different categories used at the two censuses preclude any exact comparison, but an attempt is made in Table 26 to compare the numbers engaged in the production of the main crops at the two dates. Clearly the decline in the case of the females is general. But it is the vague classes used in 1921, 'assistants in agriculture' and 'general labourers' (included under general farming) that in the main account for the steep reduction. In fact, these two categories account for a reduction of about 77,000. This strongly suggests that at least some of the decline is explicable

[1] See *West Indian Census*, 1946, vol. 1.

in terms of changing definitions of employment in agriculture adopted in 1943.

The numbers engaged in the industrial class also show an increase. In 1844 the number estimated within this class stood at

Table 26. *Comparison of types of employment occupation in 1921 and 1943*

Occupation	Male		Female	
	1921	1943	1921	1943
Cane farming	20,100	28,300	17,800	11,000
Banana cultivation	15,500	22,600	9,800	5,600
General farming	101,200	105,400	65,200	23,300
Cocoa, coconut and coffee cultivation	1,500	2,700	1,100	800
Tobacco cultivation	700	1,000	600	600
Stock raising	600	3,100	40	500
Assistants in agriculture	11,200	—	28,000	—

17,500 or about 7% of the total occupied population. By 1921 there were 73,600 persons so engaged, accounting for 17% of the total labour force. Unfortunately, it is impossible to state with assurance the precise movements in this group between 1921 and 1943 because of the changes in classification. Again it is mainly the class of 'unspecified' workers that causes the difficulty. By including these in the industrial class the latter shows an increase to 123,100; if these are excluded the industrial class in 1943 totals a much smaller figure, 94,600, which, however, is still in excess of the figure for 1921.

Participation in the working force. Changing sex composition of the working force indicates varying rates of participation for the two sexes. Rates of participation or worker rates are shown in Table 27 and in Fig. 4. The only census years for which these rates can be computed are 1891, 1911 and 1943. We consider first the rates for the males. There has, indeed, been a small over-all reduction in the proportion of males gainfully employed. Thus in 1891, 80·5% of all males over 10 were gainfully employed; this proportion declined to 78·1% in 1911 and to 72·5% in 1943. But it is important to note that the reductions in rates of employment have been marked only among the very young and the very old age-groups of

OCCUPATIONAL STATUS 93

the labour force. In the age-group 15–19, for instance, the proportion employed declined from 71·7% in 1891 to 55·4% in 1943. Similarly the proportion employed over age 65 fell from 88·3 to 68·2% over the same period. On the other hand, the proportion employed between ages 20 and 50, comprising the most productive

Table 27. *Male and female worker rates, 1891–1943*

Age	Male			Female		
	1891	1911	1943	1891	1911	1943
5–9	0·047	0·017	—	0·054	0·020	—
10–14	0·367	0·134	0·035	0·365	0·128	0·015
15–19	0·717	0·749	0·554	0·689	0·578	0·295
20–24	0·934	0·940	0·851	0·844	0·737	0·444
25–29	0·959	0·966	0·945	0·885	0·738	0·458
30–34	0·971	0·969	0·962	0·885	0·714	0·441
35–39	0·976	0·971	0·965	0·866	0·709	0·436
40–44	0·974	0·968	0·952	0·889	0·715	0·420
45–49	0·950	0·967	0·966	0·868	0·711	0·459
50–54	0·964	0·967	0·948	0·847	0·699	0·396
55–59	0·965	0·954	0·901	0·829	0·678	0·418
60–64	0·950	0·939	0·754	0·800	0·648	0·287
65 and over	0·883	0·870	0·682	0·656	0·483	0·199

Note. The rates for the males (1943) have been slightly smoothed.

section of the working population, has changed very little over the period 1891–1943. The decline in rates under age 20, and in particular the disappearance of male workers under 10 by 1943, is associated with the increased attendance at school, while the reduction in rates at advanced ages presumably indicates that more people are now in a position to retire from work, as there are now provisions for old age and retirement pension, forms of assistance virtually unknown in the island 60 years ago. The reduction in the rates of participation at lower ages, coupled with the general declines in mortality, has inevitably produced some ageing of the working force. The increase in the numbers at advanced ages is reflected in the median ages. For the three census dates, 1891, 1911 and 1943, these are 30·4, 31·7, and 33·9 respectively.

By applying the worker rates to the appropriate life-table populations the changing patterns of male occupation can be illuminated in several ways. Three useful measures can be derived in this manner, the average age at entry into the working force, the number of years a male entering the labour force at the derived average age can be expected to spend in gainful occupation, and the proportion the latter forms of the complete expectation of life of

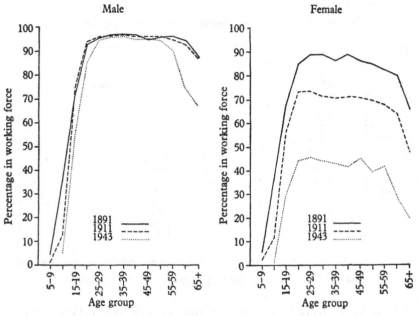

Fig. 4. Labour force participation rates, male and female, 1891, 1911, 1943

the male at the average age of entry. These measures are shown in Table 28. In 1891 the average age of entering the labour force was only 16·9 years, a figure which emphasizes the number of children employed in those days. The average age rose slightly to 18·0 in 1911, but even at this date the number of children employed was considerable. The average age at 1943, however, fully indicated the extent to which children had been eliminated from the working population; the average age had risen to 20·1. In view of the improvements in mortality witnessed, particularly since 1921, the shifts in worker rates have not adversely affected the average worker's contribution to the economy of the island. For though the average age at entry increased by 3 years between 1891 and 1943, the average time spent in the labour force also increased slightly during the same period.[1] When the average number of years spent in the labour force is related to the completed expectation of life of a male at the average age of entry into the labour force, the small gains in leisure in terms of early retirement are illustrated. In 1891 the average male about to begin earning his living could expect to spend 93% of the average length of life remaining to him in the

Table 28. *Ages of entry into labour force, years spent in labour force and expectation of life of worker at time of entry into labour force*

	1891	1911	1943
Average age at entry into labour force	16·9	18·0	20·1
Average number of years spent in labour force	36·1	36·9	37·0
Average number of years of life remaining to person entering labour force	38·7	39·2	41·8
Years in labour force as % of remaining years	93·3	94·1	88·5

labour force. This figure was nearly the same in 1911, but fell to 88% in 1943. The conclusion to be drawn is that despite greater opportunities of retirement from the labour force and higher ages of entry no fundamental alterations in the contribution of the male to the working force of the island is indicated.

In keeping with the experience of other countries, the rates of participation for the females differ basically from those for the males, as can be seen from Fig. 4. Apart from the fact that the shape of the curves for the females precludes the calculation of any simple average age of entry into the female labour force, it is important to note that there have been striking changes in the levels of female employment between 1891 and 1943. Indeed, the steeply falling proportions of women employed strongly demarcates the island from European countries where, in general, the passage of time has brought increased participation of females in the labour force. This reduction in female employment, doubtless a continuation of the break with the slave society in which every adult was potentially a 'work unit', was most marked between 1891 and 1943. At the former date, 74·5% of the females over 10 were employed.

[1] Alternative calculations for Jamaica, in which the period of economic activity is related to age 0, rather than to the average age of entry into the working force, are given in John D. Durand, 'Population structure as a factor in manpower and dependency problems of under-developed countries', *Population Bulletin of the United Nations*, No. 3 - October, 1953.

This proportion fell to 59·6% in 1911, while the still steeper decline to 34·0% in 1943 constitutes the most arresting development in the pattern of employment in the island. The reduction in the proportions employed has been considerable at all ages. For the age-group 15–19, the proportion employed fell from 68·9% in 1891 to 29·5% in 1943. Again, whereas in 1891 the proportion employed over age 65 was 65·6%, by 1943 it had been reduced to 19·9%. Similarly, in the intermediate ages drastic reductions in the proportions employed have taken place. In 1891 the worker rates between ages 20 and 60 ranged from 82·9% to 88·9%; in 1943 the range was from 39·6% to 45·9%. As the rates of participation declined, the average age of the female labour force has risen, thus paralleling the development among the males. The median ages for 1891, 1911 and 1943 were 29·6, 30·3 and 32·2 respectively.

Accompanying these changes in overall proportions of persons employed at various ages have been equally significant changes in the rates of participation in the several categories of occupations distinguished. Unfortunately, in respect of the 1943 census, these rates have to be computed on the basis of an occupational distribution in which the indefinite groups termed 'labourers (odd jobs)' looms very large, especially in the case of the males. For the present purpose these 'labourers' are included in the industrial class, a procedure which may to some degree impair strict comparability over time, especially as these 'labourers' account for 55,900 of the total 128,600 in the industrial class. The curves of employment in four major occupational classes are depicted in Figs. 5 and 6.

Considering first the rates for the males, the most striking fact we note is the fall in the proportions engaged in agriculture, especially at ages under 55. For instance, in the age-group 25–29 the proportion employed in agriculture declined from 68% in 1891 to 46% in 1943. Another important feature of male employment is the increased proportion of males of young ages employed in industrial occupations. This may be largely the result of treating the 'labourers' as part of the industrial class. Nevertheless, it seems clear that the establishment of industries in the island has resulted in the creation of a nucleus of trained workers, even if most of the men engaged in these undertakings are not sufficiently skilled to be considered industrial workers and may be more accurately described by the general term 'labourers'. There has also been increasing proportions in commercial activities at all ages over 20. Mainly as

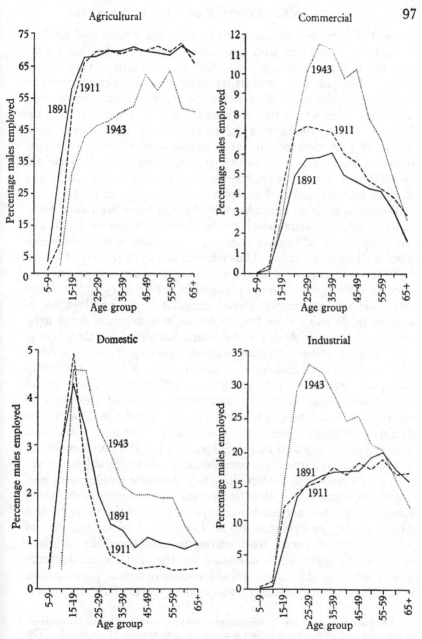

Fig. 5. Labour force participation rates by occupational class, males, 1891, 1911, 1943

a result of increased employment in domestic service and kindred occupations associated with the tourist trade there has been some increase in service occupations. But this remains a very minor avenue of employment, and its temporary nature is emphasized by the fact that the rates of participation reach a maximum at ages under 20, after which they decline steeply. In regard to the professional classes, declines in the rates of participation have been noted. But in view of the large numbers of military personnel included under this head in the past, strict comparability cannot be assured. There has been, however, an increase in the numbers of doctors, lawyers and civil servants in the island since 1921.

Withdrawals from agriculture by females have been even more marked than the withdrawals by the males. Continuous declines in the proportions of females engaged in agriculture have been evident at all ages since 1891. Thus whereas in 1891 between 62% and 64% of the females aged 30–45 were in agriculture, the proportions over the same age span in 1943 ranged from 10% to 13%. It is therefore clear that the vastly altered situation in the employment of women in Jamaica rests largely on their withdrawal from agriculture after 1921. Another important feature of female employment has been the increasing number entering domestic service. The proportion of females in domestic service in the age-group 20–24 increased from 11% in 1891 to 19% in 1943. The curve of employment emphasizes the temporary nature of this kind of occupation. The rates reach a maximum at about age 20 and then decline steeply, thus indicating that women enter domestic service at very young ages and leave it as soon as possible, presumably to secure more remunerative employment or to establish family unions. Though these factors show that domestic employment is of minor importance and that its general significance can be easily exaggerated, the increased participation in employment of this nature is also, as Professor W. A. Lewis emphasizes, evidence of growing competition for work among females, especially in urban areas where more than one-third of the females in domestic service are located.[1] Considerable increases in female participation in commercial occupations are also evident, particularly since 1921.

[1] The past and probable future course of employment in various occupations in Jamaica and in the West Indies generally is discussed in W. A. Lewis, 'The industrialization of the British West Indies', *Caribbean Economic Review*, vol. II, no. 1.

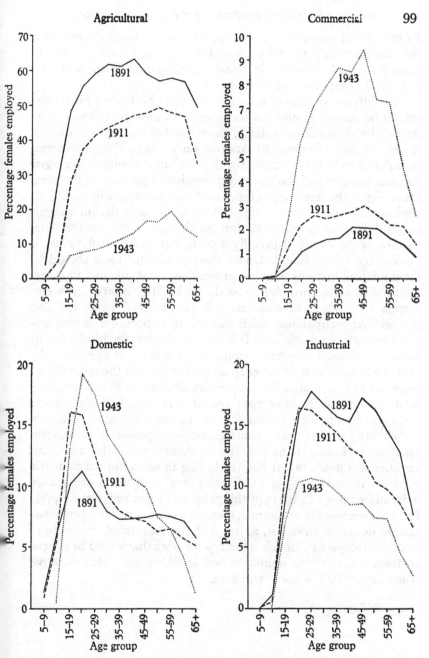

Fig. 6. Labour force participation rates by occupational class, females, 1891, 1911, 1943

In fact, at all ages the rates for 1943 are many times greater than the corresponding rates for 1891. On the other hand, there has been a marked reduction in female participation in the so-called industrial occupations at all ages.

The shifting patterns of occupation among both sexes since 1921 should be viewed against the background of growing urbanization. As will be shown in Chapter 5, there has been a considerable in-migration into Kingston–St Andrew since 1921. This movement, stimulated to a large extent by the declining demands for agri-cultural workers and the mounting population pressure in the rural areas since the cessation of external emigration, will doubtless continue on a growing scale. Urbanization and declining agri-cultural densities promise still further changes in the occupational pattern of the future, especially if more industrial development is undertaken in the island. Under these conditions there should be continued reductions in the proportions and in the numbers engaged in agriculture. So far as the males are concerned it seems certain that increased numbers will be absorbed in skilled and semi-skilled occupations which will rise in importance as employ-ment in agriculture declines. It is less certain that the displacement of females from agricultural employment will be compensated by growing employment in other spheres of activity to the same extent as in the case of males. The impressive declines in female employ-ment in agriculture after 1921 meant a profound change in the employment situation among females in the rural areas. Thus in 1921 there were 186,300 females gainfully employed in rural areas (that is, excluding Kingston and St Andrew) and by 1943 this number was down to 131,800 (including those in unpaid domestic service). In 1921 65% of the females over 10 in rural areas were gainfully employed; by 1943 this proportion was reduced to 35%. So far as females are concerned, therefore, their increased participa-tion in domestic service is, as W. A. Lewis has urged, evidence of under-employment, for this is the type of work that would be sought by female in-migrants unable to find anything more remunerative when first settled in the urban area.

CHAPTER 4

CURRENTS OF EXTERNAL MIGRATION

The fact that virtually all the inhabitants of the West Indies are descendants of immigrants of one form or another or themselves immigrants is ample evidence of the importance of immigration in the development of the region. Though as we shall see in Chapter 7 immigration was by no means the only determinant of population growth that formed a part of public policy, it was certainly the most powerful in the slave regime. Post-emancipation migration, with which this chapter is concerned, was, on the other hand, very small in size, and its effects on population growth generally negligible except for a short period 1911–21. Nevertheless, these post-emancipation migrations remain of lasting historical importance and demand some treatment in any demographic analysis of the region. Three currents of post-emancipation migration can be distinguished.

The first, though of minor importance from the standpoint of its effect on population movements in the case of Jamaica, was of considerable significance for population growth in the West Indies as a whole; this was the introduction of indenture immigrants and their families from foreign territories. It is difficult to overestimate the historical importance of this movement. Indeed, Professor W. L. Burn has, with only slight overstatement, declared, 'To write a history of immigration into the West Indies would in effect be to write the history of those colonies in the post-emancipation period.'[1] Indenture immigration raised countless social, economic and political problems, largely because of its close association with the changing fortunes of the West Indian sugar industry. It gave rise to a volume of legislation which went far beyond the mere regulation of the transfer of labourers and their families, extending to such fields as conditions of work, levels of wages, health standards and even the conditions of mating. The financing of these migrations posed weighty problems, the solution of which often involved protracted negotiations with the British Government. For the issue to be faced was a complex one. Not only had provisions to be made

[1] W. L. Burn, *Emancipation and Apprenticeship in the British West Indies*, 1937, p. 107, note.

for defraying the cost of immigration; the substantial sums involved meant in fact that the financing of the movements created major problems in taxation and general fiscal policy of the territories. These assumed importance even in Jamaica, where indenture immigration was, from the numerical standpoint, of minor significance. Moreover, the fact that the promotion of indenture immigration was one of the principal aims of the planters in the nineteenth century often precipitated grave constitutional crises. For instance, the problem of promoting immigration into Jamaica became involved in the general constitutional crises that developed in that colony in the 1840's.

Admittedly the second current of post-emancipation migration, the emigration to foreign territories, had much less historical significance. Unlike indenture migration it was completely unorganized and in no way controlled by legislation (except for some largely ineffective laws introduced by Barbados). Because no problems of financing or controlling these movements were involved the records of these migrations are far less ample than those on indenture migration. Nevertheless, from the strictly demographic standpoint, that is, so far as its effects on the growth and general composition of the island population are concerned, emigration to foreign countries proved vastly more important than indenture migration in the case of Jamaica.

The third current of post-emancipation migration was the interchange of population between the British territories. So far as Jamaica is concerned, this current can be completely ignored, as there never was any large interchange of population between it and the eastern Caribbean. The possibility of relieving distress among the 'poor whites' of Barbados by promoting emigration of some of them to Jamaica was raised in 1859, but the scheme never matured as the Barbados House of Assembly refused to vote the money necessary to defray its cost.[1] Again, in 1864 efforts were made by planters in Jamaica to start an immigration of Barbadian labourers, but this also proved abortive as Barbados refused to agree to the project. Indeed, the Governor of that island, in turning down the scheme, decried the 'very erroneous impression abroad as to the superabundance of the able-bodied agricultural labor in Barbados', and actually maintained that 'two or three considerable districts [in Barbados] have

1 *M.B.H.A.*, 1859.

been almost deserted by the people in consequence of the late emigration to Demerara'.[1]

INDENTURE IMMIGRATION

The introduction of indenture labour into Jamaica was part of the wider movement into the whole British Caribbean, and therefore it is difficult to treat it in isolation. Indentured immigration into Jamaica never attained large dimensions. The numbers of East Indians and Africans entering the island were small, while the introductions of Europeans, Chinese and Portuguese were no more than token additions to the population, of historical rather than of demographic significance. And the importance of the whole movement declines still further when the numbers that returned to their native land are taken into consideration. Nevertheless, this phase of migration cannot be ignored, and a brief analysis of its course is not only relevant to the demography of Jamaica, but also constitutes an interesting chapter in the long and chequered history of Caribbean migration.

Declines in the output of sugar throughout the West Indies during the early post-emancipation period caused great concern. It was therefore natural that attention should be directed to the factors to which this decline in output was at the time widely ascribed, the shortage of labour on the plantations. Though several factors were involved in determining the size of the plantation labour force after emancipation, two were of especial importance. The first, and by far the more important of the two, was the withdrawal of the ex-slaves from the sugar plantations in large numbers, either to settle elsewhere or to engage in some non-agricultural activity. Another factor, the precise effects of which are much more difficult to assess, was the actual movement of population, especially in the age interval 15–49, during the 1840's. It is necessary, before discussing the course of indenture immigration, to examine briefly some of the main elements involved in fixing the size of the plantation labour force and in helping to gain wide acceptance of the policy of promoting indenture immigration as a remedy for labour shortage.

The possibility that the abolition of slavery might adversely affect the labour supply of the planters did not escape the attention of the architects of abolition and others. Indeed, there was widespread

[1] Letter from Governor Walker of Barbados to Governor Eyre of Jamaica, dated 26 October 1864, *V.H.A.J.*, 1864.

fear that the Negroes would refuse to work on the plantations after securing their freedom, and the Abolition Act actually aimed at forestalling a large-scale depletion of the labour force. The Vagrancy Laws passed in some colonies in effect complemented the Abolition Act in this respect. And more drastic measures in this direction had been advocated, notably by Lord Grey.[1] It is true that the Abolition Act recognized the Negro's right to purchase his discharge from service. In practice, however, the purchase of the apprentice's unexpired term of service was not easy, and the number of persons obtaining their complete freedom by this means was very small. In the case of Jamaica, only 1385 persons applied for valuations between 1834 and 1836. Indeed, Burn claims, 'Many had no real intention of buying themselves out of apprenticeship'; in his opinion, 'being valued' was for most of the applicants 'an occasion for a holiday'.[2] The general exodus from the sugar plantations did not therefore commence in the period of apprenticeship. In fact, the masters resorted to many devices to restrain the apprentices from leaving the plantations and to exact the maximum amount of work from them.[3]

An issue raised by apprenticeship, and one particularly relevant to the question of the future labour supply, was the extent to which it aimed at or succeeded in preparing the predial workers for the 'responsibilities' of work in a free society. From this standpoint apprenticeship appeared to many as a failure, and the Negro's lack of preparation for life in a free society was advanced as a prime cause of the difficulties facing the West Indies in the 1840's. As an example of the position taken by some observers, fortified often with the wisdom of hindsight, we may take the views of John Innes. He urged that abolition was too abrupt and failed to ensure that 'slavery would safely merge into freedom. . . . The labour of the rising generation has been lost by the fatal error in the Abolition Act'.[4] This Act, it was argued, should have been preceded by and based upon the findings of a Commission of inquiry into the whole question of slavery. The slave was 'an isolated being without kindred to support or to look to for support; he was entitled to have

[1] W. L. Burn, op. cit. p. 111.
[2] W. L. Burn, op. cit. p. 278.
[3] On these points see D. G. Hall, 'The apprenticeship period in Jamaica, 1834–1838', Caribbean Quarterly, vol. 3, no. 3.
[4] John Innes, Thoughts on the Present State of the British West India Colonies and on Measures for their Improvement . . ., London, 1840, p. 8.

every want supplied. .... This was the being to be re-cast.' By first making it obligatory on him to support or assist in the support of those 'who would have been dependent on him in a state of civilization' and 'raising the wants of the negroes during apprenticeship' the slave would have been prepared for the smooth transition to freedom in a money economy. Similarly, the Committee of the House of Commons that inquired into the effects of the equalization of the sugar duties reported in 1848 that 'emancipation was carried into effect without sufficient provision having been made for providing many of the colonies with an adequate command of free labour'. And Earl Grey, whose drastic scheme for preventing the depletion of the sugar plantations of workers was rejected in 1833, affirmed in 1851, 'It was a great and unfortunate error when slavery was abolished not to place the emancipated population under regulations calculated to impose upon them the necessity of greater exertion.'[1]

It is futile to speculate on the possible outcome of measures that might have been taken to prepare the Negroes for their 'responsibilities' under freedom. The fact remains that emancipation in 1838 was followed by a movement away from the plantations. This took two forms. In the first place there was the withdrawal from the plantation areas, which reduced appreciably the numbers available for the production of sugar. Much of this movement was the result of the policy pursued by the planters, who erroneously conceived it in their interest to drive the ex-slaves from the plantations forcibly or to charge them such excessive rents that they would be compelled to leave. This, it was believed, would, by throwing people on their own resources, provide a source of cheap labour for sugar cultivation. But those thus forced away from the estates squatted on unoccupied land in places far removed from the sugar plantations, and contrary to the expectations of the planters had not to rely on employment on sugar cultivation. Others were able to purchase their own holdings with sums they had saved for this purpose during apprenticeship. The extent of this exodus from the plantations is revealed in the increasing numbers of freeholders having under 40 acres of land, which rose from 2000 in 1838 to 7800 in 1840.[2]

---

[1] Despatch from Earl Grey to Governor Sir C. E. Grey, 15 February 1851, *P.P.* 1851.
[2] The movement from the plantations is fully treated in Hugh Paget, 'The free village system in Jamaica', *Caribbean Quarterly*, vol. 1, no. 4. See also G. Cumper, 'A Modern Jamaican Sugar Estate', *Social and Economic Studies*, vol. 3, no. 2.

The second aspect of the movement away from the plantations was the seeking of occupations not connected with agriculture. As early as 1844, a substantial proportion of the population of many colonies had succeeded in establishing themselves in avenues of employment other than agriculture. Discussing Jamaica in 1861, Sewell remarked, 'Emancipation has created a middle class, . . . a class of proprietors, tax-payers and voters, whose prosperity, patriotism, happiness and comfort are bound up in the island's permanent prosperity.'[1] The first census revealed that of the 234,500 people occupied, 188,300 were in some way connected with agriculture, while about 22,800, or 10%, were returned as tradesmen, shop-keepers, merchants and kindred categories denoting a measure of respectability.

Probably because this seemed an effective means of reinforcing their arguments for increased labour supply many planters as well as others advanced as an important cause of the existing labour difficulties the supposed laziness of the Negro who, it was claimed, was disposed to work only so long as it took him to earn enough to provide for his immediate needs. Of course, indolence on the part of the slave was not unknown; it constituted a form of resistance to punishment. 'That was his usual form of resistance—passive.'[2] But the myth of the inherent laziness of the Negro gained general acceptance after emancipation and was widely used in support of immigration. For if the Negro, even when completely freed, was unwilling to work for the planters, there was no alternative but to enlist the services of some more tractable being. Picturesque variations on this theme abound in writings on the West Indies, possibly the most arresting being that of Robertson, 'The primary cause of this abhorrence of labour, is the absence of the task-masters of European society—hunger, cold and nakedness. The Children of the Sun are indisposed to continual work, to supply their few occasional wants. As emancipated Slaves, they feel their new rank as a discharge from labour altogether. . . '[3] Earl Grey's appraisal of the situation, essentially a less forthright condemnation of the attitude of the freed Negro, dwelt more directly on the effects of 'idleness' in retarding their 'advancement in civilization and

---

[1] W. G. Sewell, *The Ordeal of Free Labor in the British West Indies*, New York, 1861, p. 244.
[2] Eric Williams, *Capitalism and Slavery*, p. 202.
[3] James Robertson, op. cit.

morality'.[1] He maintained that once they were free from 'the obligation of performing gratuitous labour' the Negroes were without 'any adequate motive to work for wages for more than a small portion of their time', and 'as masters of the whole returns of their labour' they were content 'to indulge largely in the luxury of idleness'. Possibly the soundest contemporary refutation of claims of this nature came from the impartial observer, Sewell, who 'passed over with indifference . . . the lamentation of the planter that the negro won't work, because . . . the cry was an ebullition of selfish disappointment at the loss of labor'. In his view, the planters 'misjudged the negro throughout and have put too much faith in his supposed inferiority'.[2]

There were other factors which tended to gain support for a policy of immigration into the West Indies. Many saw in immigration a means of fighting the foreign slave trade by stimulating the production of British colonial sugar at prices sufficiently low to enable it to compete successfully with foreign slave-produced sugar. Large-scale introduction of foreign labour into the West Indies would 'of course, free trade prevailing, drive slave-grown produce out of the markets of the world', as it would increase the labour supply of the colonies and result in a reduction of the cost of production of West Indies sugar.[3] This view also found support from the Colonial Land and Emigration Commission, who declared in their first report, 'From the day on which sugar raised by free labour could be brought to market cheaper and of better quality than sugar produced by slaves, would date the fall of negro slavery throughout the world.'[4] The British Government was, for another reason, in favour of immigration into the West Indies. So far as this meant the immigration of liberated Africans, it offered a means of establishing permanent settlements of these people who, immediately after liberation, were accommodated in Sierra Leone and St Helena.

[1] Despatch from Earl Grey to Governor Sir C. E. Grey, 15 February, 1851, *P.P.* 1851.
[2] W. G. Sewell, op. cit. p. 130. For a modern treatment of this subject, see S. Rottenberg, 'Income and Leisure in an Underdeveloped Economy', *Journal of Political Economy*, vol. LX, no. 2.
[3] E. G. Wakefield, *A View of the Art of Colonization*, edited by James Collier, 1914, p. 449.
[4] *First General Report of Proceedings of the Colonial Land and Emigration Commission*, London, 1840. The titles of these Reports changed from time to time; they will be referred to here as *General Reports*.

As we have already seen, the fact that the first censuses of the West Indies were not taken until 1841–4 makes it difficult to say whether populations had been declining or increasing during the years immediately following emancipation. Even if there was no substantial decrease it remains doubtful whether in the light of prevailing mortality any colony experienced population increase on a scale that would counteract the general exodus of plantation labour. In the case of Jamaica, even though some population increase might have been taking place at this time, the numbers employed in 1844 were less than the slave population at work in 1834.

From Africa.[1] The distinctive feature of African migration into the British West Indies was the extent to which its course and magnitude were directed by the fortunes of slavery and the Atlantic slave trade during the nineteenth century. The majority of the immigrants were African slaves freed by the British Navy in the course of its campaign against the Atlantic slave trade. Though several attempts were made to induce a large-scale immigration of settled Africans from Sierra Leone and the Kroo Coast, immigrants of this type were of small importance in the whole scheme of African immigration.

The British Government sanctioned African immigration into the West Indies in 1840, and apparently the first groups to reach Jamaica consisted of 266 persons landed from the *Hector*. At this time East Indian immigration was not permitted and European immigration had failed to fulfil the hopes of many planters, so that the Africans were for some years the only source of additional labour available to the West Indies. At this time also the slave trade to Brazil was at a high level, so that ample numbers of captured and liberated slaves were being made available for transportation to the West Indies. Further government support for African immigration into these colonies came in 1842 with the acceptance of the recommendations of two Select Committees of the House of Commons: the Select Committee on the West Coast of Africa, and the Select Committee that inquired into the state of the West Indies. The former Committee considered that the West Indian requirements of immigrant labour could be adequately supplied from the African coast, that the transfer of settled and liberated Africans to

---

[1] The following paragraphs on African immigration are based on G. W. Roberts, 'Immigration of Africans into the British Caribbean', *Population Studies*, vol. VII, no. 3.

the West Indies 'would be of the highest advantage, it would be the greatest blessing to make such an exchange'; and that a well-controlled scheme for such immigration would in no way lead to the inauguration of a slave trade in a new guise—a fear of many who disagreed with the development of large-scale African immigration into the West Indies. The other Select Committee, viewing the diminished supply of labour as the main obstacle to West Indian progress, urged, 'One obvious and most desirable mode of endeavouring to compensate for this diminished supply of labour is to promote the immigration of a fresh labouring population to such an extent as to create competition for employment.'

These recommendations assured full official sponsorship of African immigration into the West Indies, though to prevent the impression that the British Government was taking advantage of the slave trade to recruit labourers for the colonies, it was specifically stated that this immigration had to be 'scanty or suspected'. The essence of the plan of African emigration was summed up thus in a despatch from Lord Stanley: 'The plan adopted . . . is in substance that without interfering with the bounty system already in force in the colonies, no emigration from the coast of Africa shall be permitted, or at least paid for by public funds, except through the intervention of His Majesty's Government. . . .' The meagre results yielded by the movement led the West Indian planters to urge a relaxation of the controls under which emigration was permitted as, it was claimed, these tended to retard emigration. With this the British Government complied, relaxing the restriction on the sex ratio, discontinuing the practice of maintaining liberated slaves at Government expense for periods of 6 months and even advancing small sums to emigrants before they embarked. But even these measures had little effect on the numbers entering the West Indies. In the late 1840's the British Government, as a means of aiding the West Indian colonies which were in economic difficulties after the removal of preferences on their sugar in the British market, decided to defray the full cost of the movement. This, together with the great increase in the Brazilian slave trade and the consequent rise in the numbers of slaves liberated, gave an appreciable impetus to the African immigration which reached its peak in the years 1848 and 1849. Still the numbers involved in the case of Jamaica were very small—1900 in 1848 and 1100 in 1849. With the cessation of the Brazilian slave trade in 1852, supplies of liberated Africans diminished and the African emigration ceased to be of much importance, even though the British Government continued to finance the movement. In the late 1850's when, following an expansion of the Cuban slave trade, the numbers of liberated slaves landed at Sierra Leone and St Helena increased, efforts were made to resume African immigration on a sizeable scale. But in Jamaica, as in the other West Indian colonies, the numbers secured were negligible. A further blow was dealt to the movement by the Treaty of Washington (1862) which assured Anglo-American cooperation in the suppression of the Atlantic slave trade, while the abolition of slavery in the United States meant the final doom of the African immigration.

In terms of numbers involved, African immigration into Jamaica was very small. Before the official sanction of the movement in 1840, small numbers of Africans were landed in Jamaica from time to time, but no complete records of these are available. In any event, these entries were no more than accidental and, being no part of the organized movement that began in 1841, will not be discussed here. Between 1841 and 1867 the numbers introduced into Jamaica were about 10,000. In only 3 years were the numbers introduced in excess of 1000, in 1842, 1848 and 1849. Of the total introduced, 4490 came from St Helena and 3826 from Sierra Leone and the Kroo Coast. It is of interest to compare the numbers entering Jamaica with the numbers entering other colonies. The following are the comparative figures of African immigration:

Jamaica	10,000 (28%)
British Guiana	13,970 (39%)
Trinidad	8,390 (23%)
Grenada	1,540 (4%)
St Vincent	1,040 (3%)
St Lucia	730 (2%)
St Kitts	460 (1%)

The accounts of the *General Reports of the Emigration Commissioners* show that some of the African immigrants returned to their native land in accordance with the original arrangements under which they emigrated, but it is impossible to state exactly the numbers who returned.[1]

[1] On this point see R. R. Kuczynski, *Demographic Survey of the British Colonial Empire*, vol. I, p. 141.

The present writer erred in giving numbers of emigrants returning ('Immigration of Africans into the British Caribbean', *Population Studies*, vol. VII, no 3.). Due to a misinterpretation of the heading of Appendix 19 of the *Twenty-eighth Report of the Emigration Commissioners*, the figures in this Appendix were taken to

Clearly the effects of African immigration on population growth in the island were negligible. This can be seen by examining the relevant census data for 1861, by which time most of the immigration had already taken place. Only 18,200 of the total population of 441,300 were born outside the island, a fact which emphasizes the small contribution of immigration to population growth in the island. Natives of Africa formed the main groups of foreign-born population (10,500), but as they constituted only 2% of the population, it is evident that their contribution to population growth was of no consequence.

From India.[1] From the inception of the movement of East Indians into the West Indies the conditions under which they were recruited and transported were carefully controlled by the Government of India. Emigration under indenture had been in progress to Mauritius long before it began to the West Indies. Indeed, it was the experience of the indenture system in Mauritius that led to the introduction of Indians into the West Indies.[2] The law under which indenture emigration to these colonies commenced had its origin in the recommendations of the report of the Law Commissioners on the legal control of emigration (1836). It was here pointed out that there were no laws in being except 'those already made to prevent undue advantage being taken of the simplicity and ignorance of these persons'. Following the recommendation of the Commissioners, Act V of 1837 was passed. Under this Act no emigrant making a contract of service to be performed outside 'said territories' was to be received on any vessel without a permit from a Government officer. Before the issue of such a permit the emigrant and the person with whom he contracted had to appear before the officer with the relevant contract, which had to specify the nature and terms of service as well as the wages. Contracts were not to exceed 5 years, and there had to be a provision for re-conveyance

refer only to Africans returning to their native land. Further examination of the data, however, indicates that this Appendix covers mainly, if not wholly, the movements of East Indians. Neither the records of the Emigration Commissioners nor of the immigration authorities in the West Indies give any comprehensive statistics of Africans who returned to their native land.

[1] This brief account of immigration of East Indians is based largely on J. Geoghegan, *Note on Emigration from India*, Calcutta, 1873, the *General Reports* and other official publications. Other discussions of the subject are given in D. Kondapi, *A History of Indians Overseas*, D. Nath, *A History of Indians in British Guiana*, London, 1950, and Kingsley Davis, *The Population of India and Pakistan*.

[2] J. Geoghegan, op. cit.

of the emigrant to the port from which he embarked. This Act, which applied to Calcutta only, was soon replaced by another and more extensive one, Act XXXII of 1837, applying to the whole territory of the East India Company. It was under this Act that the first group of Indians left for British Guiana in 1838.

Immigration of East Indians into the British West Indies owed its origin to Mr John Gladstone, who, doubting whether the ex-slaves would be willing 'to continue their services on the termination of their apprenticeship', approached a firm responsible for transporting many Indians to Mauritius, asking that his plantations in British Guiana be supplied with Indian labourers.[1] An Order in Council of 12 July 1837 sanctioned the introduction of Indians into British Guiana under contracts of indenture for 5 years. In 1838 the movement began when 414 emigrants left for British Guiana. Strong opposition greeted this new movement, especially after many allegations of ill-treatment of the immigrants reached England. Lord Brougham described the Order in Council under which the immigration began as establishing 'what would become a slave trade', while Scoble and others added their protests.[2] The complaints of abuse led to an investigation of the charges. Most of the complaints proved exaggerated, but there was sufficient evidence of hardships among the immigrants to prevent a resumption of the movement.

Indeed as soon as complaints of ill-treatment of emigrants reached India the Government (on 11 July 1838) directed the withholding of permits of emigration to the West Indies.[3] An effective check was put on the whole movement by the Order in Council of 7 September 1838, which, in establishing the relationship between master and servant in the colonies, provided that no contract of service was valid unless it was made 'within the limits and upon the land of the colony in which the same is to be performed'[4]. Two important steps on the part of the Indian Government followed. In the first place a Committee was appointed to examine the whole subject of emigration. Secondly, Act XIV of 1839 was passed, which prohibited the making of contracts with Indians for working in any British or foreign territory outside the provinces of the East India Company.

---

[1] John Scoble, *A Brief Exposure of the Deplorable Conditions of the Hill Coolies in British Guiana and Mauritius*, London, 1840.
[2] 'Emigrant', *Indian Emigration*, Oxford, 1924.   [3] J. Geoghegan, op. cit.
[4] This Order in Council is given in *M.B.H.A.*, 1843.

The Calcutta Committee on Emigration submitted its Report in October 1840, signed by only three of the six members. It detected fraudulent systems of recruitment and the plundering of labourers' wages, though it found little evidence of abuses on the voyages.[1] The Report condemned the whole system of emigration, considered that no legislation could prevent abuses and recommended its absolute prohibition. Unless this was done the Government would be forced to exercise complete control over all movements, which would necessitate heavy expenditure. One member recorded a minute of complete dissent, while another apparently took no part in the proceedings. What Geoghegan calls 'a valuable document' was submitted by the sixth member, J. P. Grant. Though admitting the abuses, Grant stressed the direct and indirect advantages of emigration and was of the opinion that emigration should continue under Government regulations. He advocated, among other things, the fixing of certain ports of emigration, at which Protectors were to be stationed, the guarantee of return passages and the framing of rules designed to curb past abuses. By a curious set of circumstances this was the course later taken by the Indian Government.

The Government of India remained undecided on its emigration policy, awaiting the impending debate on the subject in the House of Commons.[2] The latter proved inconclusive and the settlement of policy devolved on the Colonial Office and the Government of India. In 1842 the Colonial Office considered it no longer necessary to restrict emigration to Mauritius, and in an Order in Council of that year appropriate regulations were put forward for resuming emigration to that colony. Following this, the Court of Directors (in March 1842) left the whole question to the Government of India, merely urging them 'to prevent a project intended to promote the advantage of certain classes of the people of India, by allowing them free command of their labour, being perverted to their injury'. Accordingly the resumption of emigration to Mauritius was provided for in Act XV of 1842, later amended by Act XXI of 1843.

In 1843 Lord Stanley admitted that the supply of labourers reaching the West Indies from St Helena and Sierra Leone was inadequate, and found it 'difficult to find grounds for resisting a

---

[1] J. Geoghegan, op. cit.
[2] Ibid.

remission of the prohibition' on immigration.[1] In the same year he urged the Government of India to reconsider its ban on emigration to the West Indies, pointing out the critical position then being faced by these territories. The Government of India was, however, reluctant to permit the resumption of this movement, showing that it differed materially from that to Mauritius. The distances were much greater and communications more infrequent, while the number of colonies competing for immigrants might create difficulties. Partly to meet the latter objection, the British Government suggested that for the time being the resumption be limited to Jamaica, British Guiana and Trinidad, and, anticipating the acceptance of this by the Indian Government, appointed the Mauritius Immigration Agent at Calcutta to act for the West Indies.[2] By Act XXI of 1844 emigration from Calcutta, Madras and Bombay to Jamaica, British Guiana and Trinidad was legalized. Modifications followed in Act XXV of 1845 (which made the rules applicable to Mauritius also operative in emigration to the West Indies) and Act VIII of 1847.

The necessary laws were passed in the importing colonies, incorporating provisions for financing the movement and appropriate rules of procedure. The whole scheme was to be under public officers.[3] In the interest of economy points of departure were limited to Calcutta, Madras and Bombay. At each of these ports there were to be stationed a Protector, appointed by the Government of India, and an Immigration Agent, appointed by the receiving colony. Though every inducement would be made to wives to accompany their husbands, no restriction in regard to the proportion of sexes was to be made. Gratuities of 15 rupees could be made to emigrants before departure, these being later repaid by the receiving colony. The departure of vessels was limited to the period October to March to prevent the passage round the Cape of Good Hope during the winter months.

Emigration under these regulations commenced in 1845, and in the first season (1845–6) 816 immigrants reached British Guiana, 261 reached Jamaica and 225 reached Trinidad. Thus was started the introduction of East Indians into Jamaica. This phase of the movement was, however, of short duration, coming to an end in

---

[1] Letter from Stanley to Peel, 27 November 1843, given in K. N. Bell and W. P. Morrell, *Select Documents on British Colonial Policy, 1830–1860*, Oxford, 1928.
[2] J. Geoghegan, op. cit.   [3] *Fifth General Report*, London, 1845.

1847 after only 4551 persons were introduced. These early immigrants entered the island under contract to work on the plantation for one year, at the expiration of which they could renew the contract or seek other employment. It was claimed that 'the weaker portion of the immigrants fell into ill-health and becoming incapable of labour took to begging and thus eked out a miserable existence, living on alms and travelling from one place to another, so that the whole country was covered with beggars. . . . The parochial institutions for the relief of the poor became overcrowded and the parochial funds were unable to bear the extra strain on them.'[1]

Two factors militated against the continuance of immigration into the West Indies in the late 1840's. In the first place high mortality among the immigrant population, especially the Portuguese, and high mortality among immigrants on the long voyage from India raised doubts in many places as to the desirability of continuing indenture immigration. In fact, following a report on mortality among various racial groups in British Guiana, the Secretary of State recommended that the bounty on immigration be limited to the movement from Africa, as African labourers, deemed to possess 'superior capabilities', showed by far the lowest mortality.[2] In the second place, conditions developed during this period which made the planters unwilling to introduce labourers. The removal of protection on colonial sugar embittered the colonists. Claiming that the civil list was framed on the assumption that protection against slave-grown sugar would continue, the British Guiana legislature moved a drastic reduction in the civil list of 1848. Similar action was taken by the Jamaica legislature, which proposed a reduction in all salaries and refused to vote supplies.

Whereas in other colonies the economic difficulties and, in the case of British Guiana, the constitutional crisis, were soon resolved and Indian immigration resumed in 1850 with financial assistance from the British Government, the complex constitutional and financial situation that developed in Jamaica resulted in a complete halt in the movement until 1858. Nevertheless, Indian immigration into other West Indian colonies increased and some smaller islands which hitherto had not received these immigrants were brought into the scheme.

[1] D. W. D. Comins, *Note on Emigration from the East Indies to Jamaica*, Calcutta, 1893.
[2] *Eighth General Report*, 1848.

The small immigration into Grenada and St Lucia was of note only because the Act sanctioning it (Act XXXI of 1855) contained the important new clause that emigration would be allowed to proceed only if the laws made in the receiving colonies met with the approval of the Government of India. This principle was soon after incorporated into the whole system of indenture emigration under Act XIX of 1856.

The severe mortality on the voyage during the 1856–7 season, when a loss of over 17% was experienced among emigrants from Calcutta, led to inquiries both in the receiving colonies and in India into the facilities provided on emigration vessels.[1] Special attention was directed to the fact that losses on the route from Calcutta greatly exceeded the losses on the route from Madras. Dr Mouat, who examined the question on behalf of the Bengal Government, advanced many recommendations for reducing mortality on the voyage: more careful separation of the sick from the healthy, the reduction in the proportion of women and children embarked to 25%, the discouraging of emigration of pregnant and nursing mothers, and the employment of medical officers with a knowledge of Indian diseases and of Indian languages.

Dr Mouat's recommendations were eventually incorporated, with modifications, into the new body of regulations drawn up by a Committee appointed by the Government of Bengal in 1860 to investigate the whole question of indenture immigration.[2] To obviate the difficulty of 'procuring the services of competent and respectable surgeons at Calcutta' the employment of surgeons from the Australian emigration service was advocated. However, there were 'moral objections' to reducing the proportions of women and children among emigrants, and this proposal of Dr Mouat was rejected.

In 1858 provisions were made for the resumption of immigration into Jamaica, the cost to be defrayed by a loan raised in three instalments and repayable within 15 years. The period of indenture was extended to two years, and the law also included more stringent regulations regarding lodgings, food standards, and other matters affecting the welfare of the immigrants. This scheme was also of but short duration, lasting up to 1863, and under it the number of immigrants brought in was only 4644. Again there was dissatisfac-

---

[1] *Eighteenth General Report*, 1858, and *Nineteenth General Report*, 1859.
[2] J. Geoghegan, op. cit., and *Twenty-second General Report*, 1862.

tion with the quality of the immigrants and a Committee was appointed in 1867 to report on indenture immigration. The Committee reported that 'well-founded complaints were made that many of those imported were much emaciated, of very low physical powers and unaccustomed to agricultural labour'.[1] 'Allotment of such persons to the planters as labourers had caused much dissatisfaction and tended greatly to deter them from making further application.' Several alterations in the system were introduced on the Committee's recommendations in a law passed in 1869. The period of indenture was further increased to 5 years, employers had to supply immigrants with daily rations, while the level of wages was also fixed. About this time also, an important change in the method of financing immigration was made. Hitherto this was a general charge on the revenue of the island, being financed by a special fund, but from 1867 up to 1873 indentured immigration was almost wholly financed by the planters. In 1873 the right to the return passage was commuted to a bounty after the Secretary of State had ruled that 'the general revenue may properly be charged to a limited extent with such payments as may be made to Indian immigrants in lieu of return passages, in order to induce them to remain in the colony as free settlers'.[2]

The passing of Act XIII of 1864 constituted an important landmark in the history of Indian emigration. It was primarily a consolidating Act, repealing all previous laws on emigration to British colonies.[3] It declared emigration lawful to the following places: Mauritius, Seychelles, Natal, Jamaica, British Guiana, Trinidad, St Lucia, St Vincent, St Kitts and the Dutch island of St Croix. Permission to initiate emigration schemes to other territories might be given on the passing in the receiving country of measures of protection deemed adequate by the Indian Government. Calcutta, Madras and Bombay were, as before, the only ports of authorized emigration. There was not much that was new in the procedure for recruiting which this Act laid down, but it is necessary to outline the process, which, with slight modifications, continued throughout the course of the movement. Emigrants could be recruited only by recruiters holding licences direct from the

[1] D. W. D. Comins, op. cit.
[2] Despatch from Earl Kimberley to Sir J. P. Grant, 25 June 1873, *V.H.A.J.*, 1873.
[3] J. Geoghegan, op. cit., and *Twenty-fifth General Report*, 1865.

Protector of Emigrants and countersigned by the magistrate. The recruit had to appear before the magistrate of a district or before the Protector if recruited in a town. If the examination of the recruit showed that he understood, and was prepared to abide by, the terms of the agreement, the magistrate or the Protector registered his name, his father's name, his age, his village, the depot to which he was to go, the rates of wages and the period of service. The would-be emigrant was then furnished with a copy of the entry in the register. 'The vaccinator in the pay of the Agency up-country' was then supposed to vaccinate the recruit before the latter left his district.[1] He was then taken without delay to the appropriate depot and there washed and medically examined. If he had not had smallpox and had 'escaped the attention of the vaccinator', he was immediately vaccinated. If the medical examination showed the recruit fit the Immigration Agent was informed, but if he was rejected on medical grounds the Protector then directed the Agent to pay the recruit a sum sufficient to enable him to return to his district. The recruit was not supposed to be kept in the depot longer than 30 days, though this rule could not always be strictly observed.

Other important clauses of the Act fixed the emigration season at from 31 July to 16 March, authorized the Governor General in Council to control the proportion of women and children to be embarked, and dealt with the defraying of expenses incurred by the Indian Government in connexion with indenture emigration.

The 1864–5 season again witnessed extremely high mortality on the West Indian routes, losses of between 22% and 50% being recorded when 'a bilious intermittent fever of a typhoid character had become epidemic'.[2] As before, inquiries both in the West Indies and in India were held. The former yielded little fundamental information on the cause of the high mortality. The Sanitary Commissioners of Bengal considered the disease 'a peculiarly deadly fever known for some years in India, where at Agra, Meerut, Lahore and Saugar it had committed great ravages, but was apparently little known elsewhere'. Cases of this fever had been noted among inmates of the British Guiana depot earlier in the year. 'On board each ship in which this fatal disease afterwards

---

[1] Annual Report of the Government Emigration Agent for Trinidad, 1875, *Trinidad Royal Gazette*, 1875, and J. Geoghegan, op. cit.

[2] *Twenty-sixth General Report*, 1866, and J. Geoghegan, op. cit.

appeared there were emigrants who had in them at the time of embarkation elements of the disease which only required to be brought to maturity.'[1] The after-effects of the cyclone which struck Calcutta in the previous year were also advanced as contributing factors. But the recurrence of this fatal disease continued to cause severe mortality, and in 1871 Dr Cunningham, Sanitary Commissioner with the Indian Government, drew up special regulations to combat it.

In the early 1870's the Emigration Commissioners reported great difficulty in securing recruits in India. In the first place the necessity for maintaining the stated sex ratio among migrants made it impossible to obtain the required numbers 'without resorting to a low and impure class' of woman.[2] But 'the moral importance of procuring a reasonable proportion of women in emigration is so great that no relaxation of this rule can be admitted', it was claimed. Another deterrent to emigration was the favourable rice crop in India during these years. Again, the prevailing regulations, under which the countersigning of recruiter's licences was left to the magistrates, tended to reduce the number of recruiters. For many of the magistrates, in the belief that emigration to the West Indies would not benefit the Indians, refused to countersign the recruiters' licences.[3] In fact, the Governor of Trinidad claimed to discern 'a spirit of hostility on the part of the magistrates and Public Officers in India to emigration to the West Indies'.[4] It is possible that recruiting was hampered by the policy of certain Indian officials, acting on the view, held wrongly or rightly, that the position of the labourer was not bettered by emigrating to the West Indies; but, as will be seen presently, there were other factors tending to reduce the number of Indians leaving for the West Indies.

Another consolidating Act, No. VII of 1871, brought no substantial alterations, but soon after the passing of this Act the entire supervision of emigration in India was transferred to the Department of Agriculture, Revenue and Commerce. Another important change that took place about this time concerned the supervision of the movement in England. With the transfer of the administration of the Passengers Act to the Board of Trade (35 and

[1] *Twenty-seventh General Report*, 1867.
[2] *Thirty-first General Report*, 1871.
[3] *Thirty-second General Report*, 1872.
[4] Message from Governor of Trinidad to Legislative Council, 21 March 1872, *Trinidad Royal Gazette*, 1872.

36 Vict. c. 73), most of the administrative functions previously delegated to the Emigration Commissioners now fell to the Colonial Office. This also meant that after 1873 there was no annual comprehensive survey of indenture immigration such as the Reports of the Colonial Land and Emigration Commissioners provided since 1840. However, in the case of Indian immigration, which was virtually the only indenture immigration then in being, this gap was partly filled as the *Annual Reports on Emigration from the Port of Calcutta to British and Foreign Colonies* became available from 1875.

The condition of Indian immigrants in British Guiana was the subject of an extensive inquiry in 1870 by a Royal Commission, the Report of which, while not fundamentally affecting the course of immigration, is notable because of the abuses it disclosed and the recommendations it made for alleviating some of the harsher features of indenture. The Commission, consisting of W. E. Frere, Sir George Young and Charles Mitchell, was appointed following the receipt by the Secretary of State of a lengthy letter from G. W. Des Vœux, a former magistrate in British Guiana, alleging that 'a very widespread discontent and disaffection [existed] throughout the immigrant population', largely attributable to the inadequacies of medical services and the partial attitude of the stipendiary magistrates. The Report of the Commission is valuable because of its thorough examination of the indenture movement into British Guiana and the status of the individual under indenture. Though it established the fact that many of the particular allegations of Des Vœux were groundless, it exposed many abuses. It appeared that the 5-year reindenture period was severe. The immigrants had little confidence in the laws, which from the records of convictions and prosecutions seemed mostly administered in the interest of the employers. It also appeared that the efforts of the Immigration Agent General and his staff on behalf of the immigrants were inadequate.

Among the recommendations of the Commission were the following: the clarification of the position of the Immigrant Agent as a representative of the immigrant population; better means of examining wages; improvement in the hospital services; the appointment of medical officers by the Government and not by the planters. On the whole Des Vœux's particular allegations of neglect and ill-treatment were not substantiated; 'there was more kindness and good treatment of them than he is disposed to allow.' However, the Indian Government, in the words of Geoghegan, owed him 'nothing but gratitude, for he was the means of bringing to light the state of things which the main body of the Report is devoted to exposing'. In fact the Indian Government, on the strength of the Commission's Report, threatened to stop emigration to British Guiana in 1871, a step avoided only by the passing of Ordinance No. 7 of 1873.

Further changes in the method of financing immigration into Jamaica had to be made in 1877, by which time it should be noted the only immigration into the island was the small irregular supplies from India.[1] It was decided that one-third of the annual expenditure should be defrayed from general revenue and the rest from export duties and by employers. At this time, indeed, the difficulties of financing immigration were part of the wider financial embarrassment of the island. A heavy deficit developed in the immigration fund in 1879, and accordingly with the approval of the Secretary of State the whole of this debt was transferred to the general revenue of the colony. As a further relief for the planters, the hospital costs and medical charges were removed and they were required to shoulder only current costs of introducing and repatriating immigrants by means of export taxes on rum and sugar and a payment of £15 on each immigrant allotted to them. Another change introduced at this time was the cancellation of the right of commuting the return passage to a bounty.

Meanwhile the number of immigrants into Jamaica remained at an extremely low level. During 1869–76 the total introduced was only 8320. Complaints about the shortage of labour figured prominently in the report of the Royal Commission of 1882 (the Crossman Commission), which not only bewailed the 'natural indolence of the negro' but held that 'the inflow of outside capital has also been checked by the difficulty of obtaining a constant and trustworthy supply of labour'.[2] But a shortage of labour was by no means the cause of the economic difficulties facing the island in the 1880's; these stemmed largely from the severe depression into which the sugar industry was plunged. Even though immigration had to be suspended in 1885 because of these conditions, the belief in the

---

[1] Despatch from Earl of Carnarvon to Lieutenant Governor Rushworth, 14 June 1877 and despatch from Governor Musgrave to Earl of Carnarvon, 20 December 1877, *V.H.A.J.*, 1877.

[2] Quoted in Lord Olivier, *Jamaica the Blessed Island*, pp. 225 and 235.

magical powers of immigration persisted. Comins, sent to examine indenture immigration in the island in 1891, echoed the planters' views on this subject, enumerating five reasons purporting to indicate a shortage of labour: (1) the cessation of East Indian immigration coupled with the return to India of many immigrants whose period of indenture had expired; (2) the migration to Panama and elsewhere which was then commencing; (3) the expansion of public works in the island which absorbed a portion of the colony's agricultural labour; (4) the expansion of the banana cultivations which attracted many workers away from the sugar plantations (many labourers who formerly worked for wages on the large plantations were now growing their own bananas for sale); (5) 'the smattering of education given to youths, which has the tendency to disqualify them from free or other manual work, is also credited with the loss of part of the labour attainable.'[1]

Reviewing the general exodus to foreign countries at this time, the Governor considered it an advantage rather than a loss of labour to the island. But planter interests still considered labour supplies inadequate and a select committee of the Legislature was appointed to consider ways of increasing the island's labour supply. It considered a renewal of immigration justified, but put forward an experiment aimed at extending indenture to cover local workers. Law 29 of 1891 therefore provided for a system of indenture among non-Indians, whereby persons could be indentured to employers for 1 year under the terms and conditions applicable to Indians. On the termination of his period of indenture, the worker would be entitled to the sum of £2 provided by the employer.[2] Nothing came of this strange experiment as no applicants for indenture were received. Indeed, from the experience of the eastern Caribbean where all efforts to promote inter-Caribbean migration under terms of indenture failed dismally, such a scheme was foredoomed to failure. Accordingly, efforts to correct the alleged labour shortage were confined to East Indian immigration. But the token increments witnessed in succeeding years, though they may have helped some of the sugar plantations, cannot be held to have influenced the size of the labour force of the island as a whole.

Though in Jamaica as in other West Indian islands many problems loomed large as factors determining the size and course of

[1] D. W. D. Comins, op. cit.
[2] *Report of Immigration Department*, 1891.

East Indian immigration, factors having their origin outside the West Indies became increasingly important during the last quarter of the nineteenth century. By this time the only type of immigration in operation was that from India, and it was mainly forces at work in India that determined the decline in the migration of indentured labourers. The clearest expression of the decline in this type of immigration is revealed in the falling rate of recruitment. The small numbers of recruits secured engaged the attention of the Indian Government, who in 1881 appointed D. G. Pilcher and G. A. Grierson to investigate its causes. And on the basis of their report new regulations designed to make recruitment easier were passed.[1]

However, this action taken by the Indian Government did not imply any basic change in its direction of indenture migration, towards which from the inception of the movement it remained essentially neutral. This position did not at the same time blind it to the advantages that accrued to many emigrants to the West Indies. While not changing its position, therefore, the Indian Government turned more attention in the 1870's to the problem of stating more explicitly its policy in regard to indenture emigration. The position is clearly summed up by the Bengal Government thus: 'The advantages of emigration as a means of relieving the pressure for existence in crowded Indian districts need no demonstration. . . . The Lieutenant Governor is moreover satisfied that as a very general rule, the Bengal or Behar coolie materially improves his lot by emigration to the British colonies.' This, however, was not an unqualified endorsement of the movement. 'It is necessary to observe that help which the Government can accord to it [emigration] must in the nature of the case, be confined to narrow limits. In a country where the mere expression of a wish on the part of the officials is often interpreted as an order by the people, or distorted into such by designing persons for the achievement of their own ends, the Government could not, consistently with its duty to the people, undertake to lend active and direct aid to recruitment for the colonies. All that the administration can rightly do is to see that no obstruction is unnecessarily placed in the way of emigration, to encourage the dissemination among the masses of accurate information regarding the colonies and the prospects of colonial emigrants, and to facilitate the interchange of

[1] *Annual Reports on Emigration*, 1881–2.

communication between the emigrants and their relatives and friends in India. . . . The real and only serious obstacle to recruitment for the colonies has, doubtless, been and must continue to be, the objection of the people themselves to emigrate, due to cast prejudice, attachment to their homes, general ignorance of the conditions of life and prospects in the colonies and unwillingness to leave villages in season of prosperity and cheap food.'[1]

Still the whole indenture migration came more and more under critical examination in India and elsewhere, and the Government could not completely ignore the growing resentment with which Indians viewed this movement. And after 1890, increasing attention was paid to the conditions of life and work among indentured immigrants in the West Indies. In 1890 Major D. W. D. Comins was sent to the West Indies to examine the whole question of East Indian indenture. Comins surveyed the history of indenture immigration into the British colonies then receiving immigrants, the legal protection given them and their general health conditions in a series of reports published in 1891. A separate report of his dealt with the vexed question of the return passage.[2] Another body which turned its attention to West Indian immigration was the Royal Commission which reported in 1897. This Commission was mainly concerned with the serious conditions of the sugar-producing colonies at this time, and in the course of its recommendations had something to say on West Indian migration. Unlike the Committees of the 1840's, this Commission did not propose immigration as a solution of prevailing problems. It emphasized that a measure of immigration was important, but most of its recommendations bearing on this topic stressed the economic consequences of meeting numerous return passages should future depressions give rise to a large-scale return movement to India.[3] The Commission received evidence that the Jamaica Negro was an excellent labourer but it appeared 'that on some estates, though not on all, it is difficult to carry on cultivation without a proportion of indentured coolies, whose services can always be depended on'. No change in the

[1] *Annual Report on Emigration,* 1882–3.

[2] The following Reports were prepared by Comins: *Note on Emigration from the East Indies to Jamaica; Note on Emigration from the East Indies to St Lucia; Note on Emigration from India to Trinidad; Note on Emigration from India to British Guiana; Note on the Abolition of Return Passages to East Indian Immigrants from the colonies of Trinidad and British Guiana.*

[3] *Report of the West India Royal Commission,* 1897.

policy of immigration was recommended. 'The planter pays heavily for introducing labour [but] he has a strong inducement not to apply for coolies, unless he thinks they are absolutely essential to the working of his estate.'

In India the whole system of indenture emigration was now being heavily attacked. Several aspects of this disapproval emerged. It appeared incompatible with the growing national consciousness throughout India. Because of the many restrictions imposed on the indentured labourer the movement constituted a denial of freedom. What tended to make the whole system notorious was the ill-treatment of Indians in South Africa. In 1908 G. K. Gokhale moved in the Legislative Council: 'That this Council recommends that the Governor General in Council should be empowered to prohibit the recruitment of indentured labour in India for the Colony of Natal.'[1] This resolution was accepted by the Government of India, and under Act XIV of 1910 emigration to Natal was prohibited as from 1 July 1911. Then in 1912, Gokhale moved a more comprehensive proposal for complete 'prohibition of recruitment of Indian labour under contract of indenture, whether for employment at home or in any British colony'. He characterized the indenture system as 'based on fraud' and 'a great blot on the civilization of any country that tolerates it'. But the Government, having regard to the findings of the Sanderson Commission, did not support this motion.

The constant attack against the indenture system in India had in fact led the British Government to have the whole question examined by a Royal Commission, which was headed by Lord Sanderson and which reported in 1910. From evidence before it, this Commission concluded, 'It can safely be said that notwithstanding some unfortunate occurrences at times now remote, the system has in the past worked to the great benefit, not only of the colonies but equally of the main body of emigrants and does so still in the present.'[2] So far as Jamaica was concerned, it deemed immigration 'requisite for the maintenance of the sugar and fruit plantations in those portions of the island where the population of African origin cannot be depended upon for steady agricultural work'.

Another inquiry into the indenture system was set in motion by

[1] 'Emigrant', op. cit.
[2] *Report of the Committee on Emigration from India to the Crown Colonies and Protectorates*, 1910.

the Indian Government which sent a mission to the West Indies, Fiji and Surinam to examine the health and living conditions of Indian immigrants. This mission consisted of two members, J. McNeil and Chimman Lal, and they submitted their report in 1914.[1] A careful survey of the health and medical facilities available to East Indian immigrants as well as their general living conditions led them to conclude that on the whole the immigrants had improved their economic and social conditions by migrating to the West Indies. They indicated many harsh features of indenture but did not share the view of many critics of the system that it was utterly bad.

It was not only in India that support for indenture immigration into the British colonies was waning; in the West Indies as well voices were being raised against it. This can be seen in the evidence placed before the Sanderson Commission, one of the last bodies to investigate indenture. One group contended before the Commission that immigrants should come into the colonies as 'potential citizens' and 'if they come in at all that they shall come in free'.[2] The indenture system was described as 'vicious and degrading'. Another line of argument against the system was that the sugar industry, which, on this view, alone benefited from indenture, did not shoulder the entire cost of the movement. This argument was even extended to cover the financing of the hospital and general medical attention which at that time was a charge on the whole colony's revenue. Thus the Jamaica Baptist Union declared that 'it does not offer any objection to the importation of labourers *per se* though it deems it unnecessary . . . but our contention is that whoever imports labour into the island should do it at his own cost.' Of course, such evidence placed before the Commission did not go unchallenged. Possibly the opposing argument which carried most weight with the Commission was the stock contention long used to justify immigration into the West Indies. The Governor of Jamaica expressed it thus: 'The one great reason why it is necessary to import East Indian immigrants into the colony is because it is impossible . . . to obtain a regular and adequate supply of native labour. . . . There is the fact that the average Creole labourer has

[1] J. McNeil and Chimman Lal: *Report on the Conditions of Indian Immigrants in the four British Colonies: Trinidad, British Guiana or Demerara, Jamaica and Fiji*, Simla, 1914.
[2] *Report on the Committee on Emigration from India to the Crown Colonies and Protectorates*, 1910.

no desire to work for wages beyond what is necessary to supply his immediate needs.' The expression of such views at this time is ironical in view of the fact that manifestly it was the desire to secure better wages that was at that very period inducing large numbers of Jamaicans to emigrate to Panama and elsewhere.

Doubtless the arguments against immigration expressed within the West Indies had little influence on the course of the movement, but in view of the growing resentment against it in India, it could no longer continue unchecked. A favourable opportunity of checking it presented itself with the outbreak of World War I. The war made it impossible to secure ships to transport the immigrants, and in 1914 the Indian Government announced its intention of abolishing the indenture system with the object of conserving man-power for war purposes.[1] Thus under the Defence of India (Consolidated) Rules, 1915, the termination of the system was decreed, effective as from 12 March 1917, and recruitment of unskilled labour for indenture emigration was halted.[2] The disfavour with which the movement was viewed increased as a result of an examination of some of the harsher features of indenture revealed in the Report of McNeil and Chimman Lal. In 1915 the Government of India informed the Secretary of State that 'the time had come for His Majesty's Government to assent to a total abolition of the system of indenture of Indian labour in the four British colonies where it still prevails and in Surinam'.[3] To this the Secretary of State agreed, on the condition that sufficient time would be allowed for the necessary adjustment within the colonies. So far as Jamaica was concerned, its token importations of Indian labour came to an end in 1914 when 293 were secured.

After the end of the war were several discussions about resuming some form of Indian immigration into the West Indies, but all proved fruitless, and it can be said that the system of indenture immigration came to an end in 1917, though of course those entitled to return passages still maintained the right to claim repatriation to India and many exercised this right. The Immigration Act passed by the Indian Government in 1922 by its strict measures on the whole emigration movement effectively ruled out the possibility of any resumption of indenture immigration into the West Indies.

---

[1] *Annual Report on Emigration*, 1914.
[2] Ibid. 1917.   [3] 'Emigrant', op. cit.

A record of all known introductions into Jamaica, together with a brief note on the sources of this material, are given in Appendix II. It appears that the total introduced was only 36,400 during the 70 years of the movement, which gives a negligible annual average increment to the population of the island of about 500. It was only during the years 1866–76 that the rate of East Indian immigration attained appreciable dimensions; during these years 11,300 were introduced. Moreover, when those who exercised their right to the return passages are discounted the significance of the whole movement is drastically reduced. Up to 1916, 11,900 are known to have returned to India, and there were others leaving the island concerning whom no records exist, so that it is certain that the total net immigration of East Indians into Jamaica between 1845 and 1916 was less than the 24,500 which published records disclose.

A comparison of the numbers entering Jamaica with the numbers entering other West Indian colonies emphasizes the minor contribution of East Indian immigration to population growth in Jamaica. The recorded East Indian immigration into the West Indies compiled from available records can be summarized as follows:

British Guiana	238,900 (55·3%)
Trinidad	143,900 (33·3%)
Jamaica	36,400 (8·4%)
Grenada	5,900 (1·4%)
St Lucia	4,400 (1·0%)
St Vincent	2,500 (0·6%)
St Kitts	300 (0·1%)

Thus British Guiana and Trinidad absorbed the great proportion of these immigrants (89%), whereas only 8% entered Jamaica. East Indian immigration dominated the pattern of population growth in Trinidad and British Guiana, though it had only slight effects on population growth in Jamaica.

It is interesting to note briefly some of the distinguishing features of the East Indian immigrants. It is impossible to determine the occupations they pursued before their departure for the West Indies. Every effort was made to secure recruits fitted to agricultural labour, but it became increasingly difficult with the passage of time to obtain immigrants of this type, and towards the end of the nineteenth century recruiting meant no more than accepting persons who elected to emigrate; there were not sufficient recruits to exercise much choice on the basis of suitability. In regard to sex ratios, there was at the inception of the movement no fixed proportion laid down, though it was urged that every inducement should be made to wives to accompany their husbands. But later control of the sex ratios was introduced. As the whole aim of indenture immigration was to secure workers for the plantations, male adults were most eagerly sought after, but on the ground that too low a proportion of females was conducive to immoral conduct among the immigrant population, the necessity of having a minimum proportion of females was advanced. Other considerations were also involved in settling the sex ratios. For instance, it was at one time maintained that the relatively high proportion of females insisted on was a prime factor in the increased death-rates on the voyage. Because of the reluctance of females to emigrate it was often found impossible to recruit emigrants in the proportion prescribed between the sexes. Restrictions as to age were apparently never in force, though it was on several occasions claimed that a high proportion of children increased the rate of mortality on the voyage and that consequently not many infants should be included.

The detailed statistics published in the *Annual Reports on Emigration* after 1875 give breakdowns of emigrants in terms of sex and age. These of course are not numbers landed in Jamaica but the numbers actually despatched from Calcutta, and are more detailed than the statistics of emigrants landed in Jamaica. For the years 1875–1917 the numbers of East Indians leaving for Jamaica can be broken down as in Table 29. That the whole movement was both age selective and sex selective is evident from this summary. Thus 88% of the males were between 10 and 30. Only very small numbers over working age were imported, the proportion of males falling over 30 being less than 5%. The concentration of females within the span of working age is only slightly less marked, 82% of all females being between 10 and 30. For the whole period covered by these records, male emigrants were more than twice as many as females, the overall sex ratio being 2268. Though somewhat lower among children, the sex ratios are essentially high for all age-groups, thus again emphasizing the aim of the movement: to secure workers for the sugar plantations.

The growth of the East Indian population of Jamaica presents interesting parallels with the growth of the East Indian populations

Table 29.   *Emigrants leaving India for Jamaica, 1875-1917, by age
and sex**

Age-group	Male	Female
Under 2	244	249
2–9	622	433
10–19	2,082	911
20–29	7,860	3,206
30–39	518	198
40 and over	13	3
Total	11,339	5,000

* From *Annual Reports on Emigration*.

of British Guiana and Trinidad. Comparative population estimates
for the three territories together with intercensal rates of growth
appear in Table 30. In all three it is clear that the opening phases
of indenture immigration resulted in very high rates of growth
among the East Indians despite the very high rates of mortality
prevailing. Between 1851 and 1861 annual rates of growth in
British Guiana and Trinidad were 11% and 13% respectively, while
during 1861–71 the rate for Jamaica was 13%. But in no case did
these very high rates persist for long. In fact the many factors
tending to retard indenture immigration after 1871 brought about
dramatic falls in the annual rates of growth of the East Indians.
Between 1871 and 1881 annual rates of growth declined to 3·5%
in Jamaica, 5·2% in British Guiana and 5·9% in Trinidad. The
succeeding intercensal interval actually witnessed a small decline
in the East Indian population of Jamaica, and although the years
after 1891 brought a marked recovery in rates of growth, it can be
said that by the end of the nineteenth century the immigration of
East Indians was too small to have any marked effects on population growth in Jamaica or in any other West Indian territory.

From other areas.   Europeans were among the earliest indenture
labourers introduced into Jamaica, coming into the island between
1834 and 1844. The numbers involved are very small and the
whole movement remains merely of historical interest. 'They might,
for all their ultimate importance, have been dismissed in a footnote,
had their importation not raised questions between the Governor
and the Assembly.'[1] Among the earliest European arrivals were a
group of Germans who were supposed to introduce new methods of

[1] W. L. Burn, op. cit. p. 291.

Table 30. *Growth of the East Indian populations of three major West Indian territories during indenture immigration*

Year	Jamaica		British Guiana		Trinidad	
	Population	Average annual rate of growth %	Population	Average annual rate of growth %	Population	Average annual rate of growth %
1841	—	—	343	—	—	—
1851	—	—	7,670	36·44	3,993	—
1861	2,262	—	22,081	11·15	13,488	12·94
1871	7,793	13·17	48,363	8·15	27,425	7·35
1881	11,016	3·52	79,929	5·15	48,820	5·94
1891	10,805	-0·19	105,463	2·81	70,218	3·70
1901	—	—	—	—	86,383	2·10
1911	17,380	2·41	126,517	0·91	110,120	2·46
1921	18,610	0·68	124,938	-0·13	121,420	0·98

Note. These values are taken from the relevant census reports. For the early years the values cover only persons born in India, but evidently the numbers of East Indians born in the colonies were at these early dates very small. The populations entered refer to natives of India for 1861 and 1871 in the case of Jamaica; for 1841 and 1861 in the case of British Guiana; and for 1851 and 1861 in the case of Trinidad. The population entered for Jamaica for 1891 is taken from the Abstract of the 1891 census report giving particulars of the East Indian population. Elsewhere in this census report (in the tabulation of the 'colour' of the population) the number of 'coolies' is given as 10,116 (cf. Table 14 of Chapter 3).

cultivation and to have in general a favourable influence on the Negro workers. Later additions came from Scotland and Ireland. Complaints made by the Rev. W. Knibb about ill-treatment and high mortality among Irish immigrants who came to Jamaica on the *Robert Kerr*, together with a general disapproval of European indenture immigration into the West Indies, led to a cessation of the whole movement.[1] As far as can be ascertained from the records, between 1834 and 1845 the number of Europeans introduced into Jamaica under indenture was only 4100. There was, of course, in addition to this indenture movement, which is of no importance, a constant movement of population between Europe and the West Indies which, though also numerically very small, was of great significance.

Again from the standpoint of historical interest, it can be noted

[1] Despatch from Russell to Metcalfe, 30 April 1841, and evidence of Rev. W. Knibb, *Report of Select Committee appointed to enquire into the State of the Different West India Colonies*, London, 1842.

that attempts were made to secure immigrants from the United States and Canada, though the insignificant numbers involved robs the movement of demographic significance. In 1840 Jamaica appointed A. Barclay as Commissioner of Immigration, and he went on a voyage to North America and elsewhere in order to secure immigrants. He reported unfavourably on the prospects of large-scale emigration from North America.[1] In 1841 the Agent General of Immigration reported that a few immigrants from the United States (not all of whom were Negroes) were arriving in the island, but the movement never attained importance, recorded immigration from Canada and the United States between 1840 and 1845 being only about 400.

The introduction of Chinese under indenture into Jamaica is again of no more than historical interest. In 1853, Jamaica announced its willingness to promote Chinese immigration and two vessels were chartered for this purpose.[2] But because of high transport charges, only one ship was despatched with 310 immigrants. In the same year (1854) Jamaica obtained 205 Chinese from Panama at a cost of £6 per head. This experiment was a failure, for mortality was high and the Chinese proved sickly and unwilling workers. So unsatisfactory was this venture that attention was not again turned to this source of additional labour until 1884. In that year the failure to obtain East Indians led the West India Committee to apply for permission to secure Chinese labourers and 680 entered the island. Again there was general dissatisfaction over these immigrants, who proved poor workers and who unfortunately arrived at a time when sugar prices were falling. The Protector of Immigrants reported unfavourably on their conduct: 'The Chinese . . . proved intractable and preferred exorbitant demands for wages which could not be complied with.'[3]

Though on only two occasions were Chinese brought into Jamaica under indenture, there developed after 1911 a sizeable movement of free Chinese into the island. The course and exact magnitude of this movement cannot be traced, but by 1943, 12,400 Chinese were settled in the island, forming a very important section of the island's population.

1 Report on Immigration and Report of Commissioner A. Barclay, *V.H.A.J.*, 1841.
2 *Fourteenth* and *Fifteenth General Reports*, 1854 and 1855.
3 *Annual Report of the Protector of Immigrants*, 1885.

EMIGRATION TO FOREIGN TERRITORIES

The first records of large-scale emigration from Jamaica appear in the 1880's when work on the Panama Canal was started by the French. Jamaica and Barbados were the two colonies of the British Caribbean that contributed most of the labour force employed in the construction of this canal. In 1882 the Governor of Jamaica recorded that labourers were leaving the island at the rate of 1000 a month for Panama, Mexico and Yucatan. He did not view the movement with disfavour. 'While the immediate loss of any productive labour may be regarded as a misfortune to the country, it need not . . . be assumed to be an altogether unmixed evil, and it may . . . ultimately lead to beneficial results and a better appreciation and a more permanent settlement of the labour question and to such an improvement in the position of the labour class as will tend to remove from their minds any general inclination to emigrate with the view of bettering their prospects.'[1] On the whole, considerable advantage to the island was expected from the 'ebb and flow in the development of intelligence and habits of industry of our population', especially as many of the emigrants returned to Jamaica 'bringing with them money with which they arrange their affairs and aid their families'.

The year 1882 saw a prohibition of emigration to Mexico within the terms of Law 23 of 1879; but this concerned only East Indians, who were being enticed to go to work there under contracts of service.[2] It was estimated that during the migration season 1883–4 the number emigrating to Panama was as high as 24,300; but, on the other hand, there was a considerable return movement. During that season some 11,600 persons returned to Jamaica. 'It seems now a recognized course for the men after earning a fair amount to return to Jamaica to visit their families and friends and after a short time return to the isthmus to work again.'[3] The effects of these withdrawals from the labour force of the island might, however, have been felt by this time. The planters claimed that the shortage of labour precipitated by this movement, coupled with the low price of sugar on the world markets, made the continuance of cultivation on many sugar estates difficult, though in the Governor's

---

[1] *Governor's Report on the Blue Book*, 1882.
[2] *Annual Report of the Immigration Department*, 1883.
[3] *Governor's Report on the Blue Book*, 1883–4.

view the savings brought back to the island by returning emigrants meant that on balance emigration constituted a gain rather than a loss.

In 1885 there was a sharp decline in the volume of emigration to the canal; in fact, the recorded return movement from the area about equalled the recorded outward movement. Work was harder to secure, and following an unprovoked attack on a body of Jamaicans by Colombian troops at Culebra, the Government of the island issued a warning to persons on the possible dangers to be faced by emigrants to Panama.[1] By 1886 it was evident that Panama no longer exercised 'so great an attraction for the Negroes as formerly'. 'The cheaper living, the better climate, the home association and other local advantages are more than equal to the higher and somewhat nominal rate of wages earned at the isthmus.' Another factor tending to reduce the movement to Panama at this time was the attraction offered by Costa Rica, where railway construction was then in progress. Here again it was claimed that dangers attended emigration for work on this project. But neither here nor in the case of emigration to Panama was there any attempt on the part of the Jamaica Government to interfere with the labourers' 'desire to dispose of his labour wherever seems best to himself'.

The alleged hardships facing Jamaican emigrants to Panama and Costa Rica led to an investigation into conditions in these regions. It was concluded that the isthmus was 'overstocked with Jamaicans unable to obtain work', that there was a high degree of unemployment in Panama, while many who went to Costa Rica also failed to secure employment. Nevertheless, the records of emigration for the year 1886–7 showed an emigration of 10,400 to Panama and of 1200 to Costa Rica; these were, however, largely offset by the appreciable numbers returning to the island; 7100 and 60 respectively. Emigration to the isthmus continued to decline, and by the time of the suspension of work on the project in 1888, it was reduced to 3200.[2] The suspension of this work resulted in distress for many Jamaicans on the canal, and the Jamaican Government sent a medical officer to look after their interests. A substantial number was repatriated, but about 6000 elected to remain.

[1] *Governor's Report on the Blue Book*, 1885 and 1886.
[2] *Governor's Report on the Blue Book*, 1888–9.

The resumption of work on the Panama Canal by America in 1904 led to renewed emigration from Jamaica. Indeed, this undoubtedly set afoot what was to prove the most substantial emigration ever experienced by the island. The construction of the canal called for large supplies of labour; and even at the commencement of the work it was reported that a substantial proportion of the 3000 employees were from Jamaica.[1] In 1904 Secretary of War W. H. Taft visited Jamaica and conferred with the Governor on the possibility of securing 10,000 workers from the island. This the Governor was unwilling to allow unless the Panama Canal Commission 'deposited £5 per laborer with the island government to meet the burden which his leaving the island would probably throw on his parish under the poor law of the island for the support of those dependent on him', and agreed to meet the cost of repatriation on the completion of the work. The former unrealistic condition, strongly reminiscent of the devices resorted to by Barbados in the 1840's to restrict emigration to British Guiana and Trinidad, was not entertained by the Commission. In fact Secretary of War Taft was confident that Jamaica labourers would emigrate to the Canal Zone voluntarily in sufficient numbers as the facilities for assuring this existed. There were two direct shipping connexions between Jamaica and Panama and the fare was low, $5 per person. And in fact by 1904 the Collectors of Taxes were reporting 'a great exodus' of labourers to Colon.

In the Commission's Reports for 1905 and 1906 dissatisfaction with the numbers and quality of workers secured from the West Indies was expressed. In words that might have been uttered by a West Indian sugar planter the Commission declared: 'The majority of them work just long enough to get money to supply their actual bodily necessities, with the result that while the Commission is quartering and caring for about 25,000 men, the daily effective force is many thousands less.'[2] But this view did not long persist; in 1907 the Commission reported an improved efficiency of the working population and in the following year declared, 'the labor problem may be considered solved'. Most of the labourers obtained from the eastern Caribbean and from more distant parts were secured under contract, but so far as Jamaica was concerned this

[1] *First Annual Report of the Isthmian Canal Commission*, Washington, 1905.
[2] *Annual Report of the Isthmian Canal Commission for the year ending December 1, 1906*, Washington, 1907.

method of recruiting was hardly used at all. Only forty-seven labourers from Jamaica were brought to the Canal Zone under contract. Recruiting agents were apparently maintained by the Commission, but evidently their function was merely to encourage persons to emigrate to the canal and not to engage them under contracts of service.[1]

Work on the canal came to an end in 1914 and, indeed, for several years before this the demand for West Indian labourers was declining. Still the decade 1911–21 witnessed the climax of emigration from Jamaica. And an examination of the background of this movement is instructive. Undoubtedly the whole movement was a continuation of the large-scale emigration initiated by the exodus to Panama. The level of wages enjoyed by workers in the Canal Zone—ranging from 80 cents to $1·04 per day—was much higher than the rate of 1s. 6d. to which they were accustomed in Jamaica.[2] Consequently few were content to return to the island on the completion of the work at Panama, but went farther afield. 'Apparently these West Indians who have been accustomed to the high wages and higher standards of living prevalent on the Canal Zone, have not chosen to return to their homes in the island, but have chosen to seek employment on construction work in other fields. Over 2000 were sent to Honduras and thousands of others have gone to Costa Rica and Bocas del Toro.'[3] Moreover, 'Panama money', as the remittances made by workers in the Canal Zone to relatives and friends in the West Indies were sometimes known, revealed to many in Jamaica who had not emigrated the brighter prospects awaiting emigrants, thus constituting a strong stimulus to emigration.

Further, by 1911 the shipping routes on which such an outward movement could be promoted were fully established, largely as a consequence of the fruit trade. Unlike sugar, which could be stored for long periods at some central point until shipped, the banana, because of its highly perishable nature, called for swift transportation from the field to its final market. This made it desirable for the fruit growers to utilize as shipping points the ports nearest to the banana cultivations. The great fruit-growing concerns developed in the island late in the last century used many small sea-coast towns

---

[1] F. J. Haskin, *The Panama Canal*, New York, 1913, p. 156.
[2] *Annual Report of the Isthmian Canal Commission for year ending 1905.*
[3] *Annual Report of the Isthmian Canal Commission*, 1913–14.

as ports of call for the ships they operated. These lines, calling frequently at essentially rural areas, not only brought tourists to the island, but forged transportation links which made large-scale emigration possible.[1] This was most clearly exemplified in the case of Portland, which by the end of the nineteenth century had extensive banana cultivations, and its port, Port Antonio, evidently became the point of embarkation for most of the emigrants. As early as 1893 the Boston Fruit Company had a ship specially fitted out to encourage tourists to visit Port Antonio, and by the late 1890's it was recorded that every ship entering that port brought American tourists. By 1900 the United Fruit Company had three routes on which their ships travelled regularly. The 'direct' steamers plied directly between Port Antonio and the United States, while the 'Cuban' and the 'Haytian' lines carried cargoes from America to these islands before coming to Jamaica for bananas. They then proceeded directly to America. By 1910 the United Fruit Company's ships were calling weekly at Port Antonio. Similarly, banana interests in St Mary were as early as 1887 shipping fruit directly from Port Maria and Annotto Bay; the former indeed continued for some time to be a port of considerable importance for the parish, handling most of the exports and imports. Fruit companies also made St Ann's Bay an export point for the parish of St Ann in 1887, whilst towards the end of the century the Boston Fruit Company began developing Bowden as a centre of fruit exports from St Thomas. Trelawny likewise had its centre for banana export at the same time when Falmouth was pressed into service in this respect.

It is important to note that these shipping links with foreign countries were particularly strong with the United States, for it was American interests that were heavily engaged in the banana cultivation and this was also the chief market for the fruit. The growing link between Jamaica and the United States is perhaps best illustrated in the increasing proportion of the island's exports to that country. In 1878 the United Kingdom took 79% of the island's exports and the United States only 14%. But thereafter the proportion of exports to the United States increased sharply and by the end of the century exceeded 60%. The First World War notably

[1] The development of these shipping facilities is traced in summaries prepared annually by the Collectors of Taxes for the parishes and published in the *Customs and Internal Revenue Reports*, on which the present account is based.

changed the direction of the island's foreign trade, but this of itself could hardly have affected the current of emigration to the United States which earlier and more powerful forces had established. The cessation of large-scale emigration to the United States was due to the immigration restrictions imposed.

Extremely unfavourable conditions prevailing through most of the decade 1911–21 provided a powerful push to outward migration. This was indeed a most difficult period for 'the precarious banana', which suffered because of war conditions and because of a series of disastrous hurricanes. Soon after the commencement of the First World War shipping shortages appeared, and when America entered the conflict the market suffered a severe setback. But the four hurricanes that hit the island during the decade probably did more than anything else to create distress in the parishes dependent on this crop.[1] In 1912 a hurricane caused extensive damage to crops in Trelawny and St Ann, while the second in 1915 affected all the northern parishes. The third hurricane in 1916 wrought further destruction; in fact, from October 1916 to March 1917 not one stem of bananas was exported from Port Antonio. In most of the parishes banana cultivation was completely wiped out. The further havoc caused by the third successive hurricane (1917) 'very nearly proved the last straw', as one Collector of Taxes put it.[2] When in the following year the influenza pandemic hit the island the resulting distress in the rural areas was great. These factors, added to the general shortage of supplies and other hardships created by the war, led to a situation eminently suited to stimulate wide-scale emigration.

It was clearly parishes which depended largely on the banana that suffered most. Where sugar or logwood was produced in appreciable quantities the effects of the war were less noticeable and emigration generally less. True in the early stages of the war the sugar industry which was steadily gaining in importance also faced difficulties. For instance, Trelawny, which hitherto had specialized in the manufacture of high ether spirits for the German market, suffered until attention was directed to the production of 'a common, clear article, which commands ready sale at remunerative prices'. But at least up to 1920 sugar enjoyed a period of relative

---

[1] *Customs and Internal Revenue Reports.*
[2] Ibid.

prosperity. Similarly logwood, which up to 1914 was being gradually displaced in world markets by synthetic dyes of German manufacture, was during the war in strong demand and commanded record prices.

From the fragmentary records on emigration from Jamaica during the period of comparatively high emigration (1881–1921), it would appear that most of the emigration was to Panama, the United States and Cuba. And a rough attempt is made in Table 31 to break down the estimates of total net emigration given in Chapter 2 into emigration to these areas. The statistics on emigration to foreign countries, given mainly in the Registrar General's reports and in other official publications, are defective and yield no indication of the magnitude of the movement. But the data on persons born in Jamaica and living in these foreign countries afford some idea of the numbers involved. The statistics relevant to emigration to Panama are considered in detail by Kuczynski, and it is mainly his approach that is used to obtain the estimates given here.[1]

Table 31. *Estimated net emigration, 1881–1921*

Period	To United States	To Panama	To Cuba	To other areas	To all areas
1881–91 1891–1911 1911–21	} 16,000 30,000	17,000 26,000 2,000	 22,000	} 10,000 23,000	} 69,000 77,000

Note. The methods used to derive these estimates are outlined in Appendix III.

From the estimates obtained it appears that prior to 1911 Panama was the main centre of attraction for persons emigrating from Jamaica. During the 30 years 1881–1911 net emigration to Panama amounted to 43,000 or 62% of the total during this period. Net emigration to the United States, according to these estimates, amounted to 16,000 while emigration to other areas totalled 10,000. It is probable that most of the latter represented emigration to Costa Rica, though several other Latin American territories (including Cuba) probably absorbed Jamaican emigrants during these years. As is to be expected the completion of the work on the Panama Canal in 1914 brought emigration to that area to a close, and after

[1] R. R. Kuczynski, op. cit. pp. 7–9.

1911 the United States was the chief goal of emigrants from Jamaica. It is estimated that net emigration to the United States during 1911–21 amounted to 30,000. Other areas also absorbed appreciable numbers, for about 22,000 went to Cuba and 23,000 to other countries. Clearly the years after 1911 witnessed movements to many countries besides the United States and Cuba, though it is impossible to arrive at estimates of their dimension.

To sum up, it can be said that the total net emigration from the island during 1881–1921 amounted to 146,000, and of this 46,000 represent a movement to the United States, 45,000 to Panama, 22,000 at least to Cuba, and 43,000 to other areas. It is certain that emigration to the United States continued after 1921. According to the *Fifteenth Census*, the number of West Indian-born residents (exclusive of those born in Cuba) amounted to 87,700 and as close restriction of immigration was not introduced until 1924, it is probable that emigration continued between 1921 and 1924. But as the migration data for these years are just as defective as those for earlier periods no attempt can be made to consider the various currents of migration after 1921 and the whole intercensal interval 1921–43 is throughout this study taken as a unified period of small net immigration, the net gain to the island amounting to 25,800.

Another aspect of external migration which must be treated is the way it affected the various parishes. This can be studied only after 1911. From the rough estimates of net internal migrations derived in Chapter 5, natural increase and intercensal growth for each parish, estimates of net loss or gain experienced as a result of external migration during 1911–21 and 1921–43 can be obtained. These are shown in Table 32. All parishes lost population as a result of external migration during 1911–21. But clearly the greatest loss was to the three north-eastern parishes, St Mary, St Ann and Portland, which together experienced a net loss of 28,500, or 37% of the total net emigration from the island. It was these parishes, as we have seen, which suffered most as a result of the succession of hurricanes and the general distress that prevailed in the banana cultivation as a result of conditions associated with the First World War. The loss in the case of Kingston (7,500) was also appreciable. Only St Catherine and St Thomas experienced small net emigrations, 1300 and 1200 respectively. The movement was largely confined to males, but it is interesting to note that Kingston lost more females (4400) than males (3100). A similar feature is to be noted in the case of St Catherine, which on balance lost 900 females and 400 males.

Table 32.   *Estimated gain* (+) *or loss* (−) *to each parish resulting from external migration, 1911–21 and 1921–43*

Parish	1911–21			1921–43		
	Male	Female	Total	Male	Female	Total
Kingston	− 3,100	− 4,400	− 7,500	+ 1,100	− 4,300	− 3,200
St Andrew	− 2,700	− 1,500	− 4,200	+ 7,600	+ 7,900	+ 15,500
St Thomas	− 800	− 400	− 1,200	+ 400	+ 300	+ 700
Portland	− 4,400	− 2,300	− 6,700	− 900	− 800	− 1,700
St Mary	− 8,300	− 4,700	− 13,000	+ 1,300	+ 1,800	+ 3,100
St Ann	− 5,800	− 3,000	− 8,800	+ 2,000	+ 1,700	+ 3,700
Trelawny	− 2,500	− 1,800	− 4,300	+ 100	+ 100	+ 200
St James	− 2,600	− 1,300	− 3,900	+ 900	+ 500	+ 1,400
Hanover	− 1,900	− 1,400	− 3,300	− 100	+ 400	+ 300
Westmoreland	− 3,300	− 1,700	− 5,000	—	—	—
St Elizabeth	− 5,100	− 1,400	− 6,500	+ 1,400	+ 1,300	+ 2,700
Manchester	− 4,100	− 2,500	− 6,600	+ 3,800	+ 3,200	+ 7,000
Clarendon	− 3,800	− 1,000	− 4,800	+ 1,200	+ 1,200	+ 2,400
St Catherine	− 400	− 900	− 1,300	− 3,700	− 2,600	− 6,300
Total	− 48,800	− 28,300	− 77,100	+ 15,100	+ 10,700	+ 25,800

Note. These estimates are based on natural increase, population growth and net internal migration in each intercensal interval. The net internal migration estimates are derived in Chapter 5, and the estimates for years prior to 1921 are taken to refer to the years 1911–21.

The period 1921–43 was predominantly one of small net immigration, and only three parishes recorded any loss through net migration. These were St Catherine, Kingston and Portland. The net loss from Kingston, however, was due solely to a net emigration of females (4300); in the case of males there was a net gain of 1100. Net immigration at this period probably represented a return of emigrants who left the island during the preceding period of heavy emigration. And it is interesting to note that though most of the emigration prior to 1921 was from rural areas, the return movement after 1921 was a settlement in the urban area of St Andrew. Net immigration into St Andrew in fact exceeded the gains recorded by all the other parishes combined. In a sense therefore it can be said that the reduction of population growth as a result of emigration during the years before 1921 was to a degree temporary, for many of these emigrants ultimately returned to the island.

CHAPTER 5

INTERNAL MIGRATION

As has been shown in Chapter 2 there is evidence of a movement from the rural areas into Kingston as early as 1881. Indeed, the Committee which examined the conditions of the juvenile population of the island in 1880 reported: 'We find that there is a tendency amongst portions of the rural population to gravitate towards the towns and Kingston especially. The class to which we refer are moved by a desire to obtain their livelihood by other means than agricultural labour, and by the hope of that casual employment at high rates which is often to be obtained in towns.'[1] This movement gathered momentum in the early years of the twentieth century and St Andrew developed as the suburban area of the capital. The one important current of internal migration in the past, the exodus of the ex-slaves from the sugar plantations after emancipation, differed in character, direction and dimensions from the recent urbanward drift. The early movement away from the plantations was basically a redistribution of the rural population, a flight from plantation life to small villages and independent settlements and not in any way parallel to the urbanward drift of modern times.

The census of 1943 was the first at which detailed information on the population by parish of birth was tabulated. This gives breakdowns by parish of residence and length of time persons not born in a parish lived in that parish. It is impossible to gain from these data exact figures of the numbers who moved from one parish to another. But they do furnish the basis for an analysis of the approximate dimensions and of the direction of internal migration over two periods: the first prior to 1921 and the second for the period 1921–43. It is here assumed that persons not born in their parish of residence but who lived in that parish for periods not exceeding 20 years entered it during the intercensal interval 1921–43. Similarly, persons not born in a parish but who lived in that parish for periods in excess of 20 years are assumed to have migrated some time prior to 1921. It is generally assumed here that the latter category of internal migration refers in the main to the period

[1] *Report of the Commission upon the Condition of the Juvenile Population of Jamaica,* Supplement to the *Jamaica Gazette,* 4 November 1880.

1911–21. It would have been more appropriate to use 22 years instead of 20 years as the limit for differentiating between persons who migrated during 1921–43 and those who migrated prior to 1921, but the tabulations of duration of residence in a given parish are given only for under 20 years and over .20 years. A further source of error in the disposition of migrants between the two intercensal intervals is that persons leaving their parish of birth might have lived in some other parish before settling in the one in which they were enumerated in 1943.

Clearly, estimates of internal migration derived in this fashion from parish of birth data are understatements, since the population enumerated at 1943 represents, in the present context, only the survivors of persons who migrated in the two periods involved. It is therefore necessary to inquire into the margins of error to which such estimates are subject. A comparison between the age structure of Kingston and that of the island as a whole for 1943 (see Fig. 2) indicates that the great majority of the internal migration involves males aged 20–34 and females aged 20–39. But as there are also small numbers of migrants outside these age intervals, the age range within which internal migration is assumed to take place is widened to 15–49 for both sexes. Now the survivors of the population within this age span at 1911 would be those aged 47–81 in 1943, while the survivors of those aged 15–49 at 1921 would be the population aged 37–71 in 1943. From the life-tables of 1945–7 it appears that of the population aged 15–49 in 1911, 52·3% of the males and 61·1% of the females survived to 1943. In the case of the population aged 15-49 in 1921 the proportions surviving to 1943 are 72·3 and 77·8%. From these survival ratios it seems safe to assume the following limits of error for the internal migrations for 1911–21 and 1921–43:

	Male	Female
1911–21	48 to 28%	39 to 22%
1921–43	Less than 28%	Less than 22%

It may therefore be concluded that the estimates of in- and out-migration given for the first period may account for about 63% of the male movement and about 70% of the female movement. For the second period about 90% of the migration is probably accounted for. In general, the younger the migrants and the narrower the age interval within which they fall the more complete should be the coverage of migration achieved by this method. If, for instance, all the male migrants were aged 20–34 the proportions accounted for would be higher—72% in 1911–21 and 92% for 1921–43. Similarly, if all the females migrating were within the age span 20–39, about 74% of the migration during 1911–21 and 92% of that during 1921–43 should be accounted for. Because of uncertainty about the age distribution of migrants no attempt is made to apply survival ratios to the 1943 data. Other complications also make such a course inadvisable. For example, since we are taking estimates for all years before 1923 as covering the period 1911–21, this may mean that the proportions accounted for are probably more than the estimated 63% and 70%. Again, at least 26,000 of the 1943 population represent persons who had left the island before 1921 and who on returning did not always settle in the parish in which they were born. For these reasons the basic data derived from the 1943 census are offered in Tables 33–36 without any corrections but with the important qualification that though probably reflecting accurately the direction of internal migration they are merely minimum estimates. Throughout it is the direction of internal migration rather than its precise magnitude that will be stressed.

According to the 1943 census data, of the total native population of 1,211,200, the number who were not living in the parish in which they were born amounted to 253,600 or 21% of the total. Because of the small size of Jamaica and the short distances involved in moving even from one end of the island to another (its greatest length is 146 miles and its greatest breadth 51 miles), comparisons with larger countries cannot be too closely drawn. Still, if the proportion of the population living outside their parish of birth is compared with similar data for other countries there emerges evidence of appreciable mobility of the population of Jamaica. Thus the figure for the island is very close to the proportion of the United States population living outside their state of birth in 1940 (22%). A similar figure for Australia, persons living outside their state of birth in 1934, is 24%.[1]

BEFORE 1921

This period, predominantly one of strong emigration to foreign territories, was only to a small extent marked by internal migration. Still the small movements indicated here (which as already stated

[1] See Kingsley Davis, *The Population of India and Pakistan*, 1950, p. 107.

Table 33. *Minimum estimates of internal migration, male, 1911–21*

Parish of residence	Total	Kingston	St Andrew	St Thomas	Portland	St Mary	St Ann	Trelawny	St James	Hanover	Westmoreland	St Elizabeth	Manchester	Clarendon	St Catherine
											Parish of birth				
Kingston	4,945	—	472	247	233	431	567	198	253	127	277	497	618	356	669
St Andrew	2,645	653	—	108	139	254	199	68	87	47	96	215	274	173	332
St Thomas	1,344	54	194	—	210	112	67	52	59	26	66	139	143	116	106
Portland	1,628	51	199	114	—	344	184	59	100	82	79	111	126	92	87
St Mary	2,432	61	150	23	93	—	873	117	124	46	87	191	155	165	347
St Ann	680	41	2	3	7	111	—	107	18	7	20	52	142	98	72
Trelawny	897	13	1	1	4	15	165	—	91	21	29	239	293	15	10
St James	994	19	3	2	6	26	46	147	—	190	262	204	66	7	16
Hanover	498	7	2	4	3	5	14	20	73	—	277	67	15	3	8
Westmoreland	970	14	1	7	10	9	11	13	130	217	—	509	30	6	13
St Elizabeth	462	31	—	5	4	12	10	27	77	14	106	—	158	7	11
Manchester	812	56	6	—	8	8	41	67	18	6	21	430	—	128	23
Clarendon	2,363	56	22	26	8	43	405	34	16	13	32	404	1,056	—	248
St Catherine	2,528	99	119	28	33	269	353	53	66	52	95	384	466	511	—
Total	23,198	1,155	1,171	568	758	1,639	2,935	962	1,112	848	1,447	3,442	3,542	1,677	1,942
Migration balance	—	+3,790	+1,474	+776	+870	+793	-2,255	-65	-118	-350	-477	-2,980	-2,730	+686	+586

Note. These estimates are derived from Table 18 of the 1943 Census Report. The rows indicate minimum estimates of the number of in-migrants into the parishes, while the columns indicate minimum estimates of the number of out-migrants from the parishes. These estimates are given to the nearest unit not because of the degree of accuracy claimed for them (they considerably understate the true dimensions of the migration, as is indicated in the text) but because of the extremely small numbers involved.

Table 34. *Minimum estimates of internal migration, female, 1911–21* *

Parish of birth

Parish of residence	Total	Kingston	St Andrew	St Thomas	Portland	St Mary	St Ann	Trelawny	St James	Hanover	Westmoreland	St Elizabeth	Manchester	Clarendon	St Catherine
Kingston	7,871	—	647	403	328	659	785	387	419	201	436	848	972	778	1,008
St Andrew	3,984	891	—	161	169	368	322	118	133	95	188	392	413	261	473
St Thomas	790	56	163	—	167	47	25	16	50	10	34	55	41	69	57
Portland	1,239	43	152	144	—	325	122	57	44	55	46	72	70	53	56
St Mary	2,078	60	154	16	112	—	854	91	65	17	34	98	77	114	386
St Ann	758	36	4	6	10	147	—	115	11	3	14	33	173	141	65
Trelawny	880	6	2	4	2	9	203	—	116	14	24	231	250	15	4
St James	1,110	26	2	6	16	14	23	196	—	256	313	206	26	15	11
Hanover	552	4	4	4	3	4	11	16	104	—	347	43	7	1	4
Westmoreland	984	20	1	5	4	10	5	14	139	296	—	452	23	8	7
St Elizabeth	485	37	2	2	8	8	17	28	70	10	138	—	136	16	13
Manchester	1,135	67	9	1	2	13	46	69	18	7	27	688	—	160	28
Clarendon	2,388	59	13	8	7	39	401	36	12	5	22	330	1,018	—	238
St Catherine	2,392	98	157	24	35	291	365	52	61	25	59	293	338	594	—
Total	26,446	1,403	1,310	784	863	1,934	3,179	1,195	1,242	994	1,682	3,741	3,544	2,225	2,350
Migration balance	—	+6,468	+2,674	+6	+376	+144	−2,421	−315	−132	−442	−698	−3,256	−2,409	−37	+42

* See note to Table 33.

146

Table 35. *Minimum estimates of internal migration, male, 1921–43**

Parish of birth

Parish of residence	Total	Kingston	St Andrew	St Thomas	Portland	St Mary	St Ann	Trelawny	St James	Hanover	Westmoreland	St Elizabeth	Manchester	Clarendon	St Catherine
Kingston	16,360	—	1,062	943	1,048	1,999	1,853	683	960	438	1,126	1,563	1,426	1,321	1,938
St Andrew	22,436	6,427	—	815	949	2,435	1,852	570	794	417	1,040	1,772	1,651	1,349	2,365
St Thomas	6,737	462	510	—	1,089	730	392	122	244	160	321	789	561	563	794
Portland	3,897	231	226	377	—	929	324	120	150	121	190	325	276	283	345
St Mary	5,534	400	360	89	637	—	1,285	177	154	77	192	384	302	359	1,118
St Ann	2,528	265	42	25	60	503	—	262	102	47	100	141	310	376	295
Trelawny	2,556	125	20	16	27	95	433	—	399	134	240	375	563	70	59
St James	4,602	190	51	24	67	117	150	436	—	757	1,463	883	292	86	86
Hanover	1,777	83	11	24	30	35	55	73	418	—	702	209	51	43	43
West-moreland	3,056	220	20	40	48	59	81	94	452	877	—	917	93	76	79
St Elizabeth	1,664	226	23	35	52	46	63	103	240	44	336	—	339	60	97
Manchester	3,579	495	57	29	58	111	243	361	100	47	120	1,169	—	591	198
Clarendon	6,750	511	93	118	142	359	1,061	196	204	129	271	875	1,871	—	920
St Catherine	8,724	733	364	279	323	1,297	1,105	257	344	170	456	1,048	888	1,460	—
Total	90,200	10,368	2,839	2,814	4,530	8,715	8,897	3,454	4,561	3,418	6,557	10,450	8,623	6,637	8,337
Migration balance	—	+5,992	+19,597	+3,923	-633	-3,181	-6,369	-898	+41	-1,641	-3,501	-8,786	-5,044	+113	+387

* See note to Table 33.

Table 36. Minimum estimates of internal migration, female, 1921–43*

Parish of residence	Total	Parish of birth													
		Kingston	St Andrew	St Thomas	Portland	St Mary	St Ann	Trelawny	St James	Hanover	Westmoreland	St Elizabeth	Manchester	Clarendon	St Catherine
Kingston	27,112	—	1,417	1,440	1,693	3,267	2,955	1,188	1,472	727	1,806	3,001	2,387	2,578	3,181
St Andrew	31,079	7,355	—	1,060	1,496	3,424	2,761	968	1,149	685	1,620	2,887	2,389	2,130	3,155
St Thomas	5,148	492	459	—	1,237	485	209	97	116	106	196	487	314	417	533
Portland	3,589	301	248	433	—	936	302	89	116	114	150	251	168	220	261
St Mary	5,278	412	332	93	697	—	1,284	156	105	64	118	301	183	316	1,217
St Ann	2,870	271	50	21	76	570	—	300	92	39	80	152	349	536	334
Trelawny	2,430	108	26	19	26	92	530	—	422	81	130	289	568	82	57
St James	5,011	225	43	27	77	99	144	503	—	1,071	1,554	940	175	78	75
Hanover	1,760	89	13	16	23	27	30	50	408	—	839	167	41	30	27
West-moreland	3,074	211	32	32	47	45	55	68	414	1,023	—	929	95	54	69
St Elizabeth	1,878	242	41	28	54	43	59	149	239	76	391	—	366	90	100
Manchester	4,557	552	67	52	63	139	271	488	124	69	152	1,602	—	768	210
Clarendon	6,283	509	79	110	129	276	1,100	182	147	79	213	732	1,826	—	901
St Catherine	8,871	826	403	286	347	1,371	1,095	217	277	149	361	961	831	1,747	—
Total	108,940	11,593	3,210	3,617	5,965	10,774	10,795	4,455	5,081	4,283	7,610	12,699	9,692	9,046	10,120
Migration balance	—	+15,519	+27,869	+1,531	-2,376	-5,496	-7,925	-2,025	-70	-2,523	-4,536	-10,821	-5,135	-2,763	-1,249

* See note to Table 33.

fall short of the actual movements that took place) foreshadowed
the more powerful inter-parish migrations that were to be witnessed
between 1921 and 1943. According to the present estimates, the
total number of females who moved from one parish to another in
the period prior to 1921 amounted to 26,400 and the number of
males who migrated to 23,200. If we take these figures to refer
largely to the decade 1911–21, it appears that every year at least
5000 persons changed their parish of residence, a figure equivalent
to an annual rate of nearly 6 per 1000 of the population.

As can be seen from Table 37, there was a movement into
Kingston and, to a smaller extent, into St Andrew from most parts
of the island. These, though small in size, underline the growing
pull of the expanding urban centre. Kingston, still the main
residential area, gained to the extent of 3800 males and 6500
females. The main sources of this net in-migration were St
Catherine (1500), Manchester (1500), St Ann (1300) and St
Elizabeth (1300). The fact that the only parish which did not lose
population to Kingston was St Andrew signifies that at this early
date the growth of St Andrew as a suburban area had already
begun. In fact there was a net in-migration into this parish from all
the others except St Thomas and Portland. The total net in-
migration into St Andrew amounted to 2700 females and 1500
males. The four parishes, St Catherine, Clarendon, Manchester and
St Elizabeth, together supplied 2200 or 53% of the total in-
migration into St Andrew. Clearly this migration towards the
growing urban areas was predominantly a movement of females.
In the case of Kingston, 63% of the in-migrants were females, and
in the case of St Andrew 64% were females.

Though in-migration into Kingston and St Andrew constituted
the main feature of the small migration during these years, there
were other interesting currents to be noted. For instance, the
adjoining parishes, Portland, St Mary and St Thomas, registered
net gains of 1200, 900 and 800 respectively. No simple reason can be
advanced for these small movements. But it is probable that they
were associated with the increased banana cultivation in this area,
which constituted a notable feature of the agriculture of the island
in the last years of the nineteenth century and the early years of the
twentieth. Thus between 1891 and 1921 the proportion of cultivated
land devoted to bananas in Portland rose from 22% to 32%.[1] Over

[1] See *Reports of the Collector General.*

Table 37. *Minimum estimates of net gain or loss experienced by the several parishes as a result of internal migration, 1911–21*

Parish	Male					Female				
	Net gain (+) from or loss (−) to					Net gain (+) from or loss (−) to				
	Contiguous parishes	Kingston	St Andrew	Others	All parishes	Contiguous parishes	Kingston	St Andrew	Others	All parishes
Kingston			− 181	+ 3,971	+ 3,790			− 244	+ 6,712	+ 6,468
St Andrew	+ 171	+ 181		+ 1,122	+ 1,474	+ 545	+ 244		+ 1,885	+ 2,674
St Thomas	+ 96	− 193	+ 86	+ 787	+ 776	+ 23	− 347	+ 2	+ 328	+ 6
Portland	+ 155	− 182	+ 60	+ 837	+ 870	+ 190	− 285	− 17	+ 488	+ 376
St Mary	+ 589	− 370	− 104	+ 678	+ 793	+ 589	− 599	− 214	+ 368	+ 144
St Ann	− 1,408	− 526	− 197	− 124	− 2,255	− 1,355	− 749	− 318	− 1	− 2,421
Trelawny	+ 440	− 185	− 67	− 253	− 65	+ 392	− 381	− 116	− 210	− 315
St James	+ 432	− 234	− 84	− 232	− 118	+ 542	− 393	− 131	− 150	− 132
Hanover	+ 57	− 120	− 45	− 128	− 350	+ 101	− 197	− 91	− 53	− 442
Westmoreland	+ 211	− 263	− 95	− 330	− 477	+ 89	− 416	− 187	− 184	− 668
St Elizabeth	− 1,014	− 466	− 215	− 1,285	− 2,980	− 1,205	− 811	− 390	− 850	− 3,256
Manchester	− 882	− 562	− 268	− 1,018	− 2,730	+ 487	− 905	− 404	+ 613	− 2,409
Clarendon	+ 972	− 300	− 151	+ 165	+ 686	+ 762	− 719	− 248	+ 168	+ 37
St Catherine	+ 466	− 570	− 213	+ 903	+ 586	+ 561	− 910	− 316	+ 707	+ 42
Total	+ 171	− 3,790	− 1,474	+ 5,093		+ 545	− 6,468	− 2,674	+ 8,597	

Note. These estimates are derived from Tables 33 and 34.

the same period the increase in the case of St Mary was from 40%
to 55% and in the case of St Thomas from 11% to 26%. This
expansion of banana cultivation might have called for additional
labour. Again the in-migration into these parishes might have been
associated with the strong emigration out of the island which was
going on at the same time. For, as has already been shown, the
parishes from which large quantities of bananas were exported had
good communications with foreign countries and emigration from
them was heavy between 1911 and 1921. Consequently the net
in-migration into the banana-producing parishes might have been
a movement of labourers entering to take the place of those who
emigrated or to avail themselves of the opportunity of emigrating
from the island by means of one of the ships that called at the ports
of these parishes. In contrast to the net in-migration experienced
by Kingston and St Andrew, net in-migration into these three
parishes was predominantly male; of the total net gain they
experienced (3000) 82% was due to males. This tends to confirm
the view that the gains recorded were associated either with the
furnishing of labour for banana cultivation or with emigration to
foreign areas. The small net in-migration into St Catherine might
have been no more than the attractions exercised by the former
capital, Spanish Town, but it should also be noted that here again
an appreciable proportion of the cultivated area (22%) was in 1921
under bananas.

Out-migration was on the whole strongest from St Elizabeth
(6200), Manchester (5100) and St Ann (4700). Again it is improbable that any simple reason would adequately explain these movements. In the case of St Ann, the most significant change between
1881 and 1921 was the sharp fall in acreage under cane; in 1881
this crop accounted for 33% of the cultivated acreage, a figure
which by 1921 was down to 6%. Moreover, it is important to note
that bananas were not established on a large scale in St Ann (in
1921 only 6% of the cultivated land was under this crop), so that
there probably was a considerable surplus man-power in the parish
with the decline of the sugar industry. No comparable far-reaching
change in agricultural development marked the other two parishes,
though the fact that the cultivation of bananas never attained any
appreciable dimensions in them suggests that again there was no
new demand for agricultural workers and that consequently out-
migration was inevitable. A factor common to all these parishes of

out-migration was very high fertility, as will be shown in Chapter 8. Though they were in terms of crude densities not the most densely settled parishes in the island, their consistently high fertility rates might have meant a relatively high pressure on the land and thus have stimulated out-migration.

Another important aspect of internal migration shown in Table 37 is the movement between a given parish and the parishes contiguous to it. This not only points out secondary currents of internal migration, but also underlines the growing pull of the urban centre, which even at this early date was sufficiently strong to affect these secondary local movements. St Ann, St Elizabeth and Manchester appear as areas losing population to surrounding parishes; this, in fact, is one aspect of the strong outward movement from these parishes already dealt with. But it is important to note that in each case the greatest loss is to the contiguous parish nearest to Kingston. Thus virtually the whole loss to adjoining parishes in the case of St Ann (2800) is accounted for by net out-migration to St Mary, Clarendon, and St Catherine; net out-migration to these parishes from St Ann totalled 2600. Similarly, of a total net loss of 2200 experienced by St Elizabeth to contiguous parishes, 800 was to Manchester. Again, in the case of Manchester, it is important to note that though the net loss to surrounding parishes totalled only 1400, the net loss to the parish east of it (Clarendon) exceeded this by a considerable amount (1800). The widespread evidence that each parish suffering net losses of population to nearby areas, lost most to parishes between itself and the urban area, suggests that much of the movement towards the urban centre is not direct but takes place in stages, the migrants moving generally from one parish to another, but always getting nearer to their goal.

FROM 1921 TO 1943

Internal migration in this period assumes much greater proportions than during the preceding years. The total number of males who moved from one parish to another was, according to our minimum estimates, 90,200, while the corresponding number of females was 108,900. So that approximately one-fifth of the island's native-born population were involved in internal migration between 1921 and 1943. The increase in internal migration since 1921 is further demonstrated by the fact that 80% of all the internal

Table 38. *Minimum estimates of net gain or loss experienced by the several parishes as a result of internal migration, 1921–43*

Parish	Male					Female				
	Net gain (+) from or loss (−) to				All parishes	Net gain (+) from or loss (−) to				All parishes
	Contiguous parishes	Kingston	St Andrew	Others		Contiguous parishes	Kingston	St Andrew	Others	
Kingston	+ 5,104		− 5,365	+ 11,357	+ 5,992	+ 7,693		− 5,938	+ 21,457	+ 15,519
St Andrew	+ 712	+ 5,365		+ 9,128	+ 19,597	+ 804	+ 5,938		+ 14,238	+ 27,869
St Thomas	+ 420	− 481	− 305	+ 3,997	+ 3,923	− 565	− 948	− 601	+ 2,276	+ 1,531
Portland	+ 311	− 817	− 723	+ 1,327	+ 633	+ 321	− 1,392	− 1,248	+ 829	− 2,376
St Mary	− 2,448	− 1,599	− 2,075	+ 182	− 3,181	− 2,269	− 2,855	− 3,092	+ 130	− 5,496
St Ann	+ 608	− 1,588	− 1,810	+ 523	− 6,369	+ 369	− 2,684	− 2,711	+ 261	− 7,925
Trelawny	+ 2,030	− 558	− 550	+ 398	+ 898	+ 2,585	− 1,080	− 942	+ 372	− 2,025
St James	− 514	− 770	− 743	+ 476	+ 41	− 847	− 1,247	− 1,106	+ 302	− 70
Hanover	− 255	− 355	− 406	+ 366	− 1,641	− 418	− 638	− 672	+ 366	− 2,523
Westmoreland	− 2,326	− 906	− 1,020	− 1,320	− 3,501	− 2,615	− 1,595	− 1,588	+ 935	− 4,536
St Elizabeth	− 652	− 1,337	− 1,749	− 3,374	− 8,786	+ 98	− 2,759	− 2,846	− 2,601	− 10,821
Manchester	+ 1,425	− 931	− 1,594	− 1,867	− 5,044	+ 776	− 1,835	− 2,322	− 1,076	− 5,135
Clarendon	+ 1,529	− 810	− 1,256	+ 754	+ 113	+ 1,761	− 2,069	− 2,051	+ 581	− 2,763
St Catherine		− 1,205	− 2,001	+ 2,064	+ 387		− 2,355	− 2,752	+ 2,097	− 1,249
Total	+ 5,104	− 5,992	− 19,597	+ 20,485		+ 7,693	− 15,519	− 27,869	+ 35,695	

Note. These estimates are derived from Tables 35 and 36.

153

migration revealed by the last census occurred between 1921 and 1943.

It is at once clear from Table 38 that the most prominent feature of internal migration during this period is the enhanced movement into Kingston and St Andrew. But whereas in the past it was to Kingston that most of the in-migrants came, after 1921 St Andrew presented the greater attraction. Between 1921 and 1943, 22,400 males and 31,100 females entered St Andrew from other parishes. Since its losses to other parishes were small—2800 males and 3200 females—the net additions to the population of this parish were appreciable. In fact, the increment due to net in-migration (47,500) accounted for 64% of the total intercensal increase of St Andrew. It is important to note that the greatest movement into this parish is from Kingston (11,300); this emphasizes the shrinking importance of Kingston as the residential area of the urban centre. Next in importance as a source of additional population for St Andrew was the contiguous parish of St Mary, which supplied 5200, while the group of southern parishes from St Catherine to St Elizabeth, which added appreciably to population growth in St Andrew in the past, again appeared as prominent sources of out-migration, supplying together a net addition of 16,600.

Though Kingston continued to gain population from all parishes except St Andrew, its loss to the latter was considerable. In the case of the males for instance, its loss to St Andrew (5400) nearly equalled the gain from all parishes (6000). Net in-migration was greater in the case of the females; here the loss to St Andrew amounted to 5900, whereas the gain from all parishes totalled 15,500. Not only were in-migrants into Kingston—16,400 males and 27,100 females—less than those into St Andrew; the out-migrants were also greater in number than in the case of St Andrew —10,400 males and 11,600 females; so that the gain to Kingston resulting from internal migration was only 21,500 or less than half that experienced by St Andrew; still the contribution of migration to population growth in the city should not be underrated; it formed a sizeable proportion (46%) of the intercensal increase.

But the movements directly into St Andrew and Kingston were not the only internal shifts of importance taking place between 1921 and 1943. As in the previous period local movements, the interchange of population between contiguous parishes, were in evidence and indeed the numbers involved were greater than in the past. Actually, the patterns of local movements were largely the same as before, though the dimensions were greater. The parishes showing appreciable gains from neighbouring parishes were St James, St Catherine, St Thomas and Clarendon, while those losing population to contiguous areas were again St Elizabeth and St Ann.

St James was the only parish which did not lose heavily through out-migration, the loss (less than 100) being negligible. It is true that this parish, in common with all the others, experienced a net loss to the main urban centre (3900). But the net gain from contiguous parishes (4600) more than outweighed this. These net gains from neighbouring parishes doubtless indicate the pull exerted by the expanding Montego Bay area, which ever since the 1920's has been displacing Port Antonio (Portland) as the chief centre of the tourist industry. A consideration of the distribution of the sources of the net in-migration into St James is instructive, as it illustrates the extent to which this secondary urban attraction masks the main pull into the Kingston–St Andrew area. Most of the net in-migration was supplied by Westmoreland (2200) and St Elizabeth (1300). This constitutes an important exception to the general rule that, so far as population shifts between contiguous parishes are concerned, a parish tends to lose most to areas between itself and the main urban centre. The direction of out-migration from Hanover again bears this out. Hanover lost more to St James (1000) than to Westmoreland (400). Though the net in-migration into St James from Trelawny is smaller than that from any other contiguous parish (100), the general pattern of out-migration from Trelawny is again consistent with an appreciable pull exerted by St James. For of the four parishes adjoining Trelawny, St James is the only one which gained population from it.

Net in-migration into Clarendon and St Catherine appears to differ fundamentally from the movement into St James. Thus in the case of Clarendon, the gains are accounted for by net in-migration from St Ann (1200) and Manchester (2300), whereas there is a net loss of 1400 to St Catherine. This is not indicative of a pull comparable to that exerted by St James, but rather of an extension of the strong urban pull exerted by Kingston–St Andrew, expressing itself in inter-parish movements. The same situation appears in the case of St Catherine. Most of the population gained by this parish was from St Ann (1600) and from Clarendon (1400), the areas west and north of it; only 300 was from St Mary, the parish north-east of it. Thus both St Catherine and Clarendon gained through in-migration only because the general drift of internal migration was towards the main urban area, a drift which in part at least seems to assume the form of a movement from one parish to another but always in the direction of Kingston–St Andrew.

It is less easy to account for the movement into St Thomas. According to the present minimum estimates, net in-migration into this parish amounted to 5500 during 1921–43. The fact that an appreciable portion of the net in-migration (1500) was from the contiguous parish Portland again suggests that the movement may be merely the pull of the main urban centre expressing itself in inter-parish migration. For migration from Portland to the capital can be most easily accomplished via St Thomas. There has been no new development in St Thomas comparable to the growth of the tourist industry in St James which would attract migrants. Nor despite the fact that St Thomas now shows levels of literacy and fertility fairly close to those of the main urban centre can this pull be ascribed to incipient urban development in the parish. The only urban centre of consequence is Port Morant with a population of only 4400 in 1943. It is to be noted that agricultural activity in St Thomas increased markedly between 1921 and 1943. Thus the acreage under crops increased from 24,800 to 43,000 during this period. Similarly, the number of males engaged in agriculture rose from 9800 to 12,300 over the same period. At the same time it is doubtful whether this agricultural expansion was to any marked degree associated with in-migration. For two other parishes, St Elizabeth and Manchester, which, as has already been shown, were areas of heavy out-migration, experienced much greater agricultural expansion than St Thomas. The acreage under crops in St Elizabeth more than doubled between 1921 and 1943, rising from 11,400 to 24,100; in Manchester the increase over the same period was from 10,400 to 20,600.

Other parishes which experienced no gains from internal migration also show local migration movements strongly affected by the pull of the main urban centre. For instance, much of the loss to St Elizabeth as a result of such migration (4900) is accounted for by out-migration to Manchester (2100). Again though Manchester experienced a net loss of only 600 as a result of migration to contiguous parishes its net loss to Clarendon, the parish east of it, was much greater than this (2300). Thus the influence of the pull of the main urban centre is reflected in all the local movements affecting the southern parishes from St Elizabeth to St Catherine.

In the case of the northern parishes the pull exerted by St James has already been shown to influence strongly out-migration from Trelawny. But the effects of the attraction of Kingston–St Andrew on the movements out of St Ann are once more in evidence. For nearly the whole of the net loss to contiguous parishes (4700) is accounted for by migration to St Catherine and St Mary, all of which can be considered as a movement largely in response to the major pull of the capital.

Two distinct modes of migration towards the urban centre seem to be in operation, one which takes the migrant directly from his parish of birth to the capital, and the second, probably a slow, long-term movement involving a shift in stages, but always taking the migrant nearer to the urban centre. It is impossible to say what proportion of those who entered Kingston–St Andrew during the periods under review came directly. Though migration to the main urban area dominated internal migration after 1921 the secondary pull of St James introduced an important new phase. And this break in the drift to the chief urban district underlines the fact that any development in a given parish, such as the growth of the tourist trade in Montego Bay, which creates new demands for labour, will suffice to reduce in-migration into Kingston–St Andrew and in general to modify the pattern of urbanization in the island.

An attempt is made in Fig. 7 to present schematically the urban

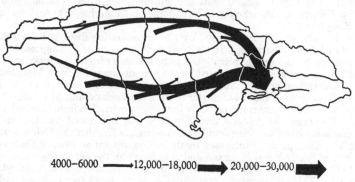

4000–6000 ⟶ 12,000–18,000 ⟹ 20,000–30,000 ⟹

Fig. 7. Main currents of internal migration towards Kingston–St Andrew
1921–1943

drift during the period 1921–43. The whole movement is here considered in terms of four currents, two major ones—a northern current and a southern current—and two minor ones representing additions to Kingston–St Andrew from Portland and St Thomas. Both of the major currents are conceived as moving eastward across the island.[1] On this basis the southern current contributes the largest share to urban growth. The dimensions of the four currents as here drawn up are as follows:

Southern current	35,000
Northern current	27,500
Portland current	4,200
St Thomas current	2,300

It is thus the southern parishes which contribute most to the urban population of the island, 51% of the total net in-migration into Kingston–St Andrew being supplied by them.

URBANIZATION AND MOUNTING POPULATION PRESSURE IN RURAL AREAS

In a sense the increased urbanization witnessed since 1921 is but one aspect of a complex of social changes that have characterized the period of demographic transition ushered in after 1921. Changes in mortality, fertility, occupational patterns and generally enhanced

[1] The construction of these currents can be illustrated by the case of the southern current and the whole procedure can be conveniently viewed in six stages as follows:

Stage I.   The current begins with net out-migration from Westmoreland to St Elizabeth, Manchester, Clarendon, St Catherine, St Andrew and Kingston.

Stage II.   The current is increased by net out-migration from St Elizabeth to Manchester, Clarendon, St Catherine, St Andrew and Kingston and is reduced by the net in-migration into St Elizabeth from Westmoreland.

Stage III.   The current is increased by the net out-migration from Manchester to Clarendon, St Catherine, St Andrew and Kingston, and reduced by the net in-migration into Manchester from Westmoreland and St Elizabeth.

Stage IV.   The current is increased by the net out-migration from Clarendon to St Catherine, St Andrew and Kingston and is reduced by the net in-migration into Clarendon from Westmoreland, St Elizabeth and Manchester.

Stage V.   The current is increased by the net out-migration from St Catherine into St Andrew and Kingston and is reduced by the net in-migration into St Catherine from Westmoreland, St Elizabeth, Manchester and Clarendon.

Stage VI.   This represents the accumulation of the previous five stages and is taken to constitute the southern current of the in-migration into Kingston–St Andrew.

social mobility of the population are all elements of this complex of changes. Though for this reason no single factor can be identified as the signal cause of growing urbanization, this movement reflects to a considerable degree the interplay between external migration and growing natural increase. In this context the drift towards the urban areas appears as an inevitable response to mounting population pressure in the rural areas, as indeed, a substitute for the external emigration which up to 1921 had to some extent curbed the mounting pressure on the land.

The migration data for 1911–21 show clearly that external migration involved predominantly rural populations. Thus of the 49,000 net male emigration 88% were from the twelve rural parishes (excluding Kingston and St Andrew), while in the case of the females 79% of the net emigration of 28,000 were of urban origin (see Chapter 4). If we assume that the same distribution of net emigration between Kingston–St Andrew and the remainder of the island prevailed since the commencement of external emigration, rough estimates of the external and internal migration balances affecting the rural areas can be made. Such estimates are presented in Table 39. The most important point revealed is the powerful effect of external emigration as a control of population growth in the rural areas prior to 1921. Whereas natural increase in 1881–91 in the rural areas was 7900 a year, the net emigration from these areas of 2100 reduced the increments to these parishes appreciably. Though the decline of emigration in the succeeding intercensal interval meant that there were appreciable increments to the rural population, the net emigration of 1900 still aided in retarding the mounting rural pressure. Clearly it was in 1911–21 that external emigration most drastically curtailed population growth in the rural areas. In fact, the annual emigration from the rural areas amounted to 6500, and the total removed in this decade was greater than the net emigration from 1881 to 1911. This considerable external emigration, coupled with the net out-migration to the urban centre, reduced the annual increment of the rural population to 2200, the lowest recorded. Though nothing in these figures supports the lasting complaint of the planters that a shortage of labour prevailed in the island in the late nineteenth century, it is manifest that external emigration constituted a very important determinant of rural population growth up to 1921.

In the light of the profoundly altered conditions after 1921 an

Table 39.   *Effects of external and internal migration on the growth of the rural population*

Period	Annual natural increase	Annual gain (+) or loss (−) through		
		Internal migration	External migration	Annual intercensal increase
1881–91	7,900	− 1,200	− 2,100	+ 4,600
1891–1911	11,000	− 800	− 1,900	+ 8,300
1911–21	10,100	− 1,400	− 6,500	+ 2,200
1921–43	14,300	− 3,100	+ 600	+ 11,800

acceleration of the rate of internal migration seemed inevitable. For although the natural increase of 14,300 between 1921 and 1943 was 42% higher than that of the preceding period, there was no complementary emigration outlets. On the contrary, there was a small net immigration, and if the situation was not relieved by the drift to the urban centre an annual increase of 15,000 would have resulted in the urban parishes. Under these conditions the numbers absorbed into Kingston–St Andrew are no more than the outflow which under past conditions might have been removed by emigration. And despite the relief afforded by out-migration to the urban centre, the annual increments to the rural population in 1921–43 (11,800) was about five times that prevailing during the period 1911–21 when emigration was at its height.

We have considered urbanization in Jamaica solely in terms of the growth of the population of the Kingston–St Andrew area. There are of course other smaller towns, but the growth of these cannot be traced, as the 1943 census was the first at which the populations of these small settlements were tabulated. The boundaries of the electoral districts, in terms of which detailed populations were given at earlier censuses, do not correspond to the boundaries of the smaller towns reported on in 1943. Still the urban returns of the 1943 census demand careful attention, as they not only suggest that urbanization was in effect no more than a considerable transfer of population from the rural areas to the chief urban centre, but also underline an important feature of the whole movement, the fact that it is divorced from industrial development. Table 40 shows the populations and the number of towns of various sizes at 1943.

Table 40.   *Urban population of Jamaica, 1943*

Size of town	No. of towns	Population	Percentage of total population
Over 1,000 and under 5,000	15	39,947	14·4
Over 5,000 and under 10,000	2	11,520	4·2
Over 10,000 and under 25,000	2	23,554	8·5
Over 25,000 and under 100,000	—	—	—
Over 100,000	1	201,911	72·9
Total	20	276,932	100·0

Note. The populations of the smaller towns are not given in the report of the 1943 census, but appear in Bulletin No. 1, dated 23 March 1943. See also *West Indian Census, 1946*, vol. 1, p. 3.

There are fifteen towns (many would perhaps be more appropriately designated villages) with populations between 1000 and 5000. The largest, Port Morant, has a population of 4400 and the smallest, Buff Bay, a population of 1200. Only two towns have populations between 5000 and 10,000: Port Antonio (5500) and May Pen (6000). Within the range 10,000–25,000 there are again only two towns: the former capital Spanish Town with a population of 12,000 and Montego Bay, the centre of the tourist industry, with 11,500. There is a large gap between the second largest town of the island, Spanish Town, and the major urban centre of Kingston–St Andrew, which has a population of 201,900. There is, in fact, no urban centre between 15,000 and 200,000. This confirms the view that urbanization in the island has meant no more than the expansion of the capital.

In view of the small size of the island, comparisons with larger populations in respect of the proportions of the populations in large cities and in respect of the size of the large cities cannot be too closely drawn. For instance, a measure frequently adopted, the proportion of the urban population in cities over 100,000, would amount to 100% in the case of Jamaica, and would in fact not be very meaningful.[1] However, it is possible to show from the available data the extent to which the urban population is concentrated in the major urban centre if we follow the earlier classification of

[1] This measure is used for instance by Kingsley Davis and Hilda Hertz, 'The World Distribution of Urbanization', *Bulletin of the International Statistical Institute*, vol. XXXIII, Pt. IV.

cities used by Davis and Casis and exclude centres below 5000.[1] In
these terms the total urban population of the island at 1943
amounts to 237,000 or 19·2% of the island population. And of the
urban population thus defined the Kingston–St Andrew area
accounts for 85%. It is interesting to note that similar high proportions of the urban population in the major towns are to be
found in other West Indian territories. Thus the four towns in
Trinidad with populations over 5000 have a total population of
203,000, and of this 159,000, or 78%, is located in the capital Port
of Spain and its environs. Even more important is the urban concentration in the major town of Barbados. There is only one settlement apart from Bridgetown which can be considered a town
and its population is only 2100. And of the total urban population of
the island (71,000) 97% is located in Bridgetown and its environs.

Despite the overwhelming importance of the major urban areas
of the population of the island, it is still relevant to consider towns
of all sizes in assessing the general urban development of the island.
An index measuring such a status has been suggested by Davis and
Casis, the unweighted average of the percentage of the population
in towns of various sizes. From such indices, calculated for the West
Indies by L. Broom, Jamaica shows an index of 17·6.[2] In view of
the difficulties of defining and in some cases of delimiting urban
centres in the West Indies, census statistics of urban population
have to be used with extreme caution and comparisons cannot be
too closely drawn. But from Broom's data, it appears that Jamaica
stands at a level of urbanization close to that of Puerto Rico and
Trinidad, the indices for these two areas being 17·8 and 16·2
respectively. Comparisons with more distant areas place Jamaica
on a level of urbanization close to that of Poland and Brazil.

As Broom rightly points out, 'Perhaps the most significant social
trend in the Caribbean today is the urbanization of agricultural
populations and the progressive concentration of people in the
major city. . . .' It is also manifest that urbanization in the West
Indies, as in many Latin American areas, has so far not been the
product of any attraction towards industrial centres, such as has
characterized the urbanization of countries that have passed

[1] Kingsley Davis and Ana Casis, 'Urbanization in Latin America', *Milbank
Memorial Fund Quarterly*, 1946.
[2] L. Broom, 'Urban Research in the British Caribbean: A Prospectus', *Social
and Economic Studies*, February 1953.

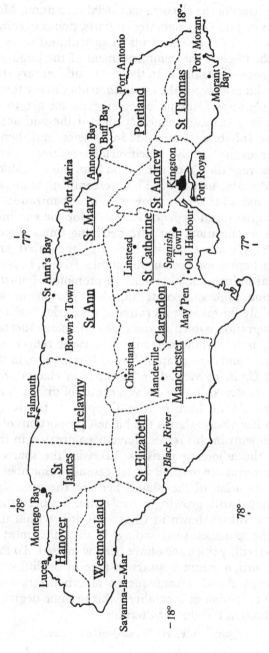

Jamaica, Parishes and Main Towns

through the phases of the classical industrial revolution. Most of the smaller towns of Jamaica originated as ports, points of communication, military stations or settlements in agricultural areas; and all were established before the commencement of the large-scale drift towards Kingston–St Andrew. In the eighteenth century the British Government did not favour the establishment of many towns in the island. The Customs Commission, ruling on an act to establish certain ports in Jamaica, assented 'provided the said act will not encourage the Inhabitants to reside in Townes, and there sett up Manufactures for the Supply of their own Necessities . . . which will not only discourage the Trade carried on from this Kingdom as well as our own Manufactures . . .'.[1] This mercantilist policy cannot be held to be responsible for the slow rate of urbanization. Its predominantly agricultural dependence made for the continuance of heavy rural concentration. But with declining importance of agriculture in the island's economy and with the rapid growth of population, urbanization became inevitable. In fact, in place of the argument that industrialization tends to promote urbanization, we have a situation, as Broom points out, where growing urbanization provides possibly the greatest incentive to industrial development.

Internal migration, which in the case of Jamaica means largely urbanization, is important not only because it reflects changing social, economic and demographic forces. It must also, in the words of Davis and Casis, be viewed as 'a source of change in its own right'. Thus construed it appears as a centre of diffusion for many social changes. As we have seen in Chapter 3, the Kingston–St Andrew area has always shown the highest proportions of literacy. And improvements in this respect have been greatest in the urban area and in the adjoining parishes. Likewise the sharp shifts in occupation patterns are to some degree causally associated with the growing urbanization of the island. Perhaps the most significant change associated with growing urbanization has been the decline in fertility. As will be shown in Chapter 8, it is within the urban areas and the parishes surrounding it that the most striking reductions in fertility since 1921 have been witnessed. To this extent therefore the urban centre appears as a point of diffusion of low fertility patterns. Another consequence of urbanization is that it facilitates the reduction of mortality ; this to some degree tends to offset the reduction it induces in fertility.

[1] Quoted in F. W. Pitman, op. cit. p. 20.

CHAPTER 6

MORTALITY

MORTALITY DURING SLAVERY AND APPRENTICESHIP

As we have already seen in Chapter 2, the cessation of the slave trade was probably followed by a reduction in the rate of natural decrease in the slave population, and this suggests that there was in fact some reduction in mortality after 1807. But in the absence of any satisfactory data, nothing definite can be said about the level of mortality in the early slave period.

Two sources of data made available early in the nineteenth century make it possible to assess very roughly the prevailing levels of mortality and the main diseases to which the population was exposed during the last years of slavery and the apprenticeship. These two sources, the study of Tulloch and Marshall and the slave registers, provide the starting-point for the mortality analysis. Despite the many limitations of these sources, they still give insights into the processes of severe wastage of human life that were such prominent features of slave society in Jamaica and in the West Indies at large at this time.

The investigations of Tulloch and Marshall, which cover the years 1817–36, constitute the most comprehensive attempt ever made to present an overall picture of health and mortality in the British West Indies.[1] However, in view of the quality of the white troops sent to the West Indies and the limitations of the military records this mortality experience cannot be taken as wholly representative of conditions prevailing throughout the region. It appears that the troops sent to the West Indies were among the worst available. And they faced conditions probably more hazardous than most of the inhabitants as the garrisons were usually located near swamps and similar notoriously unhealthy areas.[2] Moreover, there was constant movement of troops, representing replacements as well as interchanges with other commands, so that the mortality records were not the experience of a settled body of men. Again,

---

[1] Tulloch and Marshall, op. cit.
[2] L. J. Ragatz, *The Fall of the Planter Class in the British Caribbean, 1763–1833*, New York, 1928, p. 32.

the records yield data only on adult males. Indeed, the fact that records are confined to the experience of white soldiers born outside the region gives rise to the suspicion that they reveal merely the ravages of disease among Europeans who suffered so severely mainly because of their poor physique and lack of acclimatization to the region.

The inclusion of small numbers of Negroes in the military records does give comparative data for native non-Europeans, but these also are of limited reliability. For apart from the fact that, like the European soldiers, they do not include children and refer only to one sex, the numbers of troops involved are, as Tulloch and Marshall stress, extremely small. Moreover, the black troops and pioneers, to give them their full title, were never completely under medical supervision of the army. In one respect the black troops had a distinct advantage over the white troops. Unlike the latter, who were ill-selected to withstand the rigours of the West Indian climate, the black troops and pioneers were recruited through purchase 'from among the best conditioned slaves in the island'.[1] Despite these limitations, the careful analysis of the authors cannot be ignored; it does give a picture, even if in exaggerated form, of the major diseases that ravaged the West Indies during the last days of slavery and the apprenticeship.

Although their efforts were concentrated on the study of mortality they also attempted to assess the morbidity from various diseases, as this was within their terms of reference. The index of sickness used was the number of admissions into hospital per 1000 of the average strength of the troops in a given year. On this basis Jamaica appears in general more healthy than the colonies of the eastern Caribbean. The sickness rate for Jamaica was 1812 as compared with 1903 for the remainder of the Caribbean. But such indices, applicable only to white troops living in selected areas and not acclimatized to the region, cannot be accepted as adequate indications of the conditions of health in the region as a whole. The limitations of the sickness records are fully acknowledged by Tulloch and Marshall and no extensive discussion of their findings on this aspect of the subject is justified. It is also to be noted that the morbidity records among Negro troops are only fragmentary.

It is the detailed analyses of mortality rates that constitute the most interesting feature of their investigation. These take the form

[1] L. J. Ragatz, op. cit. p. 28.

of compilations and extensive discussions of annual death-rates for the several military commands. The rates used are the annual number of deaths per 1000 of the mean annual strength of the troops. These rates are calculated for twelve separate areas in the British Caribbean where troops were stationed and cover the period 1817–36. The death-rates in Jamaica and in the two eastern commands, for both Negro and white troops, are shown in Table 41.

Table 41. *Annual death-rates for troops in Jamaica and the Eastern Caribbean 1817–36**

Year	Jamaica		Eastern Caribbean	
	White troops	Negro troops	White troops	Negro troops
1817	88	45	162	46
1818	89	36	126	37
1819	294	34	83	63
1820	153	46	105	38
1821	116	42	109	40
1822	171	25	77	43
1823	65	37	49	37
1824	84	39	70	29
1825	307	18	76	35
1826	80	47	68	43
1827	224	44	85	26
1828	74	16	81	36
1829	62	28	58	46
1830	97	14	65	40
1831	133	45	69	36
1832	111	8	64	36
1833	86	15	50	37
1834	93	8	43	33
1835	75	13	57	37
1836	61	26	77	35
Average 1817–36	121·3	30	78·5	40

* From Tulloch and Marshall, op. cit.

Perhaps the most striking feature of the mortality pattern of Jamaica is the very high rates among the white troops during years of epidemics. According to Tulloch and Marshall, epidemics of fever took place in 1819, 1822, 1825 and 1827, and the death-rates for these years are 294, 171, 307 and 224 respectively. These greatly exceed the annual average for the whole period covered, which stands at 121. It is interesting that the white troops stationed in Jamaica show higher death-rates than those stationed in the Eastern

Caribbean. Death-rates among the latter average 78 during the 20 years covered by the study. Moreover, epidemics in the eastern group were much less frequent than in Jamaica.

The mortality records of the white and of the black troops are not strictly comparable. The black troops and pioneers include not only soldiers but military labourers as well, and as the latter were only to a limited degree under the medical supervision of the army, probably not all deaths occurring among them were recorded. Nevertheless, it is of interest to compare the records of the two racial groups. The outstanding difference is of course the generally lower level of mortality among the black troops. This is not unexpected in view of their presumably greater resistance to fevers than their European counterparts. In the words of Tulloch and Marshall the lower sickness and mortality rates for the black troops signify 'the superior salubrity of climate for the Negro race'. Actually the average death-rate for Negro troops in Jamaica is only 30 per 1000, or about one-quarter of the corresponding rate for the white troops. Moreover, the Negro death-rates during the periods of epidemics are not notably higher than at other times. The fact that death-rates among the black troops in Jamaica were lower than those for the eastern groups (which over the 20 years averaged 40) led Tulloch and Marshall to the conclusion that the island was 'more favourable' to the health of the Negroes than the Windwards or Leewards.

Of much greater significance is the analysis of cause of death which formed the main part of Tulloch and Marshall's work. Its importance lies not in any greater reliability of the basic material (it is, indeed, subject to the same fundamental limitations as the measures of general mortality levels already indicated), but in the information it conveys, doubtless in greatly exaggerated form, of the principal diseases that ravaged the West Indies during the period. The death-rates from various causes among white and black troops are shown in Table 42.

Of overwhelming importance among the white troops of Jamaica were fevers of various kinds. The sickness records suggest that fevers were slightly less prevalent in Jamaica than in the Windward and Leeward commands, but proved considerably more productive of fatalities. This was so especially in the years of epidemics. According to Tulloch and Marshall epidemics took the form of attacks of yellow fever, which they considered a form of remittent fever more serious than that experienced in Gibraltar. The average death-rate from all types of fevers for the period studied was 102; thus fevers accounted for more than 84% of all deaths among white troops. This was, indeed, the chief cause of death among these troops in the West Indies with the notable exception of Barbados, but in Jamaica mortality from these diseases attained levels high even by

Table 42.   *Death-rates among white and black troops from various causes, 1817–36**

Cause of death	Jamaica		Windward and Leeward Islands	
	White troops	Black troops	White troops	Black troops
All fevers	101·9	8·7	36·9	7·1
Diseases of lungs	7·5	10·3	10·4	16·5
Diseases of liver	1·0	0·4	1·8	0·9
Diseases of stomach and bowels	5·1	3·0	20·7	7·4
Diseases of brain	2·6	0·6	3·7	2·2
Dropsies	1·2	3·0	2·1	2·1
Others	2·0	4·0	2·9	3·8
Total	121·3	30·0	78·5	40·0

* From Tulloch and Marshall, op. cit.

West Indian standards. Only the small island of Tobago showed a higher death-rate from fevers (104); elsewhere rates ranged from 15 (Antigua and Montserrat) to 63 (St Lucia).

Tulloch and Marshall discussed the various forms of fever responsible for the high death-rates in Jamaica. The records of admissions and deaths due to fevers are summarized in Table 43. From this it appears that remittent fevers accounted for the greatest number of deaths as well as the highest number of admissions into hospitals. Also this group was second only to yellow fever in terms of the ratio of deaths to admissions. At the same time the weaknesses of this type of data should be emphasized. For even though the soldiers were, presumably, under close medical supervision, this did not mean that returns of the cause of deaths at military stations were very dependable. And uncertainty on the part of the surgeons in affixing cause of death was, as Tulloch and Marshall illustrate, most marked in the case of fevers. Dealing with the eastern commands they write: 'A very arbitrary distinction seems to prevail in regard to the insertion of cases under the head of yellow fever (*febris icterodes*), only one having been recorded in the medical returns since 1827; but as different opinions are entertained by medical officers in regard to the classification of fever, that which has been entered at one time as yellow fever may probably have

Table 43.   *Incidence of various kinds of fevers among white troops in Jamaica, 1817–36**

Type of fever	Admissions	Deaths	Ratio of deaths to admissions
Intermittent fevers	6,090	37	1 in 165
Remittent fevers	38,393	5,114	1 in 8
Common continued	1,971	86	1 in 23
Yellow fever (Icterodes)	20	15	1 in 1⅓
Synochus	448	1	1 in 448
Annual ratio per 1,000 mean strength	910	101·9	

* From Tulloch and Marshall, op. cit. p. 46.

been denominated remittent fever at another. The cases reported as *icterodes* must have been exceedingly fatal as about one-half died of all attacked; whereas of those attacked by remittent fevers, which may perhaps be viewed as a less aggravated form of the same disease, only one-ninth part died. These malignant fevers have of late years been principally confined to Tobago, St Lucia, Dominica and Guiana....' Dealing specifically with Jamaica they say, 'common yellow fever of the country seems ... always entered as remittent'. But whatever the true nature of the fever experienced in Jamaica, there can be no doubt about its severity, especially during periods of epidemics. These often appeared without warning and at times nearly half the troops at certain stations perished. 'Epidemics spared neither age, sex nor condition of life; the temperate and the intemperate, the prudent and the thought-less, fell victims to them in nearly equal degree.' Only immediate removal from the scene of the epidemic could arrest the progress of the disease among troops exposed to its ravages.

Other causes of death among the white troops of Jamaica were of much smaller importance. Diseases of the lungs gave rise to a death-rate of 7 per 1000 and diseases of the stomach and bowels to a rate of 5. These, it must be noted, are much lower than the corresponding rates for white troops of the eastern commands. In fact, if we exclude fevers, the level of mortality as measured by rates for the white troops is lower than that for the eastern colonies, 19 as against 42. Not only were diseases of the lungs less prevalent in Jamaica; here they were also one-third less fatal than in the eastern colonies. Again, dysentery and diarrhoea assumed 'a much more mild and tractable form' in Jamaica. Tulloch and Marshall ascribed this to 'their having twice as much fresh provisions as those in the Windward and Leeward command'.

That the general mortality from fevers may not have been so severe as the death-rates among the white troops suggest is borne out by the data for the black troops. In Jamaica the death-rate among the black troops from fevers of all kinds was only 9 or less than one-tenth of the rate for the white. This rate, incidentally, is very close to that shown by Negro troops of the Eastern Caribbean (7). In all military stations for which records are available the black troops show diseases of the lungs as the highest cause of death. In Jamaica deaths from this cause were at the rate of 10 per 1000, appreciably lower than the corresponding rate for the Windwards and Leewards (16). Other fairly important causes of death among the Negro troops of Jamaica were diseases of the stomach and bowels (3) and dropsies (3). Tulloch and Marshall sum up the position of the black troops thus: 'Climate of each colony seems to affect the constitution of the black troops in a different way; but throughout the whole diseases of the lungs and bowels are generally the most fatal. . . .'

At first sight the very high death-rates from fevers among white troops seem convincing evidence in support of the theory widely accepted throughout the nineteenth century that Europeans were subject to heavy mortality during the first few years of their life in the West Indies. However, the authors did not view the high mortality rates among white soldiers in this light, but sought to test the theory by further study of the material in terms of length of residence in the West Indies. They concluded with caution that the theory that the length of residence produces acclimatization was not proved by the data. Mortality increased with length of residence in at least as many instances as it declined.

The second source of mortality during these years—the slave registers—is again severely limited, as has already been shown, and offers only the roughest indications of the mortality position of the island. R. M. Martin calculated average death-rates over the entire period of slave registration for all West Indian colonies.[1] According to these, Jamaica (together with Nevis) has the lowest death-rate, 25 per 1000. All other rates are between 27 and 33, except for Tobago, which shows an exceptionally high rate (42). Though all these considerably understate the mortality position of the colonies, as they do not cover the experience under 3 years of age, they support the findings of Tulloch and Marshall that Jamaica enjoyed a relatively favourable mortality position during the late slave period, though, it must be noted, there is little agreement of rank-order relationship between the two series of death-rates.

A difference between the two sets of data on which Tulloch and Marshall commented is the fact that death-rates among the black soldiers were in general higher than among the slaves, except for Antigua, Grenada and Tobago. 'This is more remarkable as the mortality of the Negro slave population is calculated upon male persons of all ages; including old men and infants, sickly and healthy; whereas that of the troops is calculated upon persons in the prime of life only, which certainly shows they must be subject to some deteriorating influence from which the slave population is exempt; seeing that despite the alleged ill-treatment of the latter, a much smaller proportion is found to die than of the former, who have neither duty nor harsher treatment to undergo.'[2]

It is difficult to account for the discrepancy between the two series of death-rates. Certainly the fact that the ranks of the military personnel were being constantly recruited to replace the deaths among them reduced the reliability of rates based on annual deaths and mean annual strength of the troops. But it is equally certain that while the military reports considerably overstate mortality, the slave data considerably understate it. The only colony for which a reasonably detailed picture of slave mortality is available is British Guiana, and here the evidence of extremely heavy loss of life between 1820 and 1832 is clear. A life-table for this period shows an average length of life among slaves of approximately 23 years;

[1] R. M. Martin, op. cit. p. 20. The same rates are also reproduced in Tulloch and Marshall, op. cit.

[2] The authors are in error here on two points. Martin's rates are based on mortality among both sexes and not among males alone. Moreover, these rates do not cover the mortality of infants, as slave mortality data exclude deaths under 3 years of age.

this means a death-rate of 44 per 1000 in the life-table population.[1] Therefore even though it seems that Jamaica enjoyed a more favourable mortality level than the fever-ridden coastland of British Guiana, its death-rate would be much higher than 25. A rough estimate of the death-rate of Jamaica slaves can be made. Doubtless in the highly unfavourable conditions of the slave regime fertility was at a lower level than that for the period 1844–61, during which, according to our estimate, the birth-rate in the island stood at about 40. Even a birth-rate of 36 would have meant, on the basis of the estimated rate of natural decrease of 5 per 1000 already presented, a death-rate of at least 41. It is thus clear that the slave registers cannot be taken as giving a true picture of slave mortality, though interpreted in the literal sense as a system of population accounting they do make it possible to derive rough estimates of the level of mortality.[2]

Another important aspect of slave mortality is the cause of death. From accounts of historians a general idea of the chief diseases attacking slaves can be gained. Most accounts stress the loss of life from fevers of various kinds. It appears that the Negroes suffered less from deaths from fevers than the Europeans, but the effects of malaria and other fevers on the health and general efficiency of the former were evident. And when, after the cessation of the slave trade, great attention was being paid to the health of the slaves the regular administration of preparations to combat fevers assumed importance.   Another major cause of death among slaves was dysentery or the bloody flux. The records of the Worthy Park estate of Jamaica indicate that this was particularly severe among newly arrived slaves.[3] Yaws presented a serious problem, as, apparently, very few imported slaves escaped it. According to Renny, the disease was seldom fatal to the young but had serious effects when its victims were advanced in years.[4] Long reports that often 'a whole parcel of new Negroes, within a few weeks after they are brought on a plantation, break out altogether with this disorder';

[1] G. W. Roberts, 'A life table for a West Indian slave population', *Population Studies*, vol. v, no. 3.
[2] It is interesting to note that in the case of Grenada and Tobago, where annual returns of slave populations were made (this probably meant a smaller degree of understatement of prevailing slave mortality), death-rates were highest, 33 and 42 respectively.
[3] U. B. Phillips, 'A Jamaica Slave Plantation', *The American Historical Review*, vol. xix, no. 3, 1914.
[4] Robert Renny, *An History of Jamaica*, 1807, p. 205.

he maintained that one-third of the African slaves perished from yaws within 3 years of their arrival.[1] Smallpox also took a heavy toll of life, but Long's estimate of 70% mortality from this disease is questionable. It is of interest that vaccination was tried in Jamaica, though not systematically. Venereal diseases also attacked the slaves, not only causing deaths but generally impairing their working efficiency. Renny rated leprosy nearly as frequent as yaws. Jaw full or tetanus constituted one of the most serious causes of death among infants, 'carrying off between the fifth and the fourteenth days after birth one-fourth of all those who are born in Jamaica', claimed Renny.[2] An obscure disorder called dirt-eating in some colonies and mal d'estomac in others, and obviously rather the symptoms of a disease rather than a disease itself, evoked curious comments from writers. Renny ascribed it to 'an indulgence of the depressing passions of the mind', while Thomas Roughley considered nothing 'more horribly disgusting' than this incurable disease.[3] The number of suicides was very high also, especially among African-born slaves. Another serious cause of loss of life among slaves in Jamaica was the recurrent hurricanes. The damage these did to crops and communications throughout the island inevitably meant that many areas were isolated and exposed to disease and starvation. According to Beckford, 15,000 perished as the result of hurricanes between 1780 and 1787.[4] Probably these various diseases took the heaviest toll of the African-born slaves. Indeed, during their period of acclimatization severe losses, variously estimated at between one-fifth and one-third, were expected.

Slave registration laws of Jamaica made no provision for registration of the cause of death among slaves. This, however, was laid down in the registration law of British Guiana, which even required that medical certificates of cause of death be furnished. As a result of Robertson's thorough analysis of British Guiana slave data we have an idea of some of the major sources of loss among slaves. Table 44 shows the chief causes. Though fever probably loomed larger as a cause of death in British Guiana than in most other

---

[1] Edward Long, op. cit. vol. 2, p. 434.

[2] Robert Renny, op. cit. p. 207.

[3] According to H. Harold Scott, *A History of Tropical Medicine*, London, 1942, vol. II, p. 990, dirt-eating was 'probably due to, as well as the cause of, helminthiasis, especially ankylostomiasis'.

[4] W. Beckford, *A Descriptive Account of the Island of Jamaica*, 1790, vol. II, p. 311.

West Indian territories, it is important to note that the British Guiana record bears a general resemblance to the findings of Tulloch and Marshall. Apart from the indefinite group labelled 'aged and debility', dysentery, diarrhoea and cholic were the chief causes of death, amounting to 12% of all deaths, and thus agreeing

Table 44. *Principal causes of death among British Guiana slaves, 1829–32**

Cause of death	No. deaths	% total deaths
Aged and debility	1,338	19·1
Dysentery, diarrhoea and cholic	839	12·0
Dropsy	646	9·2
Asthma, catarrh, consumption and pleurisy	644	9·2
Fevers	567	8·1
Inflammation and mortification	311	4·4
Mal d'estomac	301	4·3
Casualties	296	4·2
Leprosy	270	3·8
Convulsions	257	3·7
Ulcers and consequent debility	249	3·5
Yaws	183	2·6
Lockjaw	183	2·6
Syphilis	70	1·0

* From James Robertson, op. cit., slightly condensed.

with the military records. Dropsy and respiratory diseases (including consumption) were about equal in importance, each group accounting for 9% of all deaths. Of the total deaths fevers accounted for 8%, while dirt-eating caused 4%. Other important causes of death were yaws, which, together with 'ulcers and consequent debility', amounted to 6% of the total. Other significant elements of mortality were leprosy, lockjaw and syphilis.

MORTALITY BETWEEN APPRENTICESHIP AND THE ESTABLISHMENT OF REGISTRATION

While it is impossible to bridge completely the demographic gap between the period covered by slave registration and the studies of Tulloch and Marshall and the period opened by the establishment of effective civil registration, rough indications of the levels and movements of mortality rates during this interregnum can still be obtained.

It will be recalled that the estimates given in Chapter 2 for the pre-registration period showed a decline in the death-rate from 32 in 1844–61, to 27 in 1861–71 and to 26 in 1871–81. These are appreciably lower than the death-rate which probably prevailed during the last years of slavery; this, it has already been suggested, could not have been less than 40. The fact that during the years 1817–29 the slave population was declining at a rate of about 5 per 1000, whereas between 1844 and 1861 the population was increasing at a rate of 8 per 1000, attests to the considerable reduction in mortality after the abolition of slavery.

Probably demographic factors were not of major importance in effecting this improvement. But one strictly demographic factor which might have contributed to the lowering of mortality was the gradual elimination of the African-born population, who were not only subject to higher mortality rates than those born in the island, but who because of their peculiar age structure would inevitably have experienced high mortality. As is evident from the age data for the British Guiana slaves, the slave populations covered by registration were heavily weighted by large proportions over age 35, the survivors of those introduced before the cessation of the slave trade in 1807. As these comparatively large groups moved through the higher age intervals they added considerably to the number of deaths and probably helped to maintain high crude death-rates. But as they declined in number the mortality of the population in general would also tend to decline. The effect of the number of aged Africans on the population of British Guiana, it should be noted, was clearly realized by James Robertson.[1] He estimated the average age of slaves at the time of entry into that colony, i.e. before 1807, at about 18½ years, whereas by 1832 the average age might have been 'nearer 60 than 50 years', and he rightly concluded that 'an excess of deaths over births must therefore continue for a considerable time to come'.

By the middle of the nineteenth century most of the African-born slaves had probably disappeared from the population of Jamaica, a factor which must have helped to some degree to reduce mortality. This is suggested by two sets of census data. In the first place the proportion of the population over age 40 in 1844, into which interval all African-born ex-slaves would presumably fall at

[1] James Robertson, op. cit.

that date, amounted to 25%, whereas the proportion declined to 22% in 1861. Moreover, it is probable that the majority of the 10,000 African-born persons shown in the 1861 census were not survivors of African-born slaves introduced before 1807 but liberated Africans brought into the island after 1841 under the scheme of African immigration.

A consideration of certain historical factors during the years 1844–61 and 1861–71 suggests that the decline in mortality arrived at from migration records and estimated births did in fact occur. The estimated death-rate of 32 for 1844–61 is influenced by the serious cholera epidemic of 1850–52, the epidemic of smallpox that followed the disappearance of the cholera, as well as by the less violent outbreaks of cholera and scarlatina that appeared in the island in 1854.[1] The attack of cholera in 1850–2 was the first ever experienced by Jamaica and was probably the greatest catastrophe ever visited upon the population. Every parish suffered, and both military and civilian populations fell victims in large numbers. In the absence of any system of registration it is impossible to state exactly the mortality from cholera during the 19 months it ravaged the island, that is, from September 1850 until March 1852. Milroy, though admitting this, concluded from his study of the course of the epidemic that the total number who died as a result of it could not have been less than between 40,000 and 50,000. However, Governor Sir Charles Grey, who challenged many parts of Milroy's report, considered this too high. He advanced the cogent reason that in Kingston and Spanish Town, where some records of cholera mortality were maintained, mortality from the disease was much below Milroy's rate.[2] Whereas the people in these towns 'were crowded in ill-drained streets' and therefore more liable to contract the disease, 'the great majority of the Negro population is thinly scattered over the interior' and therefore were less liable to be attacked; indeed, 'several districts were almost wholly exempt from the disease'. The Governor even claimed to have witnessed much more serious outbreaks of cholera in India. Still, if we accept Milroy's figures as the only ones available, and if we take the population of 1850 as approximately 400,000, then on the basis of

[1] See Gavin Milroy, op. cit., and the letter from Richard Hill to Lieutenant Governor Eyre, 2 April 1862, *V.H.A.J.*, 1862–3.
[2] Despatch from Governor Sir C. E. Grey to the Duke of Newcastle, 23 September 1853, *P.P.* 1854, vol. XLIII.

the estimated number of deaths of 224,400 during 1844–61 given in Chapter 2, it appears that, at the height of the epidemic, mortality from cholera was probably 71 per 1000, while the total death-rate might have reached 100. On the other hand, mortality during the years not affected by cholera might have been about 26 per 1000, that is equal to the rate estimated for 1861–71. If, as the Governor contended, mortality in the cholera period was lower than Milroy's estimates indicate, then the cholera rate would have been somewhat less than 100 and the average mortality over the other years higher than 26.

Further information on mortality in Jamaica in this pre-registration period is afforded by the writings of actuaries who studied the mortality of assured lives in Jamaica and other West Indian populations during these years. However, such studies need not give information on mortality for the total populations involved; in general, the experience on which they are based refers to select groups among whom Europeans formed a much larger proportion than in the general populations. It is probably this factor which led these actuaries to dwell so much on the mortality differentials between acclimatized and unacclimatized lives. While there might have been such a differential during slavery, it is doubtful whether it was of any importance after emancipation as the movement of Europeans to and from the West Indies declined considerably in the nineteenth century. As far as the study of total mortality is concerned, indeed, the Europeans could be largely ignored as they constituted very small proportions of the populations. The view taken here is that the actuarial studies based on the mortality of assured lives cover mainly the experience of upper classes, those sufficiently cultured and opulent to invest in insurance and composed mostly of people long settled in the colonies. On this view the experience thus revealed would be much more favourable than the mortality of the overall populations.

An early attempt was made by J. Marshall of the Jamaica Mutual Life Assurance Company to assess the level of mortality in the island.[1] Unfortunately, the rates used in his study are wholly arbitrary, as he clearly indicates: 'In the absence of any large collection of facts and observations in Jamaica, from which such a

---

[1] J. Marshall, 'On the rate of Mortality amongst Europeans and their Descendants residing in the Island of Jamaica', *Journal of the Institute of Actuaries*, vol. 5, no. IV, 1854.

[life] table could have been deduced, he has necessarily been compelled to proceed upon probability and approximation.' According to this life-table the average length of life among Europeans in Jamaica at this time was about 20 years. Such an extremely low value for a group who presumably represented the upper class and those least susceptible to many diseases that attack the lower classes is questionable. Indeed, it means that Europeans in Jamaica in the middle of the nineteenth century were on the whole less favourably placed than were the slaves in British Guiana in the last years of slavery, for here the average length of life was nearly 23 years. Marshall's assumption of a death-rate of 5% among Europeans in Jamaica is not acceptable and, as will be seen further on, his values of expectation of life are much lower than those founded on actuarial experience. A comment of Marshall's which seems strange in the light of the method of construction is that 'between birth and the age of 10 . . . owing to the mildness of epidemical disease among children in a tropical climate, the mortality cannot be deemed higher in Jamaica than in the mother country, according to the Northampton observation . . .'.

In assessing the life table of Marshall it is worth noting that his approach was coloured by the distinction made between Europeans born in Jamaica and those born outside the island. The importance of this distinction lies in the differential mortality it was usually taken to imply. Severe mortality as a consequence of not being acclimatized was, as has already been shown, fully borne out in the records of mortality among white troops. Indeed, it went back further, to the days of the slave trade when planters counted on a loss of about one-third of their newly acquired slaves in the process of 'seasoning'. 'The special extra risk said to await newcomers', as S. C. Thompson expressed it, engaged the attention of actuaries interested in the West Indies, as to prevent losses in insurance, it was claimed, premiums had to be adjusted in order to bear this extra risk.[1] But it also figured prominently in general discussions of mortality in the region, particularly in the case of the mortality of indentured workers during their first years of industrial residence.

Thompson developed a distinction between mortality in the East Indies and that in the West Indies based on the real or supposed

[1] S. C. Thompson, 'Address to the Members of the Actuarial Society of Edinburgh', *Journal of the Institute of Actuaries*, October 1886, p. 181.

differential inherent in the processes of acclimatization in the two regions. In the case of the East Indies 'the possessor of a vigorous constitution fresh from Europe—once he has learned the ordinary rules of living, which he can do in a very few weeks—has for some years, if he respects them, a favourable prospect of life, as compared, I mean, with that of Europeans in India as a body'. But later the European immigrant 'succumbs to the climate', and the longer he lives in the East the more enfeebled he becomes. Thompson went further and claimed that the immigrant's descendants 'became a puny race' and might wholly die out unless they had the benefit of a change to a temperate climate or 'a fresh infusion of European blood'. On the other hand, an entirely different situation faced the newly arrived European immigrants in the West Indies. Here the time of greatest danger was before 'their constitutions became adapted to the climate'. After this period of acclimatization 'the prospects of longevity become favourable'. This distinction between acclimatized and unacclimatized Europeans in the West Indies as regards mortality was certainly relevant in past periods when the numbers of Europeans living in the region were much higher and when probably an appreciable proportion of this group had recently arrived in the colonies. But with the decline of the numbers of Europeans and the decreased movement between Europe and the West Indies the distinction became pointless. In any event its significance is easily exaggerated because even in early periods of the nineteenth century Europeans formed only small proportions of the total population. For instance, the first census of Jamaica showed that only 4% of the island population were white.

Another paper on mortality in the West Indies during the pre-registration period is that of John Stott.[1] His data, based on the experience of the Scottish Amicable Life Assurance Society in these colonies, showed Jamaica to be the healthiest of the colonies. Unfortunately, the lives assured are mostly over 25 years, and for this reason the rates certainly understate the mortality position of the populations. The death-rates in terms of deaths per 1000 of the totals at risk are as follows:

[1] John Stott, 'On the death-rate among assured lives in the West Indies, being the experience of the Scottish Amicable Life Assurance Society during thirty years, 1846–76', *Journal of the Institute of Actuaries and Assurance Magazine*, October 1878.

Jamaica	20·4
Trinidad	27·6
Other islands	22·5
British Guiana	31·9

Stott also commented on 'the very important difference between acclimatized and unacclimatized lives', advancing further evidence of greater mortality among the latter. An interesting part of this paper is the analysis of cause of death. The percentage distributions are given in summary form in Table 45 for the total 139 deaths studied as well as for those dying in Jamaica. Zymotic diseases accounted for most deaths (37%) in the case of Jamaica, and here fever was the chief element. The effect of the cholera epidemic of 1850–2 is seen in the fact that '35% of the deaths under zymotic diseases in the Jamaica Mutual are explained to be due to the Asiatic Cholera epidemic in 1850–2'. However, such evidence on

Table 45. *Principal causes of death among West Indian members of the Scottish Amicable Society**

Cause of death	West Indies		Jamaica alone, % deaths
	No. deaths	% of deaths under each disease	
Zymotic diseases	42	30·2	37·5
Diseases of uncertain seat	7	5·0	5·0
Phthisis	11	7·9	3·8
Diseases of nervous system	25	18·0	16·2
Heart disease	16	11·5	10·0
Diseases of respiratory organs	10	7·2	6·2
Diseases of digestive organs	14	10·1	15·0
Diseases of urinary organs	6	4·3	2·5
Others	8	5·8	3·8
Total deaths	139	100·0	

* Condensed from Table XV of John Stott, op. cit.

cause of death must be accepted with the utmost reserve. The deaths occurred before the introduction of vital registration, and as medical certification was not in force it is doubtful whether even insurance companies could have secured reliable records of cause of death.

The last paper on pre-registration mortality to be discussed here is by far the most exhaustive.[1] And even though only 16% of the 1292 deaths on which it is based occurred in Jamaica it is none the-less relevant to this discussion, presenting as it does a detailed analysis of mortality of a select group of the West Indian popula-tions. The authors stress the diversity of sources of the basic data and the comparatively small numbers involved in each individual section, but were convinced that, 'There is, in the main, a fair correspondence in the general character of the results, which, considering the extent of the combined data, entitles them to some confidence as representing, as nearly as can at present be ascertained, the mortality generally prevailing among assured lives in the West Indies.' In constructing life tables from these data assumptions had to be made with respect to mortality under 30 and over 70: 'Below age 30 and above age 70, the facts being few at these periods of life, the observed rate of mortality was purposely exaggerated.' The expectation of life from the table for Barbados, 'which gives an indication of the rate of mortality in a tropical climate under the most favourable circumstances', as well as those for the combined experience of the West Indian colonies are compared with a few early West Indian experiences and with the values from Marshall's arbitrarily constructed life table in Table 46. It appears that the mortality level of the special groups of West Indians covered by insurance over 1840–82 was somewhat higher than that for the total Jamaica population at 1891. Thus the expectation of life at age 25 for the aggregate West Indian experience was, according to the life table of Hardy and Rothery, 31·4 years, or 4·2 years lower than the corresponding value for the females of Jamaica at 1891. In fact, the mortality of the assured lives was about equal to that shown by the British Guiana population for the period 1880–2, which, as can be seen, was appreciably higher than that for Jamaica at the same time. It is also evident from these data that not much significance can be put on Marshall's values, which are obviously based on greatly exaggerated death-rates.

Comparisons between the mortality experience based on assured lives, which in the case of the analysis of Hardy and Rothery, covers a period of more than 40 years, and the mortality based on vital statistics and census populations for 1890–2 cannot be too

[1] George F. Hardy and Howard J. Rothery, 'On the mortality of assured lives in the West Indies (chiefly Barbados)', *Journal of the Institute of Actuaries*, July 1888, pp. 161 et seq.

Table 46. *Comparison of expectation of life according to life tables of Hardy and Rothery, and Marshall, and life tables of British Guiana and Jamaica*

| Age | Tables of Hardy and Rothery | | Marshall's table | British Guiana, 1880–82 Both sexes | Jamaica, 1889–92, Females |
	Barbados	Aggregate West Indies			
0	—	—	19·8	28·4	38·3
5	—	—	30·7	37·1	48·8
10	—	—	28·6	37·7	46·0
15	38·7	—	25·4	—	42·4
20	35·5	—	23·0	30·5	38·7
25	32·3	31·4	21·4	—	35·6
30	29·2	—	19·8	27·9	32·6
35	26·2	25·1	18·2	—	29·5
40	23·2	—	16·7	23·9	26·4
45	20·3	19·3	15·4	—	23·2
50	17·3	—	14·2	19·2	20·1

closely drawn. But it seems certain that in view of the long period over which the former extends it is influenced by the heavy mortality associated with the epidemics of the mid-nineteenth century. This probably accounts for the fact that the expectation of life revealed by the tables of Hardy and Rothery is somewhat lower than that of 1890–2. Under these circumstances, the small differences noted are consistent with the view that no striking improvement in mortality has taken place between 1861 and 1921.

MORTALITY IN THE PERIOD OF CIVIL REGISTRATION

Death-rates. The average number of deaths at all ages and for three broad age intervals during successive 5-year periods since the commencement of registration appear in Table 47. Though there has been a steady rise in the number of deaths up to the early 1920's, especially high mortalities were experienced during 1906–10 and 1916–20. In the former period this was occasioned by the destruction of Kingston by fire and earthquake in 1907; during this year the number of deaths (25,100) was the second highest ever recorded. High mortality during 1916–20 was associated with a number of factors: the depressed economic conditions connected with the First World War, a series of disastrous hurricanes and small outbreaks of alastrim. But by far the most important factor was the influenza pandemic of 1918, which was mainly responsible for the record number of deaths during that year—29,600, higher than for any other year. The period after 1921 witnessed a steady decline in the number of deaths, which by 1946–50 stood at 17,500, or slightly more than it was at the beginning of the century.

Table 47. *Average number of deaths*

Period	Deaths all ages	0–4 years		15–34 years		65 years and over	
		No. of deaths	% deaths all ages	No. of deaths	% deaths all ages	No. of deaths	% deaths all ages
1881–5	13,476	5,380	39·9	2,127	15·8	2,177	16·2
1886–90	14,760	6,182	41·9	2,275	15·4	2,231	15·1
1891–5	14,670	6,253	42·6	2,294	15·6	2,217	15·1
1896–1900	16,082	7,218	44·9	2,416	15·0	2,322	14·4
1901–5	17,899	8,162	45·6	2,616	14·6	2,491	13·9
1906–10	20,415	9,082	44·5	3,180	15·6	2,424	11·9
1911–15	19,380	8,675	44·8	2,921	15·1	2,791	14·4
1916–20	23,101	8,919	38·6	3,778	16·4	3,635	15·7
1921–5	20,793	8,765	42·2	3,395	16·3	3,092	14·9
1926–30	18,706	7,934	42·4	2,950	15·8	3,074	16·4
1931–5	19,409	7,114	36·7	3,108	16·0	3,816	19·7
1936–40	18,403	6,181	33·6	2,811	15·3	4,023	21·9
1941–5	18,054	5,588	30·9	2,479	13·7	4,610	25·5
1946–50	17,451	5,422	31·1	2,167	12·4	4,608	26·4

The shrinking proportion of deaths under 5 emphasizes an important feature of mortality which will be analysed in detail presently, the appreciable declines in infant and child mortality. Another significant feature is the growing number of deaths over 65. Thus in 1921–5 the number of deaths over 65 averaged 3100 or 15% of the total; by 1946–50 this number had risen to 4600 or 26% of the total. This development follows not only from the declining mortality at younger ages; it is also associated with basic changes in the age structure of the population to which the declines in fertility after 1921 have contributed in some degree.

As will be seen from Table 48, the death-rates during the early period of civil registration were somewhat lower than those estimated for the immediate pre-registration period. The fall in mortality implied in this movement is not improbable in view of the measures which, as will be shown in Chapter 7, were introduced after 1867 in order to improve health conditions. Still in view of the under-registration in the early years of registration and the errors to which the pre-registration estimates are subject the small movements cannot be taken as marking any decisive changes in the level of mortality, and the years after 1881 can appropriately be taken as an extension of the period 1861–81. On these terms the period 1861–1921 represents one of relatively high and unchanging mortality. The mortality records of the island can therefore be divided into three phases. The first, covering 1844–61, was marked by epidemics and higher levels of mortality than the second, which extends from 1861 to 1921 and shows no basic shifts in mortality levels, though there is some suggestion of an improvement in the last quarter of the nineteenth century. The third phase, beginning in 1921, is one in which Jamaica, in common with other West Indian territories, gained effective control over an ever-widening range of diseases.

Table 48. *Comparative crude death-rates for Jamaica, Trinidad, British Guiana and Barbados*

Period	Jamaica		Trinidad		British Guiana		Barbados	
	Rate	Index	Rate	Index	Rate	Index	Rate	Index
1879–83	23·6	101·3	—	—	30·7	86·7	—	—
1889–93	23·3	100·0	26·4	100·0	35·4	100·0	26·4	100·0
1899–1903	—	—	25·0	94·7	—	—	—	—
1909–13	23·3	100·0	24·8	93·9	29·7	83·9	27·7	104·9
1919–23	24·8	106·4	23·9	90·5	31·0	87·6	35·0	132·6
1929–33	—	—	19·0	72·0	22·7	64·1	—	—
1941–5	14·6	62·7	14·2	53·8	17·0	48·0	16·7	63·2

Note. With the exception of the first rate for Barbados, which is based on deaths for 1891–3, these rates are calculated from the average numbers of deaths over each 5-year period and the relevant census population. In the case of British Guiana the aborigines are excluded. For the last period the rates for Trinidad, British Guiana and Barbados cover the years 1944–8.

Table 48 also makes it possible to compare mortality in Jamaica with mortality in three other Caribbean populations. Two facts of significance are to be noted. In the first place, the relatively high mortality before 1921 is a feature Jamaica shares with other territories; it is also clear that Barbados, British Guiana and Trinidad also failed to record any decline in death-rates prior to

1921, and that British Guiana and Barbados, like Jamaica, experienced an increase in mortality during the years 1918–23. The pronounced reduction in mortality in all four populations after 1921 emphasizes the significance of this census year as a new phase in the Caribbean as a whole. In the second place it emerges from this table that up to 1921 Jamaica shows a more favourable mortality level than the other colonies, with much lower death-rates than British Guiana and Barbados, though its advantage over Trinidad is not substantial.

The annual death-rates calculated on the revised intercensal population estimates provide a picture of the annual movements from 1879. These rates are shown in Fig. 8, which also gives

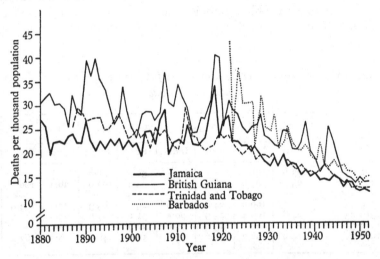

Fig. 8. Death-rates for four West Indian colonies

comparable rates for British Guiana, Trinidad and Barbados. It is clear that with rare exceptions mortality experienced in Jamaica has not been excessively high. On only one occasion, during the influenza year of 1918, has the death-rate exceeded 30; in that year a rate of 34·2 is returned, the highest ever recorded in the island. Fairly high rates also occur in 1890 and 1891, when the rates stand at 28·2 and 28·3 respectively. But in general mortality rates, even before the period of declining mortality, rarely reach 25. The lowest rate so far recorded—11·8 in 1950—is about one-half that prevailing 30 years previously and about one-third of the rate of a century ago.

The detailed examination of the trend in mortality in Jamaica during registration can best be made in quinquennial periods. Quinquennial death-rates and infant mortality rates are shown in Table 49. Here the absence of any downward trend up to 1921 is again marked. The highest rate is that for the years 1916–20 (26·9) and the lowest that for 1891–5 (21·9). There is a steady decline in death-rates after 1921. The rate for 1921–5 stands at 23·5 and by 1946–50 a decline to 12·9 has been achieved, that is, a 45% reduction from the level prevailing 25 years earlier.

Table 49. *Average death-rates and infant mortality rates*

Period	Death-rates per 1000 population	Infant mortality, rates per 1000 live births		
		Overall rates	Legitimate rates	Illegitimate rates
1881–5	22·89	158·4	—	—
1886–90	24·03	174·4	—	—
1891–5	21·91	171·1	—	—
1896–1900	22·02	174·8	137·4	196·9
1901–5	22·74	174·1	137·5	197·5
1906–10	24·95	192·0	139·8	219·7
1911–15	22·75	179·2	—	—
1916–20	26·89	174·2	—	—
1921–5	23·51	176·1	—	—
1926–30	19·56	160·0	—	—
1931–5	18·26	142·7	—	—
1936–40	16·10	122·1	93·4	129·3
1941–5	14·54	99·3	74·3	110·4
1946–50	12·94	85·5	59·2	97·0

The same general movement emerges from the series of infant mortality rates available since the establishment of civil registration. Before 1921 infant mortality rates remained high, fluctuating appreciably from year to year but showing no general downward movement. Rates exceeded 200 only in 1890 and 1907 while the lowest was that of 1882 (141). From the quinquennial rates of Table 49 it appears that infant mortality rates have been mostly in excess of 170, with the highest average falling in 1906–10 (192) and the lowest in 1881–5 (158). Since 1921 there has been a steady and marked reduction, and by 1946–50 the rate has declined to 85·5, that is, half what it was about 25 years previously.

The course of infant mortality in Jamaica over the period of registration is depicted in Fig. 9, which also shows the course of

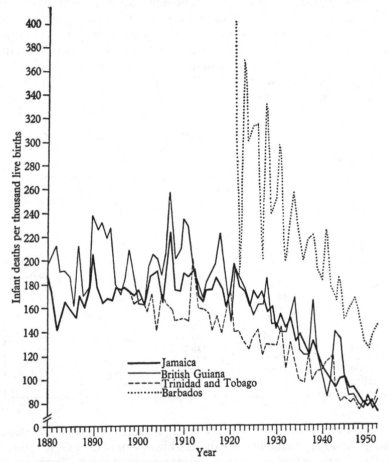

Fig. 9. Infant mortality rates for four West Indian colonies

infant mortality in three other West Indian populations. In the past infant mortality in Jamaica was somewhere between the levels of Trinidad and British Guiana, but much lower than the extremely high level shown by Barbados. It is clear that since 1921 conspicuous declines have been experienced in all populations, and in fact infant mortality, as in the case of mortality in general, is tending to a more or less common level throughout the West Indies.

Such information as is available on mortality among legitimate and illegitimate infants also appears in Table 49. The illegitimate rate is manifestly higher than the legitimate at all times. The excess varies from 38% to 64%, and has obviously widened considerably within recent years. Up to 1941–5 the decline in the two categories is about equal—46% for the legitimate and 44% for the illegitimate. The next 5-year period, however, witnesses a significant fall in the legitimate rate—from 74·3 to 59·2 or by one-fifth. During the same time the decline in the illegitimate rate has been much smaller— from 110·4 to 97·0 or by 12%, so that now the illegitimate rate is about 64% higher than the legitimate, a much higher difference than previously in evidence. The fact that the declines in the rate of mortality among the illegitimate infants, which constitute nearly 70% of all infants born, have been comparatively small during the past 10 years is of considerable significance for the decline in mortality among children in general. For the slowing down of the rate of decline in illegitimate mortality may indeed extend to the age interval 1–5 where improvements in recent years have been relatively low.

Further data for the study of differentials between legitimate and illegitimate infant mortality appear in Table 50. It is at once clear that in both sexes death-rates for all periods of infancy are higher for the illegitimate. The excess of the illegitimate over the legitimate rates is lowest in the case of the neonatal mortality. Death-rates under 1 month show an excess of illegitimate over legitimate rates of 50% in the case of males and 38% in the case of females. At higher ages the differences are more marked among the female rates. Thus the female illegitimate rates range from 12·5 (1–2 months) to 15·4 (6–8 months), whereas the corresponding legitimate rates range from 6·4 (1–2 months) to 8·0 (6–8 months). In fact, the female illegitimate mortalities are in general nearly twice as high as the legitimate. The differences in the case of the males, though smaller, are still considerable. Here the excess of the illegitimate over the legitimate ranges from 99% (1–2 months) to 56% (9–11 months).

Again, the differences between neonatal mortality and mortality over one month are greater for the legitimate than for the illegitimate infants. This is most clearly seen when the rates are expressed as

Table 50. *Analysis of infant mortality, 1951*

Age in months	Males				Females			
	Death-rates per 1000		Indices of death-rates		Death-rates per 1000		Indices of death-rates	
	Legitimate	Illegitimate	Legitimate	Illegitimate	Legitimate	Illegitimate	Legitimate	Illegitimate
0	25·25	37·80	100·0	100·0	22·06	30·47	100·0	100·0
1–2	7·18	14·29	28·4	37·8	6·42	12·51	29·1	41·1
3–5	10·64	17·52	42·1	46·3	7·47	15·08	33·9	49·5
6–8	9·89	16·64	39·2	44·0	7·96	15·37	36·1	50·4
9–11	9·56	14·88	37·9	39·4	7·44	14·85	33·7	48·7

Note. These rates are the numbers dying in the age interval per 1000 alive at the beginning of the age interval.

indices, with the rate under 1 month taken as 100. Thus whereas in the case of the males the legitimate rate at 1–2 months is only 28% of the corresponding death-rate under 1 month, the illegitimate rate at 1–2 months is 38% of the corresponding neonatal rate. A similar differential appears in the case of the female rates. From these differentials it appears that much of the higher mortality of the illegitimate infants is traceable to the relatively high rates to which they are subject at ages over 2 months.

The foregoing differences between legitimate and illegitimate infant mortality can be more succinctly presented in terms of the analysis developed by M. Bourgeois-Pichat.[1] This provides for the division of infant mortality into endogenous and exogenous components, the former referring to mortality traceable to factors inherent in the infant, and the latter to mortality traceable largely to environmental factors. By means of the relationship established by this analysis (that the infant deaths over 1 month per 1000 live births, increased by 25%, affords a close approximation to the rate of exogenous infant mortality) it is seen that the endogenous infant mortality among the illegitimate infants (19·8) exceeds that among the legitimate infants (15·7) by 26%. The difference is much greater for the exogenous rates, which amount to 71·5 among the illegitimate and 40·1 among the legitimate, the excess in this case being 78%. This emphasizes convincingly the importance of environ-

[1] J. Bourgeois-Pichat, 'An Analysis of Infant Mortality', *Population Bulletin*, no. 2, October 1952.

mental factors in determining the differentials between legitimate and illegitimate infant mortality. Further evidence of the environmental basis of the differential is afforded by the association between overall infant mortality rates for the several parishes and a measure of socio-economic status. Thus when the level of rent in each parish is taken as a measure of socio-economic status, there is a significant negative correlation ($r = -0.63$) between infant mortality and levels of rent.[1]

Mortality by parish. In general the movements in death-rates for the parishes parallel those for the island as a whole, showing no special changes before 1921 and appreciable, though varying, declines after that year. A feature of importance revealed by parish mortality is the unfavourable position of the Kingston–St Andrew

Table 51.   *Death-rates by parish, 1881–1943*

Parish	1881	1891	1911	1921	1943
Kingston–St Andrew	31·1	26·2	28·7	31·2	14·6
St Thomas	26·8	26·9	24·8	25·9	14·8
Portland	23·7	23·9	23·6	25·5	15·4
St Mary	23·8	22·7	20·5	22·3	14·3
St Ann	19·5	20·5	18·4	21·5	12·7
Trelawny	25·1	27·5	23·4	26·9	15·3
St James	23·0	21·7	23·3	24·3	16·6
Hanover	23·7	24·9	25·2	26·4	14·7
Westmoreland	22·1	23·4	23·5	23·2	14·7
St Elizabeth	20·5	22·5	20·4	21·8	15·3
Manchester	17·6	20·1	18·8	20·9	12·2
Clarendon	21·9	21·7	23·7	23·5	14·0
St Catherine	25·1	26·9	25·9	26·5	16·2

Note. The rates are based on the 5-year periods centred on each census date and on the appropriate census population.

area before the period of mortality decline.[2] In 1881, 1911 and 1921 urban death-rates were well above those for the rural districts, being 31·1, 28·7 and 31·2 respectively (see Table 51). However, it was in the urban area that the most conspicuous reductions in

[1] G. W. Roberts, 'A note on mortality in Jamaica', *Population Studies*, vol. IV, no. 1, June 1950.

[2] A single death-rate for Kingston and St Andrew is used here because, whereas in each of the other parishes the proportion of deaths registered to the total deaths throughout the island accredited to that parish approaches 100%, in Kingston it amounts to 139% and in St Andrew to 86%. By combining the two an overall urban death-rate of much greater validity is secured.

mortality were effected after 1921; the death-rate for this area declined from 31·2 in 1921 to 14·6 in 1943, or by 53%. On the other hand, it is of interest that two parishes, Manchester and St Ann, have consistently shown relatively low rates of mortality. In 1881, 1891, 1921 and 1943 rates were lowest in Manchester— 17·6, 20·1, 20·9 and 12·2 respectively, while in 1911 it was St Ann that showed the lowest rate (18·4).

Infant mortality rates also indicate that before 1921 movements in levels of mortality in the parishes were negligible, and once again emphasize that up to 1921 the urban mortality was very high (see Table 52). Between 1881 and 1921 infant mortality rates in Kingston–St Andrew ranged from 188 to 234, whereas in the parish of Manchester the corresponding range was from 130 to 152. Up to 1921 infant mortality rates were in all parishes in excess of 100, while in many rates above 200 were recorded. By 1943, however, infant mortality rates were under 100 in five parishes. Again it is of importance that the reduction effected between 1921 and 1943 was greatest in the urban area, where the rate was reduced from 199 to 86 or by 57%. In 1921 infant mortality rates

Table 52. *Infant mortality rates by parish, 1881–1943*

Parish	1881	1891	1911	1921	1943
Kingston–St Andrew	187·6	198·5	234·4	198·8	85·6
St Thomas	165·5	183·4	176·4	187·0	107·6
Portland	169·4	171·5	190·3	193·0	105·9
St Mary	153·1	173·2	173·6	163·7	93·8
St Ann	149·9	167·3	140·8	145·6	86·6
Trelawny	215·9	229·0	200·2	212·9	112·2
St James	200·8	201·9	206·8	199·5	134·2
Hanover	182·7	204·0	196·3	196·5	107·1
Westmoreland	171·6	178·4	191·8	173·5	118·3
St Elizabeth	161·6	165·2	163·0	165·5	108·3
Manchester	130·5	151·6	146·3	145·4	74·8
Clarendon	164·4	156·8	175·7	155·1	92·0
St Catherine	182·1	187·1	191·7	182·2	106·6

Note. The rates are based on infant deaths and births for 5-year periods centred on each census date.

ranged from 213 (Trelawny) to 145 (Manchester), as compared with a range of from 134 (St James) to 75 (Manchester) in 1943.

No clear-cut division of the island into areas of broadly equal levels, such as will be shown in Chapter 8 emerges in the case of fertility, can be made on the basis of parish mortality rates. But Manchester and St Ann have, for reasons which are not at present clear, experienced relatively low rates throughout the period for which records are available, whereas the urban area up to 1921 remained in a relatively unfavourable position. However, due evidently to the more effective application of measures of public health and sanitary control in the urban area after 1921, it is here that the greatest progress in the control of mortality has been achieved.

Life-tables.[1] The progress made in the control of mortality is more convincingly shown in the various values of a series of life-tables for the periods, 1880–2, 1890–2, 1910–12, 1920–2, 1945–7 and 1950–2. The average length of life or the expectation of life at birth provides the most convenient way of summarizing the experience over these six periods. These, together with comparable data for some other Caribbean populations, are given in Table 53. In 1880–1 average length of life stands at 37·0 for males and 39·8 for females. The decline during the next decade to 36·7 years for males and 38·3 for females may not be of much significance in view of the possible under-registration of deaths during the first years of registration. A small improvement to 39·0 (males) and 41·4 (females) is recorded in 1910–12, but this improvement is cancelled in the succeeding period of extremely high mortality. In fact the average length of life in 1920–1—35·9 years for males and 38·2 for females—is the lowest during the period covered by registration in Jamaica. By 1945–6 considerable improvements have been recorded and 15·4 years have been added to the average length of life for the males and 16·4 years for the females. The succeeding 5-year period also witnesses further conspicuous improvements in the mortality position of the island, the average length of life for males in 1950–2

[1] Most of the life-tables on which this section is based are given in *West Indian Census Bulletins*, nos. 9, 10 and 12. The most recent tables used (1950–2) are constructed from mid-year population estimates of the Registrar Generals for 1951 and from deaths for 1950–2. With the exception of those for British Guiana, which are given in G. W. Roberts, 'Some observations on the population of British Guiana', *Population Studies*, vol. II, no. 2, the earlier life-tables used and not appearing in the *Census Bulletins* have been specially prepared for this study. It should be noted that the first life-tables published for Jamaica were abridged tables for 1921 and 1943 prepared by S. B. Chambers and appearing in the 1943 *Census of Jamaica*. These show mortality levels virtually identical with those of the tables of 1920–2 and 1945–7. Further discussions of the Jamaica life-tables are given in G. W. Roberts, 'A note on mortality in Jamaica', *Population Studies*, vol. IV, no. 1.

being 55·7 and that for females 58·9. The additions to the expectation of life at birth during the 30 years following 1921 have therefore
been considerable, 19·8 years for males and 20·7 for females. As
can also be seen, Jamaica has at least before 1921 shown a mortality
position consistently more favourable than that for any other
population of the West Indies. In the more recent period of steep
mortality declines the differences have narrowed; indeed, the
significant feature has been the movement throughout the Caribbean
towards a more or less common level of mortality. For instance, the
average length of life for the four major colonies showed a much
wider variation in 1921 than in 1951. In 1921 the male values
ranged from 37·6 (Trinidad) to 28·5 (Barbados), while the female
values ranged from 40·1 (Trinidad) to 31·9 (Barbados). By 1951,
however, the highest male value—56·3 for Trinidad—was only 3·1
years above the lowest value for that sex—53·2 years (British
Guiana). A similar reduction in the variations among colonies
appears in the case of the females; here the highest value—58·9
years (Jamaica)—was only 2·6 years above the lowest value—56·3
(British Guiana). It is also clear that the summary improvements
registered in the average length of life are very close in the case of
Jamaica, Trinidad and British Guiana, between 18 and 21 years.
The improvement in Barbados is much higher, however—24·9
years for the males and 26·1 for the females; this is due mainly to
the very significant declines in infant mortality recorded for
Barbados from the extremely high level prevailing in 1921. The
above broad comparisons in terms of average length of life are
useful in so far as they suggest the approach to a common level of
mortality, but they mask the existence of some contrasts among
the mortality patterns of the several colonies which only detailed
considerations of specific age mortalities can reveal.

A sensitive measure of mortality movements for the lower ages
may be taken as the age at which the life-table cohort is reduced by
one-quarter. Such values calculated from the various life-tables for
Jamaica appear in Table 54. It is evident that up to 1921 the
position remains almost unchanged. And if we exclude the experience of 1879–82, which may be somewhat unreliable, the age at
which the cohort is reduced by one-quarter remains under 4 years
for at least 40 years. However, the age rises steeply between 1921
and 1946—from 2·5 to 34·7 in the case of the males and from 3·3
to 35·0 in the case of the females. Still further improvements are

Table 53. *Average length of life in years for four West Indian populations, 1870–2 to 1950–2*

Population	Sex	1870–2	1880–2	1890–2	1901–2	1910–12	1920–2	1930–2	1945–7	1950–2	Gains 1921–51
Jamaica	M.	—	37·02	36·74	—	39·04	35·89	—	51·25	55·73	19·84
	F.	—	39·80	38·30	—	41·41	38·20	—	54·58	58·89	20·69
Trinidad	M.	—	—	33·7	36·73	38·97	37·59	44·51	52·98	56·31	18·72
	F.	—	—	36·2	38·75	40·95	40·11	46·95	56·03	58·45	18·34
British Guiana	M.	26·1	28·4	22·8	—	29·9	33·5	40·3	49·32	53·15	19·7
	F.			26·9	—	32·4	35·8	42·6	52·05	56·28	20·5
Barbados	M.	—	—	—	—	28·7	28·5	—	49·17	53·41	24·9
	F.	—	—	—	—	32·5	31·9	—	52·94	58·00	26·1

recorded after 1946; and in 1951 the age at which the life-table cohort is reduced by one-quarter stands at 43·4 (males) and 45·9 (females). These values summarize adequately the conspicuous declines in mortality among infancy, childhood and adolescence between 1921 and 1951.

Table 54. *Age in years at which cohort is reduced by one-quarter*

Period	Male	Female
1879–82	3·41	4·39
1889–92	2·25	2·85
1910–12	2·10	3·75
1920–2	2·52	3·26
1945–7	34·73	35·04
1950–2	43·36	45·90

The life-table also affords a means of depicting the declines in mortality at advanced ages. The numbers surviving to age 65 out of 100,000 live-born furnish adequate measures for this purpose and are presented in Table 55. In 1881 22% of the males survived to age 65. During the ensuing 40 years this moves only slightly, in keeping with the patterns previously indicated, and stands at 21% in 1921. Subsequent improvements have been striking; in 1946 40% of the males attained age 65, and the corresponding figure for 1951 was more than twice that of 1881—49%. The corresponding values for the females, though at a generally higher level than the males, follow a similar pattern. In 1881 the percentage of females surviving to age 65 is 28. No appreciable change is noted during the succeeding 40 years and by 1921 the percentage is 27. The increase in later years is striking: 49% survive to age 65 in 1946 and 56% in 1951. The main implication of this improvement in mortality is not so much in the increased rate of population growth it signifies (survivors to age 65 are actually over the child-bearing age), but increasing proportions of persons surviving to age 65 presage a growing importance of dependents over working age.

Details of the movements of age-specific mortalities can best be examined from the life-table death-rates available, which are presented in Table 56. There are, indeed, some small changes in death-rates between 1881 and 1921, but the significant aspect of the rates during these years is their confirmation of the phenomenon repeatedly stressed in this chapter that prior to 1921 no downward trend in mortality was in evidence. On the other hand, these death-rates demonstrate more clearly than any other measures the extent

Table 55. *Numbers surviving to age 65 out of 100,000 born alive*

Period	Male	Female
1879–82	21,914	27,760
1889–92	23,161	27,753
1910–12	26,993	31,444
1920–2	21,056	26,859
1945–7	39,988	48,534
1950–2	49,287	55,945

of the mortality gains witnessed since 1921. It is convenient to discuss the declines in three phases, over the whole period of mortality decline from 1921 to 1951, as well as over the two sub-divisions of this period, 1921–46 and 1946–51. Because of the unequal length of these two subdivisions it is better to reduce the percentage rates of decline for 1921–46 and 1946–51 to an annual basis, as is done in Table 57.

In the case of the males, mortality declines between 1921 and 1951 have been in excess of 50% at all age intervals between 0 and 50. Infant mortality, which in the past so greatly retarded population growth in the island, shows a decline of 55%, from 187 to 84. Percentage declines attain greater dimensions as we proceed to higher ages and are at a maximum at the age interval 15–20, where a decline of 74% is recorded. Rates of decline are in excess of 70% between ages 10 and 35. And though at ages over 50 less impressive reductions are in evidence, even within the interval 70–75 a reduction of 14% is recorded. In the case of the females, mortality declines, though also very considerable, reveal a pattern essentially different from that which characterizes the males. Up to age 15 the percentage reductions are in excess of those shown by the males. In fact, the reduction for the interval 10–15 (75%) is the highest experienced by the females. Presumably the fact that between ages 15 and 45 declines in female mortalities are appreciably lower than those shown by the males is associated with the extra hazards facing women during these years as a consequence of child-bearing. Even

Table 56. Life-table death-rates ($1000\,{}_{n}q_{x}$) for Jamaica

Age interval	Male						Female					
	1881	1891	1911	1921	1946	1951	1881	1891	1911	1921	1946	1951
0–1	177·90	195·84	198·21	187·27	102·02	84·47	162·91	182·29	177·91	176·05	90·11	71·74
1–2	53·03	61·66	63·09	67·67	32·60	27·25	51·30	61·70	62·90	66·94	27·75	24·05
2–5	78·02	73·22	49·27	58·32	18·95	20·07	69·73	79·84	45·28	58·32	19·49	16·65
5–10	43·65	42·06	27·74	36·86	14·48	12·77	41·68	45·28	29·74	37·29	13·92	11·59
10–15	29·69	24·62	20·36	25·15	9·05	7·22	28·96	30·17	25·20	29·69	11·06	7·47
15–20	31·53	27·49	27·83	34·25	14·92	8·86	31·68	33·33	34·25	40·23	20·76	13·07
20–25	48·82	50·01	48·63	62·67	25·61	16·82	49·06	50·39	48·10	61·53	31·73	21·53
25–30	58·18	60·35	58·32	79·30	29·40	22·07	56·47	55·10	52·44	71·06	32·21	22·02
30–35	67·08	65·39	59·50	84·15	34·82	24·71	64·50	59·36	56·57	73·26	33·44	22·85
35–40	82·44	75·82	64·64	91·50	46·07	29·00	73·63	66·90	64·36	72·37	38·81	26·96
40–45	97·84	88·71	76·28	100·80	57·43	42·25	81·94	71·20	70·27	77·40	44·24	36·76
45–50	117·16	104·33	95·11	111·62	73·29	55·19	89·76	83·55	74·89	86·87	56·32	43·50
50–55	142·11	123·74	117·52	130·79	102·08	78·28	105·41	102·43	87·88	102·15	71·88	60·87
55–60	173·10	156·37	141·19	165·60	135·94	99·80	133·73	128·90	116·31	124·18	90·52	76·05
60–65	215·90	216·54	182·45	217·74	173·49	149·59	171·88	162·08	152·04	166·53	115·97	108·29
65–70	279·96	293·04	252·03	279·33	231·49	199·32	227·14	234·36	191·34	235·18	164·19	147·65
70–75	351·20	365·68	338·44	374·04	324·56	322·56	305·11	320·53	258·43	329·51	237·54	242·53
75–80	413·19	430·46	445·33	500·83	470·01	478·98	386·73	386·38	376·92	431·09	368·46	365·22
80–85	486·32	521·72	561·35	598·32	645·76	583·18	490·37	497·27	503·58	516·67	530·28	580·55

so, the reductions within the reproductive span are impressive, ranging from 63% to 69%. Over age 60 females show greater reductions in mortality, the reduction for the interval 70–75 (26%) being twice that experienced by the males. It remains clear that

Table 57. *Rates of decline (%) in $_nq_x$ values*

Age interval	% decline 1921–51		Annual rates of decline (%) 1921–46		Annual rates of decline (%) 1946–51	
	Male	Female	Male	Female	Male	Female
0–1	54·9	59·2	2·4	2·6	3·7	4·5
1–5	61·7	66·8	3·4	3·7	1·7	2·9
5–10	65·4	68·9	3·7	3·9	2·5	3·6
10–15	71·3	74·8	4·0	3·9	4·4	7·6
15–20	74·1	67·5	3·3	2·6	9·9	8·8
20–25	73·2	65·0	3·5	2·6	8·1	7·4
25–30	72·2	69·0	3·9	3·1	5·6	7·3
30–35	70·6	68·8	3·5	3·1	6·6	7·3
35–40	68·3	62·7	2·7	2·5	8·9	7·0
40–45	58·1	52·5	2·2	2·2	5·9	3·6
45–50	50·6	49·9	1·7	1·7	5·5	5·0
50–55	40·1	40·4	1·0	1·4	5·2	3·3
55–60	39·7	38·8	0·8	1·3	6·0	3·4
60–65	31·3	35·0	0·9	1·4	2·9	1·4
65–70	28·6	37·2	0·7	1·4	2·9	2·1
70–75	13·8	26·4	0·6	1·3	0·1	—

for both sexes the most important mortality gains have been made between the ages of 10 and 30.

The mortality reductions over 1921–46 and 1946–51 emphasize a very important feature of mortality changes in the island, that declines in recent years are much more impressive than those recorded earlier in the period of mortality control. Indeed, it can, with some qualification, be maintained that an acceleration in the rates of mortality decline has taken place since 1946. Whether this acceleration will continue, whether in fact mortality will decline more rapidly after 1951 than it did during 1946–51 must remain the subject of speculation. But even a continuation of the rates of decline prevailing during 1946–51 should suffice, within a short time, to place Jamaica securely on a level of mortality characteristic of European populations and far removed from the level currently associated with the so-called under-developed countries.

Rates of decline during 1921–46, though smaller than those of

later years, are still considerable. From infancy up to age 45
annual rates of decline among the males are in excess of 2% and
reach a maximum of 4% in the interval 10–15. Females also show
declines broadly similar, though two contrasts between the experience of the two sexes must be noted. In the first place the
declines exhibited by females during the child-bearing span are
substantially lower than the corresponding declines for the males,
a differential probably associated with the extra risks to which
women are exposed during these ages. In the second place, the
reductions shown by the females over age 50, though themselves
not large, are substantially in excess of the corresponding male
reductions.

With the exception of the mortalities between ages 1 and 10,
declines after 1946 have been much greater than those recorded
prior to 1946. This is particularly so in the case of the males, where
the highest annual rate of decline recorded for 1946–51, for the age
interval 15–20, is nearly 10%. Again, whereas prior to 1946 the
rates of decline in male mortality over 40 are meagre, those after
1946 are on a considerable scale, being between 5% and 6% for the
ages 40–60. Though females also show much more rapid rates of
decline after 1946, there are a few significant contrasts between the
experience of the two sexes. Whereas during 1921–46 only modest
mortality declines are recorded for females of child-bearing age, the
declines after 1946 are considerable. Between ages 10 and 40 the
rates of mortality decline are large and very consistent, lying
between 7% and 9%. On the other hand, females show much lower
gains than males at ages over 50. It is probable that this is associated
with the changing patterns of cause of death, which is especially
marked among females of advanced ages. For, as will be shown
presently, degenerative diseases in general, which under present
conditions of mortality control are not susceptible of marked
improvements, are rapidly becoming the foremost cause of female
deaths over age 50, and this presumably contributes somewhat to
the relatively small reductions among females at advanced ages
after 1946.

An examination of the mortality differentials between the sexes
is necessary in view of the important part played by mortality in
determining the sex composition of the population. Jamaica shows
the almost universal feature of a relatively favourable mortality
among females. The advantage of females over males can be shown

in general terms by the ratios between T_0 values for the two sexes, which are as follows for 1911 to 1951:

| 1911 | 943 | 1946 | 939 |
| 1921 | 939 | 1951 | 946 |

These ratios are not far from that shown by the United States Negro population at 1939–40 (941), but they are appreciably higher than the ratio for the United States white population at the same date (933). The fact that the advantage of the females is much smaller in the case of the Jamaica population than in the case of the United States whites is of importance in view of the relatively low sex ratio at birth shown by the former. From this it would appear that, on the basis of the sex ratios at birth, the excess of females to be expected in the population is lower than would be the case if the advantage of females over males in Jamaica was as great as that prevailing among the United States white populations.

A convenient way of comparing mortality between the sexes is in terms of the ratios between male and female q_x values. Such ratios are shown in Fig. 10 for Jamaica and Barbados, two West Indian populations similar in racial composition, as well as for the white and Negro populations of the United States. Despite some irregularities, the ratios for Barbados and Jamaica are largely similar, but they differ widely from the pattern of the United States white mortality. Whereas in the case of the United States whites the ratios are well above unity throughout the life span, the ratios for the West Indian populations over an appreciable portion of the life span fall well below unity. As these low values appear in ages under 30 it is possible that they may be associated with special risks to which females are exposed as a result of child-bearing. On the other hand, at advanced ages the ratios tend to be higher for the populations of Jamaica and Barbados. The pattern of the ratios for the two West Indian populations shows some similarity to that for the United States Negro population up to about age 30: among all three the ratio declines sharply in the early years of the child-bearing span; but at higher ages the United States Negro ratios show a curious pattern distinctly at variance with the other ratios considered here.

Probably biological (racial) factors contribute much to the pattern of mortality differentials. But if the low values for the Jamaica population between ages 10 and 30 are associated with special mortality risks incurred over these years by females, the

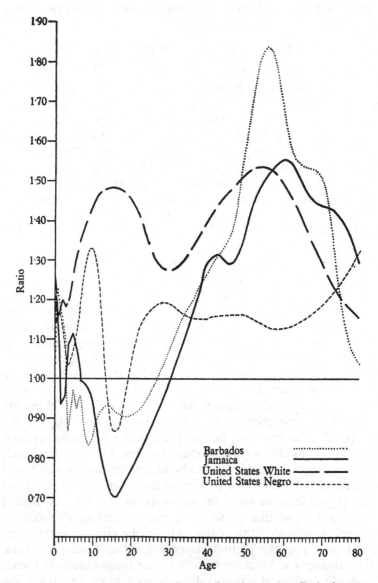

Fig. 10. Ratios of male/female $_xq_x$ values for Jamaica 1945–7; Barbados 1945–7; United States white population 1939–41; United States negro population 1939–41

The United States values are calculated from T. N. E. Greville, *United States Life Tables and Actuarial Tables, 1939-1941*, Washington, 1946.

effects of which may be reduced with the growing control of mortality, then the ratios may eventually exceed unity at all ages. The much more rapid declines in female mortality over these ages since 1946, to which attention has already been called, suggest a development of this nature. On the other hand, the growing importance of certain causes of death among females at advanced ages should tend to reduce somewhat the advantage of the females over males. It is thus possible that improvements in mortality may mean in general a greater conformity to the mortality differentials shown by the United States whites. Should this take place its main effect would probably be to emphasize further the feminine character of the population of the island.

A comparison of mortality in Jamaica with mortality in European and American populations brings out the fact that despite the considerable reductions effected since 1921, a large gap still separates the island from these countries in respect of the general control of mortality. For instance, the average length of life for the United States non-white population in 1950 (59·2 years for males and 63·2 for females) exceeds the corresponding values for the Jamaica population by 3·5 and 4·3 years respectively. More striking is the difference between the United States white population and the Jamaica population. The former, with an average length of life of 66·6 years for males and 72·4 for females, shows an advantage over the Jamaica population of 10·9 years (males) and 13·5 (females).

Comparisons between Jamaican mortality and that of England and Wales over the past 60 years are also instructive. These emphasize not only the great gap between the two populations in the past but also the rapid declines effected in Jamaica after 1921. While mortality remained virtually unchanged between 1891 and 1921 in Jamaica appreciable reductions were recorded in England. So that the advantage over the Jamaica population, in terms of the average length of life, which was between 20% and 25% in 1890, rose considerably, at 1921 being about 56% (see Table 58). At this period in fact the average length of life for the English population was about 20 years above that for the Jamaican. Steep declines in mortality effected in Jamaica closed the gap appreciably, and by 1950 the advantage over the Jamaica population was reduced to 18% in the case of males and 20% in the case of females. Thus the gap was lower than it was in 1890.

Table 58.   *Excess of average length of life for populations of England and Wales over corresponding values for the population of Jamaica*

Period of Jamaica life table	Average length of life, England and Wales*		Excess over corresponding value for Jamaica population			
			In years		% excess	
	Male	Female	Male	Female	Male	Female
1890–1	44·13	47·77	7·39	9·47	20·1	24·7
1910–12	51·50	55·35	12·46	13·94	31·9	33·7
1920–2	55·62	59·58	19·73	21·38	55·0	56·0
1945–7	66·39	71·15	15·14	16·57	29·5	30·4
1950–2	65·8	70·9	10·1	12·0	18·1	20·4

* These are the nearest to the corresponding Jamaica values. Those entered for 1890–1 are for 1891–1900, while those entered against 1945–7 are for 1948; those entered against 1950–2 are for 1951.

The average length of life for the Jamaica population in 1891 was lower than that for any value shown by the official English life tables. But by 1946, the expectancy of life at birth for the Jamaican population rose to a level shown by the English population in 1911, 51·5 years (males) and 55·4 (females). So that within the period of 25 years mortality in Jamaica was reduced to a greater degree than the reduction effected in England between 1841 and 1911. Moreover, the further declines in death-rates in Jamaica between 1946 and 1951 resulted in a rise in the average length of life to a level similar to that prevailing in England in 1921, 55·6 years (males) and 59·6 (females). Thus the gain recorded in Jamaica during 1946–51 is equivalent to that brought about in England during 1911–21, which means that after 1921 rates of mortality decline in Jamaica have exceeded comparable rates for England and Wales.

Cause of death.   High proportions of deaths not medically certified have long constituted the chief weakness of the mortality statistics of Jamaica. It is clear, however, that within recent years steady improvements in the proportions medically certified have been effected. In 1946–47 the percentage certified was 54, and this rose to 58 by 1950–1. In the case of the infants the proportions medically certified remain too low to permit any detailed analysis of cause of death, the proportion being 41%, as compared with 34% in 1946–7.

On the assumption that the overall estimates of deaths from various diseases, derived from the proportions medically certified within each age-group, constitute reliable guides to such mortalities, an analysis of cause of death for 1950–1 in terms of eight major causes has been prepared, strictly comparable to that prepared for 1946–7. From these two levels the lines of progress in the control of mortality during this 5-year period can be traced. Recent mortality statistics of Jamaica are classified according to the new World Health Organization Lists. The groups studied here have been constructed from this classification to conform as closely as possible to those analysed for 1946–7, which were classified according to the 1938 International List of causes. These deaths account for nearly 70% of the total mortality of the island.

As only 41% of all infant deaths are medically certified no detailed analysis of death-rates under one year of age can be made. Nevertheless, death-rates calculated on the assumption that the proportion of infant deaths not medically certified is the same for all causes of deaths show changes between 1946 and 1951 worth noting.[1] It is clear from Table 59 that the reason for the absence of any considerable declines in infant mortality must be ascribed to the failure to reduce mortality from diseases peculiar to infancy. These, which constitute by far the largest group of infant deaths, show a slightly higher death-rate in 1950–1 than in 1946–7, 29·8 per 1000 live births as against 27·5. Deaths of this category seem to constitute the chief obstacles to further rapid declines in infant mortality. But the fact that the ill-defined elements in this category (A 134 and A 135 of the Intermediate World Health Organization List) account for 73% of deaths peculiar to infancy emphasizes the unsatisfactory classification of cause of infant deaths and makes any detailed study of this aspect of mortality impossible. On the other hand, some improvements have been recorded in other causes of infant deaths. Death-rates from pneumonia declined from 16·7 to 12·9. Appreciable reduction was also effected in the death-rates from syphilis, from 7·2 to 5·6, while the death-rates from diarrhoea and enteritis fell slightly from 14·0 to 13·5.

It is also clear from Table 59 that mortality is in all four classes

---

[1] The method of estimating total deaths from the numbers registered and the proportions of registration in each age group was used in the mortality analysis for 1946–7. See G. W. Roberts, op. cit. This paper also gives the mortality rates for 1946–7 quoted in this section.

Table 59. *Death-rates under 1 year of age from four causes, 1946–7 and 1950–1*

Cause of death	Death-rate per 1000 live births			
	1946–7, all infants	1950–1		
		Legitimate	Illegitimate	All infants
Deaths peculiar to first year of life	27·5	21·0	33·8	29·8
Gastro-enteritis	14·0	8·5	15·8	13·5
Pneumonia	16·7	10·2	14·2	12·9
Syphilis	7·2	2·9	6·7	5·6

considered here higher among the illegitimate. This is especially marked in the case of syphilis where the illegitimate rate (nearly 7 per 1000 live births) is 2·3 times that of the legitimate rate. The very low proportion of infant deaths medically certified precludes a satisfactory analysis of cause of death in terms of endogenous and exogenous components. But it appears that in the classes of mortality where environmental or exogenous factors are of great importance—notably in the case of pneumonia and gastro-enteritis—virtually all deaths occur among infants over one month of age. Thus the mortality differentials noted are presumably almost wholly environmental in origin. On the other hand, mortality strongly determined by endogenous factors—represented here by deaths from diseases peculiar to infancy—occurs mostly in the neonatal stage. And here the difference between legitimate and illegitimate neonatal mortality rates (15 and 18 respectively) is very small. At ages over 1 month the corresponding death-rates show much larger differences, being 6 and 16 respectively. Despite the limited nature of the infant mortality analysis imposed by the low proportion of deaths medically certified, the available data suggest that the comparatively high rate for the illegitimate infants is rooted in the less satisfactory social conditions of the parents.

More detailed analyses of mortality on the basis of the whole population can be carried out with more assurance.[1] It is first of

[1] As in the previous study of mortality in Jamaica, the proportions of deaths medically certified in the various age groups are used to estimate total mortality from each group of diseases. In effect therefore the proportion of medical certification is assumed to be the same in all causes of death.

interest to compare the actual numbers of deaths attributed to the groups of causes studied at 1946–7 and 1950–1. As can be seen from Table 60 the total number of deaths from these groups of causes was less in 1950–1 than in 1946–7, 10,900 as against 11,600. In fact, with the exception of diseases of the circulatory system and cancer, deaths from all causes decreased in number. The saving of life effected by the reduction in deaths from respiratory diseases was considerable, deaths being reduced from 2130 to 1660 or by 22%. Deaths from malaria also declined appreciably, from 1040 to 730, while deaths from syphilis fell from 1050 to 850. On the other hand, the increase in deaths from the degenerative diseases was considerable. Deaths from cancer rose from 750 a year to 1070, an increase of 43% within 5 years. Diseases of the circulatory system amounted to 2350 in 1950–1, which also is an increase over the number for 1946–7 (2080).

It is clear that death-rates per 100,000 of the population from these causes declined appreciably between 1946 and 1951, the only exceptions being cancer and diseases of the circulatory system. But as a more complete picture of the improvements in mortality from these causes can be drawn by reference to the stationary populations the crude rates will not be considered in detail.

Considered in terms of the proportion of deaths in the stationary population, the mortality position in 1950–1 emphasizes the growing importance of diseases of the heart and of the circulatory system and of cancer. Whereas in 1946–7 the former group accounted for 17% of all deaths in the stationary population, the proportion from these causes had by 1950–1 risen to 21%. Even more striking is the growing importance of cancer as a cause of death. In 1946–7 5·8% of the male deaths and 6·4% of the female deaths were ascribed to cancer, but 5 years later these proportions had risen to 7·1 and 11·2% respectively. By contrast the proportion of deaths from other diseases declined appreciably, being marked in the case of respiratory diseases and diseases of the digestive system. For instance, among the females, respiratory diseases accounted for 6·5% of all deaths in 1950–1, as compared with 8·3% 5 years earlier. Similarly, the proportion of female deaths from digestive disease declined from 8·8 to 6·9%.

The changing importance of the groups of deaths considered within the short period of 5 years can best be seen by comparing the death-rates per 100,000 of the stationary population. In the case of

Table 60. *Analysis of mortality from various diseases, 1946–7 and 1950–1*

Cause of death	Actual population				Stationary population				
	Average no. of deaths		Death-rate per 100,000 population		% total deaths		Death-rate per 100,000 population		% decrease or increase in death-rates 1946–7 to 1950–1
	1946–7	1950–1	1946–7	1950–1	1946–7	1950–1	1946–7	1950–1	
Male									
Diseases of circulatory system	1020	1150	161·8	167·1	17·2	21·1	330·2	378·5	+14·6
Diseases of nervous system	600	550	95·0	79·9	8·9	9·2	171·5	165·5	−3·5
Diseases of digestive system	1090	1020	172·9	148·5	10·4	8·7	200·5	156·4	−22·0
Diseases of respiratory system	1100	860	174·3	125·4	9·4	7·3	181·4	131·6	−27·5
Cancer	330	410	53·1	59·2	5·8	7·1	111·1	126·9	+14·2
Syphilis	650	530	103·0	76·9	7·0	5·9	133·9	106·8	−20·2
Tuberculosis, all forms	610	520	97·3	75·6	6·3	5·0	120·8	89·7	−25·7
Malaria	550	400	87·8	58·7	4·5	3·3	87·4	59·1	−32·4
Female									
Diseases of circulatory system	1060	1200	161·6	165·2	17·5	21·4	319·2	363·5	+13·9
Diseases of nervous system	740	760	113·4	103·8	11·6	13·9	211·7	236·1	+11·5
Diseases of digestive system	930	820	142·2	113·3	8·8	6·9	161·1	117·9	−22·9
Diseases of respiratory system	1030	800	157·3	110·3	8·3	6·5	151·5	110·3	−27·2
Cancer	410	660	63·5	90·9	6·4	11·2	116·7	189·4	+62·3
Syphilis	400	320	60·6	44·3	3·5	3·0	63·8	50·4	−21·0
Tuberculosis, all forms	630	510	96·9	70·7	5·5	4·3	100·4	73·3	−27·0
Malaria	490	330	74·4	44·7	3·9	2·3	71·4	39·6	−44·5

the males diseases of the circulatory system comprise the main cause of death and the rates from these diseases increased from 330 to 378 or by 15%. Likewise the death-rate from cancer rose notably from 111 in 1946–7 to 127 by 1950–1, or by 14%; whereas it ranked seventh among the causes studied here in 1946–7 it was fifth most important 5 years later. In 1946–7 diseases of the digestive system were second with a death-rate of 200, but with the reduction to 156 by 1950–1 the rank of this group was third. Deaths from respiratory diseases appeared as the third most important cause of death in 1946–7, the death-rate being 181; but this rate was reduced to 132 or by 27% during the succeeding 5 years. Only small reductions in the death-rate from diseases of the nervous system were recorded, the rate falling from 171 to 166, or by 3%, so that in 1950–1 this appeared as the second highest cause of death, though in 1946–7 its rank was fourth. The reduction in the death-rate from syphilis has been striking, from 134 to 107 or by one-fifth. Appreciable reduction was also recorded in the tuberculosis death-rate, from 121 to 90 or by 26%. Though the least important of the causes studied here, malaria death-rates showed the greatest decline, from 87 to 59 or by 32%.

In general, declines in the mortality rates for females parallel those shown by the males, but there are a few differences which are noted here. Diseases of the nervous system, which at both dates constituted a greater hazard to females than to males, showed a death-rate of 236 in 1950–1, as compared with 212 at the earlier period, thus increasing by 12%. Again the increase in the female death-rate from cancer is striking, from 117 to 189 or by 62%. Finally, the female death-rate from malaria shows by far the greatest reduction, from 71 to 40 or by 44%.

An effective way of assessing the effects of various diseases on the mortality pattern of a population is to estimate the years of life lost through these diseases. In other words, if a particular group of diseases is eliminated, what may be the resulting gain in longevity? A comparison between the resulting gains at 1950–1 and those at 1946–7 reveals interesting changes. In fact, the rising importance of cancer and diseases of the circulatory system, particularly among the females, is perhaps more forcefully brought out in this analysis than in any of the measures already discussed.

The years of life lost through the eight groups of causes studied are shown in Table 61. We consider first the position of the males.

Diseases of the circulatory system, the major cause of death among the eight groups of causes, account for an appreciable lowering of the average length of life by 2·59 years or 4·6%. The corresponding loss in 1946–7 was 2·23 years. Up to age 55 the loss exceeds 2 years and for a considerable portion of the span of life is above 2·8 years. The importance of these diseases at higher ages is emphasized by the fact that the loss at age 55 is 13% of the expectation of life at this age.

Losses in expectation of life caused by diseases of the nervous system are much lower in 1950–1 than in 1946–7. The reduction in the average length of life is only 1·21 years as against 1·32 at the previous period. Up to age 35 losses exceed 1 year.

Diseases of the digestive system, which claim most victims in infancy, are responsible for the greatest reduction in the average length of life, 2·63 years, but the improvements in mortality from these diseases emerge from the fact that the corresponding loss at 1946–7 was 2·79 years. Thus the percentage reduction in average length of life has declined from 5·4 to 4·7%. Throughout the life span the years of life lost from these diseases are much less in 1950–1 than at the preceding period.

The rapidly shrinking importance of deaths from respiratory diseases is effectively demonstrated by the fact that only 2·15 years are cut off the expectation of life at birth in 1950–1, which is much lower than the loss at 1946–7 (2·85 years). In fact above age 5 not much could be added to the expectation of life by the complete elimination of these diseases, which are no longer the source of severe loss they were in the past.

Reduction in the expectation of life from deaths from cancer assumes impressive dimensions, though, as we shall see, it is among females that the disease takes its heaviest toll. The loss in average length of life (0·93 year) is only slightly greater than the corresponding figure for 1946–7 (0·73). At higher ages the loss is more important, though again the values are much smaller than those in the case of the female.

It would appear that modern methods of treatment of syphilis, which had long been 'one of the commonest diseases of Jamaica, being met with everywhere in all its various stages', have successfully brought this disease under control. Whereas in 1946–7 the loss in expectation of life from syphilis was 1·64 years at birth it has been reduced to 1·32 in 1950–1. At higher ages as well the losses in

Table 61. *Years of life lost through certain causes of death, 1950–1*

Age	Circulatory system	Nervous system	Digestive system	Respiratory system	Cancer	Syphilis	Tuberculosis	Malaria
Male								
0	2·59	1·21	2·63	2·15	0·93	1·32	1·47	1·07
1	2·82	1·24	1·69	1·95	1·01	1·02	1·56	0·90
5	2·81	1·10	1·17	0·72	0·90	0·81	1·44	0·52
10	2·83	1·09	1·07	0·63	0·91	0·80	1·43	0·46
15	2·83	1·08	1·03	0·62	0·90	0·79	1·41	0·42
20	2·81	1·08	1·02	0·60	0·91	0·78	1·34	0·39
25	2·83	1·07	0·98	0·57	0·91	0·78	1·14	0·35
35	2·78	1·04	0·83	0·50	0·89	0·71	0·68	0·24
45	2·64	0·93	0·64	0·40	0·83	0·55	0·26	0·11
55	2·27	0·74	0·32	0·22	0·57	0·29	0·01	0·00
65	1·65	0·38	0·00	0·00	0·08	0·00	0·00	0·00
Female								
0	2·94	1·89	2·42	2·28	1·88	1·08	1·69	1·12
1	3·15	1·97	1·65	1·43	2·01	0·83	1·79	0·92
5	3·16	1·87	1·21	0·84	1·96	0·62	1·72	0·63
10	3·17	1·87	1·10	0·75	1·98	0·63	1·72	0·56
15	3·15	1·88	1·04	0·73	1·98	0·63	1·68	0·53
20	3·15	1·89	1·01	0·70	2·00	0·63	1·56	0·50
25	3·15	1·90	0·95	0·66	2·02	0·61	1·24	0·46
35	3·11	1·94	0·84	0·62	2·02	0·56	0·80	0·37
45	2·96	1·91	0·70	0·55	1·85	0·44	0·47	0·28
55	2·65	1·72	0·53	0·45	1·48	0·31	0·23	0·18
65	2·29	1·47	0·37	0·31	0·93	0·15	0·06	0·05

expectation of life have shown marked declines since 1946. The loss of life from syphilis is greater among males than among females at all ages.

Some progress has been made in the control of tuberculosis, though the disease still remains an important cause of death. The loss in the average length of life was 1·68 in 1946–7, but was reduced to 1·47 by 1950–1. At ages where the disease is most active, between 15 and 30, improvements have been recorded, but its importance is still evident. The loss in expectation of life at age 20 fell from 1·76 to 1·34 in 1950–1, but this is still the second most important loss of the groups of diseases studied at age 20. In fact, the expectation of life at age 20 is reduced by 3% as a result of deaths from tuberculosis.

In recent years malaria has never been a very high cause of death in Jamaica. In fact it seems only half as important here as it is in British Guiana. Still the reductions in deaths from this disease has been hardly less marked than in the case of British Guiana. The loss in average length of life declined from 1·47 in 1946–7 to 1·07 in 1950–1. It is therefore evident that malaria is the least significant of the elements of mortality considered here; its complete elimination would add very little to the expectation of life, particularly at advanced ages.

With the exception of syphilis and diseases of the digestive system the groups of diseases considered here bear much more heavily on females than on males. This is most marked in the case of diseases of the circulatory system. Here the loss in the average length of life amounts to 2·94 years or 5%, as compared with 2·28 in 1946–7. At advanced ages the losses are very considerable, being up to age 35 in excess of 3 years. And at age 55 the loss (2·65 years) amounts to 13% of the corresponding expectation of life.

In respect of diseases of the nervous system females also show greater losses than do males. The length of life is reduced by 1·89 years or 3·2%, which is considerably higher than the corresponding loss for the males. At higher ages female losses are even more impressive. Above age 20 the losses exceed 4% of the expectation of life and at age 55 the loss (1·72 years) is nearly 9% of the expectation of life.

Cancer also appears as more important among females, the loss of life being about twice as high as in the case of the males. Deaths from cancer cut off 1·88 years from the average length of life, which is much more than the loss revealed in 1946–7 (1·02). And whereas in 1946–7 the number of years lost at age 45 was only 1·10, the loss at this age in 1950–1 had risen to 1·85. At advanced ages the loss of life due to cancer is about equal to that from diseases of the nervous system.

Despite the progress made in the control of mortality in Jamaica, the loss of life from many diseases remains high by comparison with European populations. This is well illustrated by Table 62, which compares the mortality from the eight major groups of disease studied here as well as from diseases peculiar to infancy for Jamaica and the United Kingdom. Apart from the fact that malaria, though producing a death-rate of 52 per 100,000 in Jamaica is of negligible importance in the United Kingdom, the most arresting contrast is in respect of diseases of the digestive system. The death-rate in Jamaica is nearly three times that of the United Kingdom. It is basically infant mortality that contributes most to the un-favourable position of Jamaica, for diarrhoea and enteritis, as has already been shown, are second only to diseases peculiar to infancy as a cause of death among infants. Jamaica also appears in an unfavourable light in regard to diseases peculiar to infancy; here the death-rate (100) is more than twice that for the United Kingdom (40). The differences shown for these two groups of diseases emphasize that it is chiefly infant mortality that accounts for the relatively high overall rates for Jamaica. But in respect of two groups of diseases which bear more on the population of higher ages Jamaica also stands at a much higher mortality level. Thus the Jamaica death-rate from syphilis (60) is twelve times that for the United Kingdom. The latter also has a marked advantage in the case of tuberculosis, as the death-rate for Jamaica (73) is 39% higher than that for England and Wales.

It is important to note that for a very considerable group of diseases the advantage lies in favour of Jamaica. Diseases of the respiratory system offer the most interesting case here. For the death-rate among the population of Jamaica (118) is 7% lower than the United Kingdom rate. The advantages enjoyed by Jamaica in the other cases all apply to diseases that exact their heaviest toll at advanced ages, and to this extent the unfavourable position of the United Kingdom is to be expected in view of the much younger population of Jamaica. Though disease of the circulatory system and cancer are of mounting importance in Jamaica the mortality from these diseases still remains small by comparison with European populations. The overwhelming importance of diseases of the circulatory system in the population of the United Kingdom is seen in the fact that they yield a death-rate of 368, or more than twice

Table 62. *Comparison of death-rates from certain major causes of death in Jamaica and the United Kingdom*

Cause of death	Death-rates per 100,000 of the population	
	Jamaica, 1950–1	England and Wales, 1947–8
Diseases of circulatory system	166·1	367·6
Diseases of nervous system	92·2	145·0
Diseases of digestive system	130·4	44·9
Diseases of respiratory system	117·6	126·6
Cancer	75·4	186·6
Syphilis	60·1	5·1
Tuberculosis, all forms	73·0	52·7
Malaria	51·5	0·0
Congenital malformations and debility and diseases peculiar to infancy	99·7	40·4

Note. The rates for the United Kingdom are the averages of the rates for 1947 and 1948 given in the United Nations *Demographic Yearbook, 1953.*

that of Jamaica. Mortality from cancer is also much greater in the United Kingdom, accounting for a death-rate of 187 and constituting the second highest of the causes studied here. In the case of Jamaica mortality from cancer is less than half of this (75) and it constitutes the fifth highest of the causes studied. There is also a small advantage for the Jamaica population in respect of diseases of the nervous system, the death-rate being 92 as against 145 for the United Kingdom.

The foregoing comparisons are subject to several limitations because of the great difference in age structure of the populations compared. Still two significant differences emerge. In the first place much of the unfavourable light in which Jamaica appears is due to high mortality from certain diseases which affect infants mostly—diseases peculiar to infancy, and diarrhoea and enteritis. Hence much of the future progress in the control of mortality in the island depends on the success that attends attacks on these diseases. In the second place the fact that, despite the mounting importance of cancer and diseases of the circulatory system in Jamaica, the mortalities are still much lower than in European populations, raises important questions as to the future course of mortality in view of the basic differences in age structure of the two populations. As will be shown in Chapter 9, the increase in population of advanced age in Jamaica promises to be considerable in the near future. But even this increase will not be such as to bring the death-rates from the degenerative diseases up to a level comparable to that shown by European populations within the course of one generation. It therefore seems that the reduction in mortality from diseases that affect infants primarily should result in further striking overall declines in mortality in the island.

CHAPTER 7

CHANGING PATTERNS OF REPRODUCTION

The principal method of recruiting the population of Jamaica during slavery, the slave trade, together with the policy aimed at inducing large numbers of Europeans to settle in the island, constituted in effect state-sponsored programmes for increasing the population. Similarly the whole movement of indenture immigration surveyed in Chapter 4 illustrates another aspect of policy designed to recruit the population of the island. But it was not solely in terms of immigration that policy for promoting population growth was framed. An important and entirely different line of policy developed during the later slave period had also as its aim the promotion of population growth. Since in many respects this policy conforms to what is generally termed a population policy the conditions of its development and its precise methods of operation deserve some study. Two further lines of policy that emerged in the post-emancipation period are also treated in this chapter, though, as will be seen, they were in no sense continuations of the policies in operation during slavery or in any way aspects of a formal population policy. Still the controls these bodies of legislation introduced—controls over the general health of the population and over the legal status of family unions as well as over conditions of mating—are so closely interwoven in many basic elements of the demography of the island that they also must be considered here. These several types of policies are surveyed in terms of the main periods into which the island's history can be conveniently divided for the purpose of demographic analysis.

At the outset it is necessary to note some distinctive features of the policy formulated to encourage population growth among the slaves. It has been shown by Professor D. V. Glass that most of the population policies pursued in European countries rested on certain fundamental principles, summed up conveniently in the words of the Italian jurist Filangieri: '[There is] no one government that has not reserved some prerogatives for the fathers of families, that has not granted some privileges and exemptions to those citizens who have given a certain number of children to the state;

that has not provided some express laws to increase the number of marriages.'[1] The uniqueness of the population policy that emerged in the later years of slavery lies in the absence of two of the principles enunciated by Filangieri. In the first place the chief financial rewards for reproduction accrued not to those supplying the additions to the population, the parents of the slave children, but to their owners. Moreover, such prerogatives and privileges as were accorded the slaves for adding to the population went to women and not to men; the responsibility for reproduction, it seems, was firmly and exclusively fixed upon the females. Another principle of Filangieri that did not appear in the slave population policy was the stimulation of the rate of marriage, or more exactly of mating (as will be shown, marriage among slaves was virtually impossible, even after the passing of the slave code of 1826). Mating was in no sense deemed complementary to fertility in considerations of slave reproduction.

These two departures from the principles stated by Filangieri are significant not only in the context of the uniqueness of the population policy developed during slavery; just as important, they serve to emphasize another aspect of the slave regime, the unstable nature of the slave family and in particular the weakness of the link binding the parents together. Consequently, in this survey of patterns of reproduction some examination will be made of the conditions of mating during slavery, quite apart from its association with population policy. The forms and conditions of slave mating have an important bearing on the general forms and conditions of mating now encountered in the island. As in the study of all human populations, mating demands careful attention because of its close association with fertility. Because of the diversity of existing family forms in the island the relationship between fertility and mating now attains great complexity. An examination of mating conditions during slavery should therefore help to place this whole complex relationship in proper historical perspective.

Because of the variety of legal, social and other factors determining population growth which will be taken up here, the treatment is oriented towards the components of reproduction over the several historical periods rather than towards a survey of legal measures.

---

[1] *Analysis of the Science of Legislation, from the Italian of the Chevalier Filangieri,* translated by W. Kendall, p. 28, quoted in D. V. Glass, *Population Policies and Movements in Europe,* p. 86.

It is, in fact, the changing background of fertility, mating and mortality that will be dealt with. A thorough study of such aspects of the population would involve in effect the tracing of the social history of Jamaica an undertaking beyond the scope of this chapter. For the present purpose it suffices to consider some of the relevant laws and such historical and social factors as are deemed necessary to portray the main features of the two broad historical periods in terms of which this brief survey is made: slavery and the post-emancipation era. However, before dealing with these two major periods it is instructive to note briefly a few features of the patterns of reproduction in the very early years of the island's history.

BEFORE THE CONQUEST BY THE BRITISH

The fact that the indigenous Indian population of Jamaica was completely wiped out within half a century of the settlement of the island by the Spaniards makes the pattern of reproduction of more than passing interest. Still, too much importance cannot be attached to reproduction in these early years, since the rapid disappearance of the Indians meant that their reproductive performance had no influence on later population movements in the island. In respect of social organization the indigenous inhabitants doubtless differed profoundly from the Negro slaves later established in the island. But in two respects the two populations were alike; they both seemed to experience a deep social shock on being impressed into slavery and both experienced heavy mortality when enslaved. Though we are without any information on the numbers of Indians on the island when first visited by Columbus, their rapid extermination suggests that they were not large. However, it is usual for historians of the island to ascribe this rapid extinction to the severe mortality resulting from the harsh treatment meted out by the Spaniards. In the words of Las Casas: 'I have seen them remplir les campagnes de fourches patibulaires; on which they hanged these unfortunates by thirteens in honour of the thirteen apostles; I have beheld them throw the Indian infants to their dogs; I have seen five caciques burnt alive; I have heard the Spaniards borrow the limb of an human being to feed their dogs and next day return a quarter to the lender.'[1]

But, as Kubler has argued in the case of the Indian population

[1] Quoted in G. W. Bridges, *The Annals of Jamaica*, vol. 1, note 1.

of Mexico, the policy of the Spaniards cannot be so crudely stated as a policy of extermination.[1] The frequent explanation of severe depopulation of the Indians in terms of what he aptly calls the homicidal theory of population loss tends to underrate the importance of epidemics and the more subtle elements of loss resulting from the social shock and reorientation towards a Christian, mercantilist economy. Doubtless in Jamaica, as on the mainland, Spanish occupation engendered high mortality in virtue of the severe economic exploitation it involved, exploitation aimed not so much at extermination of the inhabitants, but rather at maximum economic gain. In both areas high rates of suicide in order to escape slavery have been reported.

THE PERIOD OF SLAVERY

Some historians of the West Indies have argued that the position of the Negro slave was no worse than that of the European he displaced, the indentured servant brought from the British Isles. Though in the early days of slavery some of the harsher features of the regime were in essence little different from the disabilities under which European indentured servants worked, it is easy to exaggerate this parallel.[2] By treating the Negroes as distinct from the European population in virtue of their status as slaves, the development of their mating institutions appears more meaningful.

The social function of the Negroes, their status and conditions of life, were rigidly and unequivocally defined in social codes having the fullest sanction of the law, though it was often the master who was mainly responsible for their enforcement. Until the late eighteenth century the codes were not held to be unduly severe. Indeed, in view of the current conviction of the innate inferiority of the Negro (in the opinion of Long he was but one remove from the animals) the slave laws were to most observers perfectly justified.

Rigid ordering of the life of the slave in terms of slave codes meant that changes in the broad patterns of slave life were, to some extent, mirrored in the series of laws promulgated from time to time.

[1] George Kubler, 'Population movements in Mexico, 1520–1600', *The Hispanic American Historical Review*, vol. XXII, no. 4, 1942.
[2] Eric Williams, op. cit. p. 18, and Frank Tannenbaum, *Slave and Citizen*, p. 102.

The volume of slave law in the West Indies is vast. Here we shall consider briefly four important slave laws illustrating the changing position of the slaves and the patterns of slave reproduction at three different periods of slavery in Jamaica. The first, that of 1696, illustrates the legal position of the slave in the heyday of the sugar colonies. The second, passed in 1792, is of significance primarily because it contains the elements of a pro-natalist population policy for slaves. The third law discussed is that of 1816, the chief interest of which is its restatement, with minor variations, of the pro-natalist measures of the law of 1792. The fourth law, that of 1826, the direct outcome of the anti-slavery movement, aimed at improving the social and moral conditions of the slaves.

Though these four codes outline the legal restraints on the slave, and emphasize his status as a 'work-unit' and as the property of his master, it is to the accounts of contemporary observers, preserved in their writings or in evidence before committees, that we must in the main turn for descriptions of life in the time of slavery. Such accounts, the work of historians, travellers, medical men, planters and others, have their limitations. The several ways in which the elements of reproduction—mating, fertility and mortality—interacted against the background of slavery can be only imperfectly gleaned from these records. So strongly do individual viewpoints on several basic aspects of slavery intrude themselves in the literature that it is often difficult to distinguish between objective description and thinly disguised approval or condemnation of existing social conditions. Moreover, these accounts are not without contradictions on some important points. Above all the dearth of statistical material on every facet of life on the plantations not only makes interpretation and assessment difficult, it also means, as has already been shown, that only the crudest estimates of population movements before 1816 can be made. Despite these limitations, however, it is possible to obtain a rough picture of the chief patterns of mating and reproduction among Negroes in the days of slavery.

1696–1792. The slaves, being a group devoid of natural or civil rights, a class to whom the principles of English common law did not apply, were subject to entirely different laws from those relating to their masters. The slave code of 1696, like most of the later laws of this nature, had certain definite aims.[1] Basically it sought to prevent slave uprisings and to afford means of dealing with any

---

[1] This code is outlined in Edward Long, op. cit. vol. II, p. 485.

that might arise.[1] It effectively determined the gulf between master and slave though it did guarantee a few rights to the latter. His mobility was severely limited; he could travel or hire himself out for work only with the permission of his master. Heavy penalties were prescribed for infractions of the code. Beyond laying down what clothes he should receive from his master and certain annual holidays that should be given, it imposed very few responsibilities on the master regarding the treatment of his slaves. An interesting clause stated that pregnant female convicts should not be executed until after their delivery, evidently more a device in keeping with the preservation of property than a genuine humanitarian gesture. The code also laid down that owners should provide for the instruction of their slaves 'in the principle of Christian religion and to facilitate their conversion, and to do their utmost to fit them for baptism', but said nothing about mating among slaves.

At the time of the passing of this law there was apparently a general effort on the part of the planters to ensure reproduction among the slaves. But the only method relied on for promoting this was to ensure equality between the sexes in all purchases of slaves. In the words of Blome, 'For the encrease of the stock of Negroes they generally take as many Men as Women.'[2] The view seems to have been that as long as the sexes were present in equal numbers slaves would automatically mate and produce children. An additional advantage of such a policy, it was claimed, was the improvement in slave morale it engendered.

But the promotion of reproduction among slaves was incompatible with their legal position and the conditions of plantation life. It therefore soon gave place to a policy of relying wholly on the slave trade in order to recruit the slave population. For the expansion of the plantation economy it was essential not only to keep up existing numbers but also to ensure an increase in slave population in the face of extremely high mortality. So that by the early eighteenth century the policy of relying wholly on the slave trade for population increase was generally acknowledged. Henceforth instead of seeking to secure equal numbers of both sexes, the aim was to secure

[1] R. W. Smith, 'The legal status of Jamaican slaves before the anti-slavery movement', *Journal of Negro History*, vol. xxx, July 1945.
[2] Quoted in F. W. Pitman, 'Slavery on British West India plantations in the 18th century', *Journal of Negro History*, vol. xi, pp. 584–668, October 1926. This study, from which extensive quotations are given here, contains by far the most thorough analysis of reproduction in the West Indian slave population.

a preponderance of males. John Steward and John Wright, agents in Jamaica for the Royal African Company, declared in 1714 that in procuring slaves in Africa, 'there may be three men to one woman, no old people nor young children'.[1] At the height of the slave trade Hippesley reported that the rule was to export five or six times as many males as females from Africa. This he claimed favoured polygamy in Africa, which in turn increased the population of that continent and the source of labour for the West Indies. Thus in the words of Pitman, 'until 1807 Africa remained the cradle of the slave population'.[2]

It is true that a policy of slave reproduction meant, as was forcibly impressed upon West Indian planters after 1807, the maintenance of a high proportion of non-effective slaves (women and children) with a consequent sacrifice of labour power. But it is doubtful whether the economic loss such a policy implied contributed in any way to the abandonment of the policy of promoting slave reproduction. This change of policy followed as an inevitable consequence of the failure of the slave population to replace itself. Vital processes could not ensure the population increase required for the plantations; indeed, without continued large-scale introduction of slaves swift depopulation would have resulted. Probably any policy not explicitly based on the slave trade was foredoomed to failure. The recruiting of the slave population must also be viewed in the broader context of the economic basis of slavery which has been demonstrated by Eric Williams.[3] It was economically impossible to maintain slavery as a profitable institution unless the rapidly wasting population was constantly recruited by importations of slaves. This could easily be done at the low prices at which they could be secured. Sir James Stewart considered the cost of rearing slaves in the West Indies was greater than the cost of constant importations of Africans.[4] As M'Neill put it, slaves were considered 'as so many cattle' who would for seven or eight years 'continue to perform their dreadful tasks and then expire'.[5] In short, slaves were expendable. The issues are neatly summed up in the words of U. B. Phillips: 'The laborers were considered more as work-units than as men, women and children. Kindliness and

---

[1] Quoted in F. W. Pitman, op. cit. p. 630.   [2] Ibid. p. 628.
[3] Eric Williams, op. cit.
[4] F. W. Pitman, op. cit. p. 637.
[5] H. M'Neill, *Observations on the Treatment of the Negroes in the Island of Jamaica*, London, 1788, p. 4.

comfort, cruelty and hardships were rated at balance-sheet value; births and deaths were reckoned in profit and loss and the expense of rearing children was balanced against the cost of new Africans.'[1]

That the failure of reproduction as a means of assuring population growth among the slaves was to a large extent rooted in the prevailing levels of mortality will have been evident from the previous survey of the catalogue of diseases that ravaged the island during slavery. Still, it is interesting to note that even at this early date certain observers were conscious of the inherent disadvantages of relying on so wasteful a system of demographic balance, and were urging improvements in the conditions of slaves in order to reduce mortality. Governor Robinson of Barbados, noting that the increase in the population of that island was not commensurate with the numbers of slaves imported, attributed this to 'the want of some effectual Municipal Law to oblige their Masters to use less severity and cruelty towards them . . .'.[2] And Dr Campbell expressed the view that 'a little more humanity would have enabled Barbados to save two-thirds of the cost of the slave importation'.[3] But until these arguments were reinforced by the hard facts of changing economic conditions, there was no chance that such views would prevail and the harsh conditions of plantation life remained, a signal cause of the tremendous mortality of the period.

Nevertheless, high mortality was not simply the result of prevailing conditions of the plantation. Another factor must here be considered, one which assumes importance not only because of its effect on mortality but also because of its effects on the whole reproductive situation during slavery. This is the demoralization the Negroes experienced when brought to the West Indies under slavery. Here, in fact, we are faced with a situation similar to that noted by Kubler in the case of the Indians when they were brought under Spanish domination.

There is evidence that the general feeling of despondency that pervaded the African-born slaves profoundly influenced the patterns of reproduction during slavery, contributing much to the establishment of the extremely wasteful type of demographic balance embodied in plantation slavery and to the establishment of one concomitant of this form of balance, the unordered system of mating relations.

[1] U. B. Phillips, *American Negro Slavery*, 1928, p. 52.
[2] Quoted in F. W. Pitman, op. cit. p. 621.   [3] Ibid. p. 623.

In the first place it induced high mortality among the slaves. The general feeling of despondency manifested itself from the time the slaves boarded the vessels for transportation to the Western World. Once on board they evinced 'signs of extreme distress and despair from a feeling of their situation and regret at being torn from their friends and connections'.[1] This feeling not only expressed itself in the deep dejection into which the slaves remained plunged, but also actively in their numerous attempts to destroy themselves. Many refused food because 'they wished to die' and had to be forcibly fed on the voyage; starvation was, in fact, frequently resorted to as a means of suicide. Another important means of suicide was jumping overboard, a practice which necessitated many slavers being specially constructed to prevent this. To the same attitude on the part of the slaves was also attributed many uprisings on slave ships. Many witnesses claimed that high mortality on the Middle Passage was partly due to the 'diseased mind' of the slaves.[2] In this state they refused food and medicines when sick and expressed the desire to die. Surgeon Wilson even claimed 'that of the deaths two-thirds of those who died in his ship the primary cause was melancholy'. 'The original disorder was a fixed melancholy, and the symptoms lowness of spirit and despondency.' Another witness declared that slaves 'often fall sick sometimes owing to their crowded state but mostly to grief for being carried away from their country and friends'.

On arrival in the West Indies slaves also showed this spirit of desperation and sought to commit suicide by various means. 'It is also notorious that the Africans, when brought into the colonies, frequently destroy themselves. . . . The causes of it are described in general to be ill treatment, the desire of returning home, and the preference of death to life when in the situation of slaves . . .'.[3] So frequent were these 'acts of desperation' that the claim was made that 'the Gold Coast negroes, when driven to despair, always cut their throats, and those of the most inland country mostly hang themselves'. As further evidence of the distress of Negroes in their life as slaves, some witnesses pointed to the rejoicing that marked the funerals of slaves. 'This joy is said to proceed from the idea that

[1] *Abstract of Evidence delivered before a Select Committee of the House of Commons in the years 1790 and 1791 on the part of the Petition for the Abolition of the Slave Trade*, Cincinnati, 1855.
[2] Ibid.   [3] Ibid.

the deceased are returning home.' In this same connexion Charles Leslie wrote: 'They look on Death as a blessing. . . . They are quite transported to think their slavery is near an end and they shall revisit their happy native shore . . .'.[1] It was this belief that by death they could return to their native land that was, according to Lewis, the cause of many suicides in Jamaica. But this, he emphasizes, was characteristic only of the 'fresh' slaves.[2]

Doubtless much of the evidence of the anti-slavery interests was either specially selected to prove their contention of the harshness of slavery or deliberately distorted with this end in view. But whatever the degree of distortion or exaggeration inherent in this evidence it suffices to sustain the thesis that the will to self-extermination dominated the lives of African-born slaves. Probably the same attitudes were less widespread among the creole slaves; these were less recalcitrant, more pliable, more subservient to their masters, more organically part of the regime into which they were born because they knew no other. Indeed, observers detected a marked hostility between the creole slaves, 'the aristocrats of the Negro world', as Pitman terms them, and the African-born; as M. G. Lewis expressed it, 'they hate each other most cordially'. But probably as long as the slave trade continued the proportion of African-born slaves remained large, and for this reason the symptoms of social shock were widespread among the slaves.

The failure of the slave population to replace itself stemmed as much from levels of fertility and forms of mating as from high mortality. Low fertility was an inevitable result of the policy of relying on the slave trade and of the general conditions of slavery. Indeed, to the masters and the slaves alike high-fertility patterns were unacceptable. The function of the female slave as a 'work unit' was heavily stressed; in this capacity she was as essential to the plantation as the male slave, being required for domestic service and for the lighter operations connected with field and factory. It was even claimed by Governor Parry of Barbados, 'the labour of the females . . . in the works of the field is the same as that of men'.[3] The rearing of children impaired her function as a labourer and thus was not countenanced by the master. The

---

[1] Charles Leslie, *A New History of Jamaica*, London, 1740, p. 307.
[2] M. G. Lewis, *Journal of a West India Proprietor*, 1834, p. 100. The extent of suicides among the slaves cannot of course be demonstrated statistically.
[3] Quoted in F. W. Pitman, op. cit. p. 637.

position of the pregnant slaves, it seems, was not a happy one. In the words of Ramsay, they were 'wretches who are upbraided, cursed and ill-treated ... for being found in the condition to become mothers'.[1] A witness before the Select Committee of 1790–1 declared that 'a female slave is punished for being found pregnant'.

So far as the African-born woman was concerned it is probable that the despair into which she was thrown when impressed into slavery made her most reluctant to bear children. For, exposed to severe treatment, with the full knowledge of the hardships of pregnancy under slavery and the fate of her slave offspring who could easily be severed from her, she would doubtless employ any means to escape child-bearing. It seems that Africans were familiar with methods of fertility control before their arrival in the West Indies.[2] Such practices or modifications of them were evidently widely known among West Indian slaves. According to Long and Bryan Edwards, abortion among slave women was common, though we may disregard the reasons they advanced for its pre-valence. The 'common prostitutes', as Long termed the female slaves, took 'specifics to cause abortion, in order that they [might] continue their trade without loss of time or hindrance of business'.[3] Bryan Edwards considered the frequency of abortion as 'the usual effects of a promiscuous intercourse'.[4] Two plants found in Jamaica are mentioned by Long as used by slaves as abortifacients, the fruit of the calabash tree and the aloe plant. It is less certain that infanticide was practised during this period, but this allegation is advanced by the Central Board of Health.[5] Martin, referring to British Guiana, wrote of 'the arts to which African women resort to prevent their being mothers'.[6] At the same time this aversion to child-bearing might have been largely confined to the African-born slave; the creole slave, unhampered by the desire to regain her lost freedom or to escape slavery by death, must have been more resigned and probably less averse to child-bearing.

Mating assumes importance primarily as a factor in reproduction, but in the present context its wider aspects, and particularly its emergence from the matrix of slave society, are just as important.

[1] Quoted in F. W. Pitman, op. cit. p. 639.
[2] Norman Himes, *Medical History of Contraception*, 1936, p. 9.
[3] Edward Long, op. cit. vol. II, p. 436.
[4] Bryan Edwards, op. cit. vol. II, p. 144.
[5] *Report of the Central Board of Health*, 1852, p. 113.
[6] R. W. Martin, *History of the British Colonies*, vol. II, p. 28.

In the nature of things it is impossible to outline with any certainty the forms of mating among the non-European population. The only information bearing directly on the subject, the writings of contemporary observers, are not always rewarding. Still a consideration of some of the elements that might have influenced mating habits suggests that nothing approaching the form characteristic of European populations could have been expected under slavery.

The conditions under which slaves were captured and transported to the colonies, calculated, as we have already shown, to produce symptoms of profound social shock among them, made the continuance of African mating institutions impossible. Uprooted from various environments, each with its own marriage ceremonial and prescribed mating relationships, and thrust forcibly into a slave society where conditions of mating were of little relevance to their owners, the slaves inevitably experienced a dissolution of their traditional family forms. And in the absence of any sustained effort on the part of the slave owners to assure adequate reproduction, interest in encouraging any new, orderly relations among their slaves could at most be casual.

Certainly the most important condition affecting mating among the slaves was the absence of any legal provisions for this. In fact, it was only towards the end of slavery that ineffective provisions in this direction were incorporated into the slave codes. Contemporary observers viewed mating among slaves not as something peculiar to the slave society but as a departure from the norm of the monogamous union instituted by civil or religious rites and resting on full legal sanction. But this was an unrealistic position. For the marriage laws in the West Indies applied strictly to Europeans, not to slaves. There was apparently no law in Jamaica preventing the marriage of slaves during this period, though such laws were known elsewhere in the West Indies.[1] But the absence of provision for marriage was only part of a wider complex. For marriage as understood in European society was incompatible with slavery. A stable union of the spouses necessitates their sharing a common household; but slave law could not guarantee this. A union could be easily broken by the sale and removal of one of its members. In this connexion it is interesting to see how an eighteenth-century Jamaica slave law sought 'to prevent as far as possible the separation of different

---

[1] In British Guiana for instance there was a law against marriage among slaves. See E. A. Wallbridge, *The Demerara Martyr*, 1943, p. 68.

branches of the same family'.[1] This law made it necessary for a merchant or factor conducting the sale of newly arrived slaves to swear 'that in the sale, he has done his utmost to class and sell together mothers and their children and brothers and sisters'. Possibly a distinction was drawn between newly arrived slaves and others in regard to the separation of families at sale. For another law stated that slaves 'taken on any writ of vendition' and exposed for sale should be sold singly 'unless in cases of families in which case and no other the provost marshal . . . may set up to sale such family consisting of a man and his wife, his, her or their children'.[2] It remains clear that apart from the absence of provisions for marriage in the slave laws, the whole legal system governing slavery denied the conditions essential for the foundation of stable unions. Indeed, slavery was incompatible with stability among family unions; its continuance as a profitable economic institution necessitated the right of ready sale and purchase of individual slaves, and a system of marriage, binding slave couples together in indissoluble unions, would have made this impossible.

Another significant factor influencing patterns of mating during slavery was the relation between the female slave and her European master. To the extent that such relations gave rise to the coloured population they contributed to population growth. It has even been suggested that such unions were encouraged on the West Indian plantations 'in order that the gang of slaves might be enlarged', but this is untenable.[3] In the present context it is the legal and social position of the coloured people and the mating habits in which they were involved that are more relevant. In one sense the sexual relation between the female slave and her master can be considered as a result of the dominance of the master over his slave, even as the exaction of a right to which, presumably, his status entitled him. On the other hand, it cannot be denied that the custom of a master taking a concubine from among his slaves was perfectly in keeping with the unordered familial relations among the slaves themselves.

[1] 'An Act to repeal an Act entitled "an Act to regulate the sale of newly-imported negroes . . ."'

[2] 'An Act to prevent hawking and peddling and disposing of goods clandestinely' (8 Geo. 2, c. 6, 1735), given in John Lunan, *An Abstract of the Laws of Jamaica Relating to Slaves*, St Jago de la Vega, 1819.

[3] *Antigua and the Antiguans: a full Account of the Colony and its Inhabitants*, 1844, vol. II, p. 158.

Many European males, coming to the West Indies young and unmarried and finding themselves in complete ascendancy over the Negroes, not unnaturally sought sexual relations with slave women. This tendency was, possibly, strengthened by the shortage of European females in the colonies. But the establishment of such associations was by no means confined to the unmarried males. And writers have remarked on the curious situation where the legitimate (white) children of the head of the big house mingled with his illegitimate (coloured) offspring.[1] It was the domestic servants of the plantation household or the 'housekeepers' of the European bachelor that generally contracted such relationships. Many forms of relationships between the European male and the slave female can be distinguished. J. Ramsay emphasized their more unsavoury aspects. 'Mulatto girls, during the flower of their age, are universally sacrificed to the lust of white men; in some instances to that of their own fathers. In our town, sale of their first commerce with the other sex, at an unripe age, is an article of trade for their mothers and elder sisters; nay it is not uncommon . . . for their mistresses, chaste matrons, to hire their services and their persons to some of the numerous band of bachelors.'[2] And in the evidence before the Select Committee on the abolition of the slave trade witnesses spoke of the outdoor slaves in these terms: 'those unhappy females who have leave to go out for prostitution and are obliged to bring their owners a certain payment per week. . . . They are punished if they return without the full wages of their prostitution.' But probably such overtly mercenary connexions between Negro and White were far from universal. More generally unions between the two races were in no sense prostitution. Indeed, it was to the woman's advantage to encourage a union on a firmer footing. For a union with a white man conferred some status on her, especially when children were born. Paternity was usually acknowledged and ample provision made for the upkeep of the children. Indeed, the extent of these provisions led in 1762 to the passing of a law in Jamaica 'to prevent the inconveniences arising from exorbitant grants and devices made by white persons to Negroes and the issue of Negroes

---

[1] As M. G. Smith rightly points out in 'Some aspects of social structure in the British Caribbean about 1820', *Social and Economic Studies*, vol. 1, no. 4, the concepts of legitimacy and illegitimacy were alien to slavery. But the terms seem applicable, with some reservations, to the offspring of the slave owners.

[2] Quoted in F. W. Pitman, op. cit. p. 633.

and to restrain and limit such grants and devices'.[1] And in later years it became usual for the slave owner to grant freedom to his coloured offspring.[2] If in general the union between black and white was somewhat more than the relation between prostitute and customer, it was still a far cry from the union characteristic of European society.

As an indication of prevailing mating habits, the relationships developed among the free mulattoes themselves are probably more significant than the relations they established with their European masters. This, unfortunately, is a subject hardly touched on at all by contemporary writers. Nevertheless, the fact that we are again dealing with a class denied the rights and legal institutions of the Europeans once more rules out the possibility that unions resting on legal sanction could be widely rooted. Only a small proportion of the free coloured population enjoyed the full rights of English subjects. The more numerous group constituted a marginal class, free in the sense that they were not enslaved, but denied the rights open to Europeans. Indeed, since much of the law of the period aimed at preserving 'the distinction requisite and absolutely necessary to be kept up in this island between white persons and Negroes, their issue and offspring', severe legal disabilities attached to the free population.[3] Apparently there was no law in Jamaica prohibiting marriage among mulattoes, though the whole spirit of the law was against any union between white and non-white. Even among the lower order of whites on the plantations, the overseers, marriage was discouraged.[4] And at least one colony had a law designed to restrict marriage between white and non-white. An act passed in Montserrat made it an offence punishable by a fine of £100 for a minister to marry a white person to a Negro.[5] Since, in the words of Wesley, 'distinctions as to color were continued from the cradle to the grave', it remains doubtful whether even the class of the non-white population most intimately associated with the Europeans participated to any degree in the institution of marriage.

[1] Edward Long, op. cit. vol. II, p. 323.
[2] A. C. Carmichael, *Domestic Manners and Social Conditions of the White, Coloured and Negro Population of the West Indies*, 1834, p. 94.
[3] Quoted from a law of 1761 by C. H. Wesley, 'The emancipation of the free colored population in the British Empire', *Journal of Negro History*, vol. XIX, 1934.
[4] John Bigelow, *Jamaica in 1850: or the effects of sixteen Years of Freedom on a Slave Colony*, 1851, p. 26 (note).
[5] F. W. Pitman, *Development of the British West Indies, 1700–1763*, p. 27.

Hardly less important as a factor influencing familial patterns was the virtual exclusion of the Negro population from the Christian community of the Europeans. Opinion on the merits of conversion of the slaves varied. Some argued that acceptance of Christianity would make them more docile. For instance, Governor Robinson of Barbados urged the adoption of the French policy of widespread conversion on the ground that this would enable the priests 'to keep a strong hand over them against the Revolting or Rebelling against their Masters'.[1] But the majority of planters held differently. As Ramsay expressed it, 'Religion is not deemed necessary to qualify a slave to answer any purposes of servitude.'[2] Moreover, there was no inducement to the slave to embrace Christianity, since conversion brought no mitigation of his lot; as early as 1696 an act provided that no slave should be made free by becoming a Christian.[3]

Despite the essential validity of Long's remark that 'the mere ceremony of baptism would no more make Christians of Negroes . . . than a sound drubbing would convert an illiterate faggot maker into a regular physician', it remains true that the widespread extension of this rite to the slave population would have indicated a spread of Christianity among them.[4] In contrast to the situation in the case of marriage, as early as 1707 the established church gave permission to the clergy 'to instruct all free persons of colour and slaves who may be willing to be baptised and informed in the tenets of Christian religion', and, as already seen, baptism was provided for in the slave code of 1696.[5] However, practical difficulties in the performance of baptism and other Christian rites made their general extension outside the ranks of Europeans almost impossible. Above all the cost of performing these ceremonies was, as has already been shown in Chapter 1, very high. At the rates prevailing the ceremonial Christian rites for a slave family of six (male and female partners and four children) would be at least £5 and might conceivably exceed £10. And these charges fell on the slave owners; they could not be collected from the slave. Certainly the expenditure of such sums in connexion with the conversion of his

[1] Quoted in F. W. Pitman, 'Slavery on British West India Plantations in the 18th century', op. cit. p. 621.
[2] Ibid. p. 661.
[3] Cap. 11, 1696, 'An Act for the better order and Government of Slaves'.
[4] Edward Long, op. cit. vol. II, p. 428.
[5] F. W. Pitman, op. cit. p. 659.

slaves would not have been countenanced by the planter, especially as he was averse to the spread of Christianity among them. It is in any event questionable whether the small number of clergy in the island during slavery was adequate to perform the large number of baptisms and marriages the widespread conversion of slaves would have entailed. This, of course, does not mean that baptisms and marriages among slaves were unknown. Indeed, small numbers of baptisms, marriages and even burials were performed, as is evident from the entries in the parish registers of many colonies. In the case of Jamaica registrations of this type appear as early as 1666. But clearly the conditions of slavery made the general adoption of marriage as a method of initiating a family union among slaves impracticable.

One strictly demographic factor which must have influenced powerfully both the forms and the frequency of mating during this period of slavery was the imbalance between the sexes. Most historians treated this imbalance as of importance because, they claimed, it reduced fertility. Its effects on fertility, however, were secondary compared with its implications for mating. The policy of importing an abundance of male slaves doubtless resulted in very low sex ratios before 1807, but no reliable indication of sex composition is available before registration. Registration data for Jamaica give no indication of the sex composition of African-born slaves. But the British Guiana slave data tabulated by Robertson show that in 1817 (that is, 10 years after the cessation of the slave trade) the sex ratio among African-born slaves in British Guiana was 1728, as compared with 953 for those born in British Guiana. It is therefore probable that at the height of the slave trade about two-thirds of the adult population of the West Indies were males. Moreover, the custom of Europeans taking concubines from among the female slaves further reduced the number of females available for mating. Such a sex composition in fact powerfully reinforced the social factors making for the establishment of unstable forms of mating, and would have tended to extend the promiscuous relationships of the type described by Long and others.

Descriptions of prevailing family forms by contemporary writers are not always objective and are in most cases vague, which is not unexpected in view of the absence of prescribed legal forms of mating during slavery. Manifestly the operation of factors such as those outlined above precluded the widespread development among slaves of mating patterns characteristic of European populations.[1] But many assessments of the situation based on this departure from European norms are unrealistic. Perhaps the most extreme of this line of comment is the pietistic utterance of Phillippo: 'Every estate on the island—every negro hut was a common brothel; every female a prostitute and every man a libertine.'[2]

Long maintained, 'They are married (*in their way*) to a husband or wife, *pro tempore*, or have other connections, in almost every parish throughout the island, so that one of them perhaps has six or more husbands or wives in several places.'[3] His conviction of the tenuous nature of familial relationships among slaves is further emphasized by the statement that the slaves seemed 'to hold filial obedience in much higher estimation than conjugal fidelity'. But the attempted explanation of the alleged primacy of the filial bond suggests that the relations were not so unordered as he at times contends. '. . . of the whole number of wives or husbands one only is the object of steady establishment; the rest although called wives are only a sort of occasional concubine or drudge whose assistance the husband claims in the culture of his land, sale of his produce and so on; rendering to them reciprocal acts of friendship when they are in want.' Bryan Edward's picture of slave mating is equally indefinite.[4] According to him, polygamy was 'very generally adopted among the Negroes in the West Indies', and only someone 'utterly ignorant of their manners, propensities and superstitions' would urge marriage as a 'remedy' for this. 'It is reckoned in Jamaica on a moderate computation that not less than 10,000 of such as are called Head Negroes (artificers and others) possess from two to four wives. This partial appropriation of the women creates a still greater proportion of single men, and produces all the mischiefs which are necessarily attached to the system of polygamy.' A similar picture is conveyed in the words of J. Ramsay: 'A man may have what wives he pleaseth, and either of them may break the yoke at their caprice.'[5] These and similar accounts suggest that the

[1] These as well as other aspects of slavery in the West Indies have been treated in many recent papers in *Social and Economic Studies*. See especially, G. E. Cumper, 'A modern Jamaican sugar estate', vol. 3, no. 1; M. G. Smith, 'Some aspects of social structure in the British Caribbean about 1820', vol. 1, no. 4, and 'Slavery and emancipation in two societies', vol. 3, nos. 3 and 4.
[2] J. M. Phillippo, *Jamaica: Its Past and Present State*, 1843, p. 218.
[3] Edward Long, op. cit. vol. II, p. 414.
[4] Bryan Edwards, op. cit. vol. II, p. 143.
[5] Quoted in F. W. Pitman, op. cit. p. 633.

initiation of a union among slaves was not marked by any ceremony or rite, but it is an over-simplification to contend, in the words of Long, that the Negro laughed at the idea of matrimony. Nor is it at all definite that unions assumed among slaves were lightly dissolved. Indeed, the dissolution of a union necessitated the performance of a ceremony described by Phillippo. A cotta or circular pad formed from the leaf of the plantain tree was divided, each member of the union taking half. Claims Phillippo, 'Regarding the circle as a symbol of Eternity and the ring of perpetual love and fidelity, it was a ceremony that certainly did not inaptly express their eternal disunion'.[1] The fact that the dissolution of a slave union was accompanied by such a rite suggests, however, that the union was not always casual or irregular.

1792–1807. The commencement of this period is fixed at 1792 merely for convenience. This year witnessed the passing of an act consolidating many measures the necessity of which had been realized by observers long before. In fact the new policy to which this act gave full official sanction emerged but slowly.

Several factors induced the adoption of new means of recruiting slave populations. Mounting competition from other sugar producing areas resulted in a fall in the price of sugar and economic loss throughout the region.[2] It was widely held that only a reduction in the imports of slaves could bring about the fall in output of sugar, which, under existing circumstances, seemed imperative. Again, the tenor of the Parliamentary debates on the trade in the 1790's was for a reduction in slave imports.[3] In these debates the merits of stimulating slave reproduction were carefully weighed against the disadvantages of the slave trade. Of those supporting abolition some claimed that it should be gradual, lest the plantation working force be depleted. Others contended that mortality was high only among African-born slaves and that if imports were discouraged the slave population as a whole would replace itself. Further, anti-slavery interests urged that the only way of securing improvements in the treatment of slaves was to abolish the trade. Once more, slave uprisings in Santo Domingo led many to question the wisdom of having large aggregations of African-born slaves in the British colonies.

The rising cost of slaves which resulted from the growing com-

---

[1] J. M. Phillippo, op. cit. p. 219.   [2] Eric Williams, op. cit. p. 149.
[3] Thomas Clarkson, *History of the Rise, Progress and Accomplishment of the Abolition of the African Slave Trade. . .* , London, 1839, Chapter XXVII.

petition from the American mainland reinforced the strong opposition to the trade in the West Indies and in England, 'The value, or more properly speaking, the original price of the Negro, has in the course of thirty years risen upwards of one third; the proprietor is therefore led to view the Negro property as an object of great concern, and consequently is disposed to preserve it by every prudent method.'[1] Again, Long, after showing that between 1761 and 1768 the average annual 'dead loss' among slaves was about 3000, concluded that 'the annual loss in value' resulting from this might be about £105,000, 'a most astonishing sum'.[2] Long rated the cost of slaves as 'the most chargeable article attending these estates and the true source of distress under which their owners suffer'. It was apparently this view which led him to urge an amelioration of the lot of the Negroes, 'the sinews of West-India property', so that they might forget 'the very idea of slavery'.[3] It was thus being realized on all sides that slaves were no longer cheap and expendable.

The new attitude towards the slave found full expression in the ordinance of 1792, which consolidated many acts purporting to ameliorate the conditions of the slaves. As we have already urged, however, the significance of this code lies not so much in the easier conditions it sought to introduce but in the pro-natalist policy for slaves it patently expounded.[4] It remains questionable whether the efforts of the slave owners to secure slaves at bargain prices were any more successful than the policies pursued by some European Governments in the present century with the express purpose of buying babies at bargain prices; but the code of 1792 none the less is an interesting document in the history of population policy.[5]

Clearly it was a necessary first condition of any such policy that adequate records of the numbers of slaves should be kept. This the code sought to ensure. At the end of each year owners had to submit to the justices of the vestry and parishes particulars of their slaves, such as their numbers and the number of births and deaths during the year, under a penalty of a fine of £50. At the same time doctors in attendance at the plantations had to submit returns of deaths among the slaves.

Such returns formed the basis of the plan to stimulate fertility,

[1] H. M'Neill, op. cit. p. 5.   [2] Edward Long, op. cit. vol. II, p. 431.
[3] Edward Long, op. cit. vol. II, p. 437 and p. 502.
[4] This law is given in Bryan Edwards, op. cit. vol. II, p. 151 and in Robert Renny, op. cit. p. 245. An earlier version was passed in 1788.
[5] Cf. D. V. Glass, *Population Policies and Movements*, 1940, p. 371.

which in the first instance took the form of substantial monetary grants to the slave owners and their staff. 'And in case it shall appear to the satisfaction of the justices and vestry from the returns of the owner ... that there has been a natural increase in the number of slaves on any such plantation ... the overseer shall be entitled to receive from the owner or proprietor ... the sum of £3 for every slave born on such plantations. ...' Further, the owner was entitled to a reduction from his public taxes in respect of such payments to overseers. The act aimed at making child-bearing more attractive to the female slave : 'In order that further encouragement may be given to the increase and protection of slave infants ... every female slave who shall have six children living shall be exempted from hard labour in the field or otherwise, and the owner ... of every such female slave shall be exempted from all manner of taxes for such female slave, anything in the act commonly called poll tax law or any other of the tax laws of this island ... to the contrary notwithstanding, and a deduction shall be made for all such female slaves from the taxes of such owners. ...' Adequate proof had to be submitted by the owners that both the mother and her child were living and that the mother was indeed 'exempted from all manner of field or other hard labour and is provided with the means of an easy and comfortable maintenance'.

The law thus aimed at stimulating fertility by affording the slave owner tax remission in respect of births to his slaves and more directly by making child-bearing less burdensome on the female slave. But these efforts were not limited to activities of this nature, which, in a sense, may be construed as state aids to reproduction. The planters themselves offered incentives to their female slaves. On Worthy Park estate for instance, 'upon the birth of each child the mother was given a Scotch rug and a silver dollar'.[1]

The code aimed at population control not merely by promoting births; it also sought to prevent slaves from leaving the island. Any slave convicted of aiding another to escape was liable to death, and a similar penalty was incurred by any slave caught in the act of escaping from the island. A penalty of £100 was also incurred by any white person assisting in such an escape. The penalties prescribed for the murder of a slave might have been an expression of growing humanitarianism, but they might equally well have

---

[1] U. B. Phillips, 'A Jamaica slave plantation,' *The American Historical Review*, vol. XIX, no. 3.

indicated a further attempt to protect the life of the slave in the interests of population increase.

Apparently there were no complementary legal measures aimed at the reduction of mortality, though the general advisability of action along these lines was realized. Indeed, under the conditions of slavery it was probably only by means of the efforts of the individual slave owners that improvements in health conditions could be achieved. Legal measures to ensure this even if passed could under the conditions of plantation life hardly be enforced. Nevertheless, greater care of the sick was taken, according to contemporary writers, towards the end of the eighteenth century. At the first sign of illness the slave was, according to M'Neill, 'exempted from all labour' and committed to the plantation hospital or hot house where he was well fed and treated. 'Poverty, want and affliction are by no means the concomitants here of a sick bed.'[1]

The efforts to stimulate fertility among the slaves thus differed in several respects from the basic principles enunciated by Filangieri. In the first place the monetary rewards provided by the state accrued not to those actually supposed to add to the population of the island, the slave parents, but to their owners. In the second place such prerogatives and privileges as were meted out to the slaves went only to the mothers. The fact that fathers did not receive any consideration emphasizes the secondary role assigned to mating, both as a factor in population increase and in the general scheme of plantation life.[2] The small attention paid to mating at this period is well illustrated in the records of Worthy Park, studied by U. B. Phillips. In the 1790's, by which time efforts to recruit the slave population by natural increase were firmly established, records of deaths and births were maintained with care, but 'matings were listed in the records only in connection with the birth of a child'.[3] In the light of this it is not surprising that another of Filangieri's principles—action to increase the rate of mating—was also absent from these policies.

[1] H. M'Neill, op. cit. p. 8.
[2] The complete disregard of the role of the male in reproduction by the framers of slave population policy offers an interesting contrast to the disregard of the role of the female in the writings of Malthus. As Norman E. Himes points out, Malthus writes as if women played no part in the reproduction of populations. See Francis Place, *Illustrations and Proofs of the Principles of Population*, edited by Norman E. Himes, London, 1930, Appendix A, p. 285.
[3] U. B. Phillips, op. cit.

After 1807. With the cessation of the slave trade in 1807 the planters had to redouble their efforts to stimulate reproduction among the slaves. 'The planter has thus been placed under the necessity of rearing all the slaves requisite for the cultivation of his land.'[1] No new legislation was passed to ensure this. The same measures provided by the Act of 1792 remained in force, being restated with minor variations in 'An Act for the subsistence, clothing and better regulations and government of slaves . . .', passed in 1816.[2] In fact, the whole campaign was based on a more diligent application of the methods which were introduced towards the end of the eighteenth century. More encouragement was offered to slave women to bear children and greater care was taken of the young and of the sick. In the words of J. Stewart, 'Labour is now mild; the slaves are better fed, clothed and lodged when sick, experience kinder attention and are more amply supplied with the necessary comforts; and above all the breeding women are carefully attended to and receive every necessary indulgence and assistance.'[3]

Though there was still no legislation passed specifically designed to improve the health of the slaves, the planters were fully alive to the urgency of action along these lines, and the provision of better medical facilities on the plantations became general. The Central Board of Health, in comparing conditions in the island in the mid-nineteenth century with conditions in the later years of slavery, seemed to rate the medical facilities provided too highly. But their summary of those provisions is worth quoting: 'During the period of slavery every estate was yearly and periodically attended by a medical man. On every estate there was a hospital or "hot house", to which was attached a dispensary, or place containing a collection of necessary drugs; besides this a person, generally known as the "hot house doctor", was selected to attend and take care of the sick (the hot house doctor on most estates was an intelligent man, who had been for some time apprenticed to some regular medical practitioner) to administer the medicines and to perform the minor operations in surgery, such as bleeding, tooth

---

[1] *Report of the Select Committee on the Commercial State of the West India Colonies*, 1832.

[2] *Slave Law of Jamaica: with Proceedings and Documents relative thereto*, London, 1828.

[3] J. Stewart, *View of the Past and Present State of the Island of Jamaica*, Edinburgh, 1823, p. 230.

drawing, dressing sores, etc., etc., under the direction of the medical attendant. Besides this every estate had also its "yaw hut", or hospital, and its "grandy", an old woman who took care of the pregnant females before, during and after their confinement. She also superintended the infants and young children on the estate.'[1] Though these methods often assured that 'the sufferer was restored to health unscathed', the Central Board admitted that due to the absence of a sound sanitary system the measures in force failed to provide any effective control of mortality, and during periods of epidemics they proved powerless to stem the spread of disease.

Each planter introduced his own plans for increasing fertility among his slaves. For instance, M. G. Lewis gave each of his female slaves one dollar for 'every infant which should be brought to the overseer alive and well on the fourteenth day'. Each mother also received 'a scarlet girdle' which entitled her on feast days and holidays to special privileges. For each child born a medal was affixed to the girdle and 'precedence is to follow the greater the number of medals'.[2]

Efforts to encourage child-bearing and to reduce infant mortality were general throughout the West Indies, and probably the most thorough account of the form these took appears in the evidence of John Innes before the Select Committee on the Commercial State of the West Indies (1832). He recounted in great detail the methods adopted on the Katz estates in British Guiana designed to stimulate the reproduction of slaves, and some of these are worth noting here. As in Jamaica, managers of plantations received bonuses for increases in their slave populations, and various inducements were held out to women to bear more children. Pregnant women ceased to do regular field work about the fifth month of pregnancy and were put to do 'trifling light jobs about the building'. About 14 days before delivery they were sent to 'a woman . . . who keeps an establishment as a midwife'. On their return to the plantation, the mothers went to their own homes and devoted 'the whole of their attention to the child', and for some months after continued to be relieved of field labour. It often happened that before the expiration of this period of light activity some of the women 'are again in a state of pregnancy and do no field work for two years'. Innes also outlined the medical and hospital facilities provided on the Katz

[1] *Report of the Central Board of Health*, 1852, p. 262.
[2] M. G. Lewis, op. cit. p. 125.

estates, where hospitals were supplied 'with every description of medicines of the best quality'.

Some contemporary observers denied that the treatment of slaves at this period had improved to the extent suggested by Innes. For instance, the Rev. John Smith claimed that 'much more care and attention are commonly bestowed on the horses, cattle, etc., than on the Negroes', while he characterized the estate hospitals of British Guiana as dirty 'charnel-houses', without ventilation or beds.[1] Still, as the planters were fully aware that the only way of maintaining their stock of slaves was by forcing up the level of reproduction, it can be accepted that genuine efforts were made to improve health conditions among slaves.

Even if the elaborate evidence of Innes placed the health schemes of the Katz plantations in an unduly favourable light and were not completely representative of the practices followed elsewhere, they probably illustrate the chief aims of the new policy. It is clear that the major attention was centred on the mother and child rather than on the family union as a whole. The male partner did not share in the incentives to reproduction. But relief from heavy work and even a measure of ease were assured the pregnant woman in the attempts of the masters to reduce the hazards of child-bearing and to induce her to bear more children. Even if, as seems certain, the success of the measures to curtail the loss of life among infants was at best meagre, the new policy, tending to strengthen the bond between mother and child and to assure the woman a less arduous life, might have helped to rid her of the lurking fear of child-bearing which must have beset her in the earlier and harsher times of slavery, and made her less averse to rearing a family.

In the heyday of the slave trade, the slave was pre-eminently a work unit, and labour on the plantation took precedence over reproduction. The cessation of the trade did not radically transform this pattern; the chief task of the plantation was still to extract the maximum amount of work from the available labour force. But an important condition was imposed; the supply of labour had to be maintained, and this the new policy sought to ensure.

Forced to rely solely on reproduction to recruit their slaves, the planters were naturally led to consider closely the economics of slave rearing. The Select Committee of 1832, from evidence placed

[1] E. A. Wallbridge, op. cit. pp. 57 and 65.

before it, concluded that the cost of rearing a slave up to age 14, when he could be employed in field labour, was £87, and this expense, it was claimed, added an extra 15s. 10d. per cwt. to the cost of production of sugar. A further implication of having to rear his own slaves was that it forced upon the slave owner the necessity 'of maintaining in addition to his effective male slaves, and a limited number of non-effective persons, the large number of women and children whose existence on the particular colony, if not on the particular estate, this necessity occasioned'. This second factor, it was claimed, added also to the cost of production of sugar. The planters were fully aware of the roles of high mortality and low fertility in the forcing up the cost of rearing slaves. Evidently on economic, if not on humanitarian grounds, they were convinced that the new policy was inevitable.

Apart from the privileges accorded to the slaves with the immediate aim of stimulating reproduction, there were other changes in the general conditions of slaves which conceivably might also have had some bearing on slave fertility, by contributing some-what to the stability of slave unions.[1] In the first place the idea that the slave belonged to the plantation rather than to the master became widely accepted. Indeed, long before the end of the slave trade it was apparent that many were beginning to view the West Indian Negro as a serf rather than as a slave. In 1784 Ramsay declared, 'All plantation slaves as at present is the custom in Antigua should be considered as fixed to the freehold that they may not be sold or carried away wantonly at pleasure.'[2] And John Blagrove in his famous will referred to his slaves as 'my loving people, denominated and recognized by law as, and being in fact my slaves in Jamaica, but more estimated and considered by me and my family as tenants for life attached to the soil'.[3] If slave owners acted in general accord with such interpretations of the status of the slave during this period, the chances of the slave family being broken by the sale of one of its members were probably reduced.

There were other signs of the gradual change in the status of the slave. At this time appeared the 'task system' of plantation

---

[1] It is here assumed that the phenomenon which will be discussed in Chapter 8 in terms of recent fertility data, namely that instability of family union tends to result in relatively low levels of fertility, was also a feature of slave populations.

[2] Quoted in F. W. Pitman, op. cit. p. 616.

[3] F. Cundall, *Historic Jamaica*, London, 1915, p. 293.

work. In regard to the rights of property as well the slave's position improved notably. In terms of a useful distinction of slave rights advanced by M. G. Smith, rights *in rem* and rights *in personam*, rights *in personam* were being considerably expanded, both as a matter of custom and in legal formulations.[1] By the nineteenth century most slaves in Jamaica had some form of livestock and some even had slaves of their own, and laws had to be passed to restrict this. Indeed, Beckford was so convinced of the slave's right to property that he urged 'his hut should be his castle'. Moreover, the passing of laws declaring slaves to be real estate rather than chattel, and often explicitly designed to prevent the depopulation of the plantations by the recovery of legacies left by testators, tended greatly to reduce the chances of slave unions being broken.[2] All these factors conduced to the development of firmer unions and thus probably tended to stimulate fertility.

The last slave code to be discussed here, that of 1826 (it was actually passed after 1826 and only after considerable resistance), was at once the most idealistic of the four and the least acceptable to the slave owners.[3] It was essentially an ameliorating measure and was passed largely at the prompting of the anti-slavery interests in England. It ranged over a wide field of slave activity, being expressly aimed at promoting 'their religious and moral instruction and by means whereof their general comfort and happiness may be increased, as far as is consistent with due order and subordination and the well-being of the colony'.

At first sight the provision for the solemnization of marriages among slaves which the ordinance contained seems highly significant; this was, in fact, the first time such provisions were made in the island for slave marriages. But actually it was largely ineffective. For though the performance of the ceremony was made free of charge, two conditions imposed robbed the measure of much of its value. Slave marriage was permissible only if the clergy of the parish was satisfied that the couple 'possess proper knowledge of the nature and obligations of such a contract and produce also proof that they have the sanction of their owner . . . for the performance of the ceremony'. Even if, as was apparently the case, the owner

---

[1] M. G. Smith, op. cit.

[2] For instance, according to a law passed in St Vincent in December 1825, 'All slaves . . . are hereby declared to be real estate and not chattel . . .'

[3] A summary of this law is given in B. M. Senior, *Jamaica, as it was, as it is, and as it may be*, London, 1835.

would willingly give such permission it remains doubtful whether the small body of the clergy in Jamaica at the time would have undertaken such extensive additional duties without receiving corresponding increases in remuneration. It is important to note that such provisions for marriage had no connexion with the plans to promote the natural increase of the slave population. This emphasizes the secondary role assigned to slave mating even at a period when every slave owner was directing his attention towards the increase of slave reproduction. The rates and conditions of mating, it seems, were not regarded as associated with the general problems of population growth. Marriage crept into the law, apparently, with the sole object of improving the 'moral' conditions of the slaves.

Probably more significant than the marriage clause as a factor making for stable unions was the clause that slaves taken under any writ of vendition by the provost marshal should be sold in families and not separated. True, levies could still be made on individual slaves; but the operation of such a measure would undoubtedly enhance the stability of the family unit more than the ineffective marriage provisions. The law also guaranteed property rights among slaves, though cattle belonging to them could be kept on any person's land only with the latter's permission. Moreover, the slave's right to receive bequests and legacies was recognized. The law also restated, though in slightly different form, the generous treatment of mothers. Mothers with six children or adopted children were exempted from hard labour and owners of such slaves were relieved of taxes on them.

Again there is a dearth of reliable information on the forms of mating during these years, but there is nothing to suggest that the improvements in the position of the slaves after 1807 and the provision of the 1826 law wrought any revolution in the forms and frequency of mating among slaves. So far as these were governed by the increase of marriage, the attitude of the clergy must be weighed. There is evidence that even before the passing of the laws for marriage among slaves some missionaries tried to develop a form of union sanctioned by the church, though not resting on a legal basis. 'The members of the congregation [Moravians] entering into wedlock were solemnly joined at their Christian meetings. They gave each other the hand of promised faithfulness; the church owned their union, but the law disregarded them.'[1] But the great body of

[1] Richard Hill, *The Lights and Shadows of Jamaica History*, 1859, p. 73.

the established clergy was not eager to extend marriage among the slaves, a factor of importance since it was only through the church that marriages could be contracted. Nowhere is this attitude more bluntly expressed than in the answer one clergyman gave to a question on the number of marriages performed in the island of Grenada. 'The legal solemnization of marriage between slaves in this island is a thing unheard of and, if I might presume to offer my sentiments, would, in their present state of imperfect civilization, lead to no beneficial result. Their affection for each other, if affection it can be called, is capricious and short-lived; restraint would hasten its extinction and unity without harmony is mutual torment.'[1]

That the conditions of mating at this time remained as indefinite as during the period before 1807 is suggested by descriptions of some writers. For instance, the Rev. Smith, writing about slavery in British Guiana in 1822, said: '. . . scarcely one planter in a hundred pays the least attention to the household concerns of the Negroes under his management. When their attachments take place, which is at an early age, they feel no delicacy in declaring their passion for each other. The man will simply ask the woman whether she will live with him as his wife? and the woman often puts the same question to the man. An answer being given in the affirmative, all is soon settled, and the contract almost immediately consummated. . . . They have no ceremonies for these occasions, except those of drinking and dancing, and these, especially the latter, are frequently dispensed with . . . sometimes "the first morning that dawns on the marriage witnesses also its virtual dissolution".'[2] However, Mrs Carmichael noted that unions, established usually when the female was 16 or 17 years old and the male perhaps a year older, necessitated the building of a new house and apparently some sort of ceremony grew up around this.[3] The construction of the house was a cooperative effort, and to those lending assistance the owner or manager shared out rum and sugar. The work was made the occasion for merrymaking, and the couple often gave a supper when they took possession of the new house.

It seems that if a slave couple did decide to enter into formal

[1] Papers and returns pursuant to an address relating to the Slave Populations of Dominica, Grenada . . ., *P.P.* 1823.
[2] E. A. Wallbridge, *The Demerara Martyr*, Georgetown, 1943, pp. 65–6.
[3] A. C. Carmichael, op. cit. p. 131.

marriage the necessary permission from the owner would be granted. When the written permission to marry was presented to the minister a day was fixed for the marriage. 'Friends are invited, and the usual ceremonies and festivities conclude the business.'[1] At these occasions the master or the mistress usually contributed to the celebrations 'if the individuals are worthy of their indulgence'. But such marriages were not frequent. Senior suggested that the reluctance to enter into formal marriage was mainly on the woman's part. She feared that 'if they were tied together not to be separated he [the husband] would beat and ill-treat her'.

Though no detailed statistical assessment of the effects of the new policy can be made, it appears that it did have some success. It is clear that it did to some degree reduce the wastage of human life which was fully compatible with slavery so long as slaves could be easily and cheaply imported, but which if allowed to proceed unchecked after 1807 would have resulted in early and drastic depletion of the plantation working force. It has already been noted in Chapter 2 that the rate of natural decrease revealed during slave registration in Jamaica (5 per 1000) was much lower than the rates which probably prevailed during the eighteenth century. The precise contributions of changes in mortality and changes in fertility to this improvement are not easily measured.

It seems certain that reductions in natural decrease from over 20 per 1000 to 5 per 1000 must have been accompanied by declines in mortality, though it is impossible to gauge the magnitude of the decline. This reduction in natural decrease is again compatible with a rise in fertility, though once more no definite rates can be given. If the aversion to child-bearing which we have imputed to African-born slaves did in fact induce them to resort to abortion and to other means of controlling fertility to a greater degree than their Creole counterparts, then the reduction in the proportion of African-born slaves that resulted from the cessation of the slave trade might have contributed to a rise in fertility. This would have been more likely to occur if with the cessation of the trade the preponderance of adult males was also reduced. Registration data for Jamaican slaves, though limited in scope, tend to support the assumption of a rise in fertility. If the Jamaica slave population of 1817 had shown an age and sex composition similar to that of the British Guiana slaves it is doubtful whether any measures could have

[1] B. M. Senior, op. cit. p. 41.

succeeded in forcing up fertility. Indeed, Robertson, who in his interesting analysis of the British Guiana data all but arrived at a satisfactory measure of reproduction, rightly concluded that a population so heavily weighted with persons of advanced age 'could not long support itself'. However, it seems that the composition of the Jamaica slave population differed appreciably from that of British Guiana. Whereas the latter, with a sex ratio of 1311 and with 64% of its total concentrated within the age range 10–40, remained heavily influenced by the slave trade 10 years after it had ceased, it seems that by 1817 the Jamaica slave population showed very little traces of the effects of past immigration, for here the overall sex ratio was 1003. This strongly suggests that, unlike the population of British Guiana, which in 1817 showed only 45% of the slaves as Creole, the slave population of Jamaica consisted mostly of Creoles. If this was indeed the case, the Jamaican population was of a sex and age composition much more favourable to reproduction than that which probably prevailed during the years of the slave trade. Though changes of this nature suggest that a rise in fertility did take place, some planters in Jamaica evidently held that their efforts in this direction achieved little. For instance, M. G. Lewis, on whose plantation considerable efforts were made to increase fertility, doubted whether his policy was successful. Reflecting on a miscarriage of one of his slaves (evidently a frequent occurrence), he was led to remark that there were never 'above eight women on the breeding list out of more than one hundred and fifty females'. Despite the fact that the slaves were well clothed and fed and relieved of work when pregnant, and despite 'their being treated with all possible care and indulgence, rewarded for bringing children and therefore anxious to have them, how they manage so ill I know not, but somehow or other certainly the children do not come'.[1]

Though the stimulating of the rate of mating formed no part of the new policy, it is possible that the redress of the imbalance between the sexes did have some effect on the rate of mating. Such an increase in the frequency of mating would also tend to increase fertility. It is possible also that the removal of the excess of males tended to reduce the frequency of promiscuous associations that obtained in the eighteenth century when, presumably, there was a pronounced shortage of females. Of course such changes, if they did

[1] M. G. Lewis, op. cit. p. 380.

in fact occur, would not necessarily mean the establishment of any essentially new form of family union.

Whatever the success that attended the population policy of the slave period, it remains clear that the planters' deep concern over labour shortage throughout the nineteenth century dates from the cessation of the slave trade, which halted the ready acquisition of labour from overseas on which the sugar industry was built. As is revealed in the Report of the Select Committee of 1832, the planters persisted in ascribing all their difficulties to the termination of the slave trade.

POST-EMANCIPATION PERIOD

When the master–slave relationship was destroyed by the abolition of slavery, the basis of the population policy developed since the later eighteenth century was removed. The rescinding of the slave laws meant in the first place that the efforts to increase fertility, both by the tax remissions afforded the planters by the state and the privileges granted by the planters directly to their female slaves, were no longer practicable. This, however, did not mean that fertility declined. On the contrary, we have seen in Chapter 2 that the level of fertility in the immediate post-emancipation period was very high. Indeed, despite the claims of the Central Board of Health that abortion 'together with its allied crime infanticide' continued and that the prolonged nursing of infants was done in order to escape child-bearing, there is ground for assuming that fertility rose after emancipation.[1] In the second place emancipation brought to an end the hospitalization services of the plantations and the attendant health programmes.

Thus in terms of the central theme of this chapter—reproduction —the altered social relationships meant that there were no longer any measures in force for influencing directly the levels of fertility or mortality. Henceforth the only measure taken by the Government that directly influenced the movements of the population was the sponsoring of indenture immigration; but this was basically more a means for securing abundant supplies of cheap labour than a formal population policy. Nevertheless, state action did have some bearing on population movements, for the legislation establishing the medical services and aiming at the improvement of health conditions in general formed the background of the mortality

[1] *Report of the Central Board of Health*, 1952, pp. 113–15.

declines that ultimately appeared. The important question of regularizing mating habits, which was largely ignored during slavery, was early legislated for and came under review on several occasions; but as far as can be seen this legislation had no influence on population growth in Jamaica. The two aspects of reproduction that will be surveyed in this section therefore are mating conditions and mortality control.

Conditions of mating. In contrast to the slave period when despite the attention directed towards increasing fertility there was no effort to affect mating or in any way to regularize the legal status of slave unions, the post-emancipation period witnessed a keen interest in the introduction of measures calculated to extend the institution of marriage throughout the West Indies and to regularize the course of mating in general. These efforts of course were in no way associated with influencing the level of fertility; they were wholly concerned with the moral uplift of the ex-slaves.

In the fluid situation immediately following 1834 the validity of many marriages solemnized during slavery became questionable. In fact, the rescinding of the laws on which slavery rested meant that the whole basis of the West Indian family as a legal institution was rendered more obscure.

Proposals for resolving all doubts as to the legal status of existing unions and for assuring a ready means of contracting unions resting on full legal sanction formed a major portion of Lord Glenelg's despatch of 15 September 1838, which considered 'the enactments of laws to meet the new state of society consequent on the termination of the system established by the Act for the Abolition of Slavery'.[1] The marriage law was one of the seven major categories of legislation listed by Lord Glenelg as 'calculated to meet the new exigencies of society'. This despatch enclosed a model marriage ordinance, an Order in Council, dated 7 September 1838, the main provisions of which will now be noted. Its preamble states: '... since the abolition of slavery ... marriage laws ... have been found inappropriate to the altered condition thereof and inadequate to the increased desire for lawful matrimony therein.' A uniform procedure of marriage by publication of banns was made, the cost of the ceremony being fixed at 4s. Permanent records of such marriages had to be kept by the Island Secretary. As the

---

[1] *Extracts from Papers printed by Order of the House of Commons, 1839, relative to the West Indies,* London, 1840, p. 341.

number of clergy was too small to perform all marriages among slaves if such a procedure was adopted to initiate all family unions, the Governor was given power to appoint additional marriage officers. The legislation also aimed at settling the legality of certain 'marriages contracted and solemnized previous to the abolition of slavery... between slaves... and since the abolition of slavery between apprentices and other persons..., by members of the Christian religion other than clergy of the United Church of England and Ireland'. Such marriages were in fact declared to be 'good, valid and effective...'. Probably the most realistic section of the act was the one which sought to invest the countless unions among slaves not initiated by formal marriage with a full measure of legality. 'And whereas in consequence of imperfect instruction in the Christian religion, and from other causes, many marriages *de facto* have taken place between persons one or both of whom were in the condition of slavery, but which marriages *de facto* have never been sanctioned by any public ceremony or formally registered... and it is expedient that provision should be made for enabling such persons to conform upon their children the benefit of children born in lawful wedlock... it is further ordered that it shall be lawful for all persons having contracts of marriage as last aforesaid at any time within one year after coming into operation of this order, duly to solemnize the marriage ceremony before any clergyman of the established church or in any other manner authorised by this order; and every person so recognizing a previous marriage *de facto* shall at the same time make and sign the following declaration....' In the declaration the parties affirmed that they on a given date and in a given place 'intermarried with each other and... had issue of the said marriage'.

Couched in objective terms contrasting strikingly with most of the literature on mating among West Indian Negroes, and viewing the existing unions as marriages *de facto* rather than as promiscuous relationships, Lord Glenelg's despatch seemed eminently suited to solve the basic problems. But the legislation that sought to put it into effect in Jamaica—the marriage law of 11 April 1840—failed to induce couples already in states of marriage *de facto* to have these legalized by processes of law; nor did it induce those about to establish family unions to do so by means of the prescribed measures. Moreover, the law of 1840 which gave effect to Lord Glenelg's recommendations evoked the anger of the non-conformist ministers, who claimed that it was a discriminatory law aimed at suppressing missionary activities in the island.[1] The discriminatory clauses were, however, removed by an amending act of 22 December 1840.

The transition from slavery to freedom introduced two significant changes in the Negro family. In the first place it provided a ready means of initiating such unions; in place of the indefinite way in which such unions were established during slavery a legally sanctioned mode of doing so was established. The fact that only a fraction of the unions after emancipation were initiated in this way meant in effect that for the first time in the history of the West Indies two distinct types of family unions could be distinguished, those resting on legal sanction and those not so based. Thus legitimacy-illegitimacy became a meaningful dichotomy among the Negro elements of the population. In the second place, in so far as it was no longer possible to disrupt a family union by the sale or removal of one of its members as was possible in slavery, the stability of the union, regardless of its legal status, rested solely on the attitude and action of the couple.

Some observers erroneously claimed that the new marriage laws greatly stimulated marriage in the island. 'Marriages are happily beginning to be very common, and it is thought a disgrace to live otherwise than in honourable marriage life', declared John Candler.[2] Another observer, claiming to discern a similar development, also remarked on two features, which, as will be seen in the following chapter, still constitute important aspects of mating in the island—the late age at which marriage tends to occur and the fact that it often merely legally cements unions long in being. 'Marriage is more common, but none marry until they have lived some time together; and the man generally lives with two or three women before he marries and leaves them with children and he seldom does anything to maintain them.'[3] But as the marriage laws of 1840 set up only a rudimentary system of marriage registration, it is impossible to gauge the level of marriage. Still it soon became evident that the great majority of the population continued to initiate unions without any reference to the marriage laws. The

[1] E. Bean Underhill, *The West Indies: Their Social and Religious Condition*, London, 1861, p. 218.
[2] *Extracts from the Journal of John Candler, whilst Travelling in Jamaica*, London, 1840, p. 14.
[3] Robert Paterson, *Remarks on the Present State of Cultivation in Jamaica: the Habits of the Peasantry;* . . ., Edinburgh, 1843, p. 14.

Jamaican Richard Hill, one of the few who viewed the conditions of mating more as the outcome of historical factors than as evidence of the immoral propensities of the population, declared: 'This island, from one end to the other, is strewn with wives without husbands, and children without paternity.'[1] Underhill, writing in 1861 claimed: 'Outside the non-conformist communities, the neglect of marriage is almost universal.' And the Report of the Sanitary Committee of the Royal Society of Arts claimed: 'The crowds of bastard children that are brought to the churches of the Establishment for baptism show how sadly the marriage ordinance is neglected and the multitudes that are still living in the sin of open and unblushing fornication.'[2] It was even suggested that the clergy refuse to baptize illegitimate children. This, however, posed a dilemma, for the Roman Catholic church freely baptized children, legitimate or not, and the refusal of any denomination to do so might lead to a widespread conversion to Catholicism.

Illegitimacy and kindred problems came to the fore with greater prominence following the discussions provoked by the letter addressed by Dr Underhill to Secretary of State Cardwell in 1865. Governor Eyre, possibly seeking to discredit Underhill, sent circular questionnaires to ministers throughout the island about the general conditions of the people. Replies generally rated illegitimacy high among the causes of social distress and many suggested bastardy laws as a remedy.[3] As subsequent history has shown, bastardy laws have proved no more effective in reducing the level of illegitimacy than any other measure aimed at the same result. But one practical difficulty to be encountered in the enforcement of bastardy laws was pointed out by Eyre: 'I believe it [a bastardy law] would be extremely useful, but hitherto it has been found impracticable to get such a law enacted, on the alleged ground that the females of the country are so abandoned that they would swear children to any person for the purpose of extorting money.' Moreover, it was claimed that 'many of those who would be called upon to pass the law would themselves come under its operation.'[4]

Consideration of the illegitimacy question formed part of the wider discussions that began after the disturbances of 1865. In his

[1] Richard Hill, *Lights and Shadows of Jamaica History*, Kingston, 1859, p. 63.
[2] E. B. Underhill, op. cit. p. 188.
[3] This correspondence is printed in *Parliamentary Papers Relative to the Affairs of Jamaica*, 1866.
[4] Ibid.

despatch to Governor Sir J. P. Grant, the Earl of Carnarvon stated: 'A prevalence of concubinage amounting almost to a national habit will suggest to you the question whether ... the charge for the support and education of bastards should not be thrown upon the putative fathers.'[1] But 'the difficulty of ascertaining paternity and the little credit to be given to the oath, of the mothers' were acknowledged obstacles to the passing of such laws. However, the first bastardy law was passed in 1869. Law 31 of 1869 was designed to compel parents to support their children, whether legitimate or not, but proved a dead letter because it had no provision for fixing paternity.

The Commission which examined the conditions of the juvenile population of the island in 1880 also dwelt on the problem of illegitimacy.[2] It was held that much of the distress among the population of Kingston was due to the fact that many women had been deserted by the men who cohabited with them. Soon after this Commission reported the Bastardy Law (No. 2 of 1881) was passed. This compelled the father of an illegitimate child to contribute towards its support, and provided satisfactory proof of paternity was forthcoming an affiliation order would be issued for the payment of 5s. a week by the putative father for the child's support.

With the introduction of civil registration, and the effective registration of marriage provided by Law 15 of 1879, figures on the extent of illegitimacy and on the low rate of marriage became available for the first time. The high rates of illegitimacy proved of the utmost interest to the first Registrar General, S. P. Smeeton. He devoted long sections of his annual reports to discussions of 'this Hydra-headed evil', and advanced many suggestions for its eradication. And one writer even invoked a category termed moral statistics in dealing with the phenomenon of high illegitimacy in Jamaica.[3] The breakdown of infant mortality into legitimate and illegitimate groups by the Registrar General showed very clearly that mortality was higher among the latter. This attracted the attention of the medical authorities of the island, and thus

[1] Despatch from Earl of Carnarvon to Governor Sir J. P. Grant, 1 August, 1866, *Parliamentary Papers Relative to the Affairs of Jamaica*, 1866.

[2] *Report of the Commission upon the Condition of the Juvenile Population of Jamaica*, Supplement to the *Jamaica Gazette*, 4 November 1880.

[3] F. L. Hoffman, 'The Negro in the West Indies', *Publications of the American Statistical Association*, vol. IV, 1894–5, p. 181.

illegitimacy came under close scrutiny, both for its supposedly moral implications and for its supposed effect on the mortality of the island.

It was Smeeton's view that the registering of the name of the father of the illegitimate child 'while it would not directly meet the evil of neglect of marriage, it might act with an educational effect in placing responsibility upon the fathers, instead of refusing their offered testimony and so more than suggesting to them their present irresponsibility'.[1] The practice of entering the father's name on the register in the case of illegitimate children had been followed by Smeeton for some time during 1880–1, though this was not in keeping with the registration law.

A committee appointed by the Governor in 1904 to inquire into the laws of marriage and civil registration found it necessary to consider also the question of illegitimacy. Its report stated: 'Some men consider even in cases in which they are faithful to one woman that the woman behaves better and works more satisfactorily when she is not a wife, and some women consider that the man is more faithful to the woman and treats her better when she is a concubine.'[2] Replies to a questionnaire sent out by the committee showed that many held the view that 'our women invariably prefer marriage to concubinage and that it is only the men who prefer the irregular connections'. In many cases, it was reported, couples postponed marriage until they could afford it, one factor tending to delay marriage being the costly displays associated with these ceremonies. The committee realized that legislation could do little to increase the rate of formal marriage, but held that 'the cultivation of a higher moral tone and a better public opinion among all and especially the lower classes' might reduce illegitimacy. In any case the high rate of illegitimacy was no indication of 'hidden immorality among the mass of our people'. The registration of paternity in the case of illegitimate children, which the committee considered, was not held to be a solution to the illegitimacy problem, though it was suggested that greater facilities for the registration of fathers on a voluntary basis should be made.

The most recent examination of illegitimacy has been by a

[1] *Annual Report of the Registrar General*, 1890–1.
[2] *Report of Commission appointed by Sir A. Hemming to enquire into the the working of the Marriage and Registration Law . . .*, Supplement to the *Jamaica Gazette*, 28 July 1904.

committee which reported in 1941. This committee sought to ascribe many social evils to illegitimacy, such as the low standard of life, unemployment and 'the listless unhealthy condition of many of the children'. But one is tempted to comment that cause and effect are confounded in many of the committee's pronouncements. Among its numerous recommendations was one for a consideration of 'legitimatizing cases of permanent concubinage by some form of common law marriage', the first time such a realistic proposal had been made in the island.[1]

Manifestly no extensive efforts have been made to control the high rate of illegitimacy and its concomitant, the low marriage rate, despite the many official investigations of the situation and the critical comments of many observers. Indeed, most official reports realized the futility of legislating against institutions which, though not in conformity with the basic marriage law of the land, are by no means beyond custom. And apart from the provision of the marriage laws of 1840, the only direct action taken with the intention of influencing the situation has been the passing of various bastardy laws. As will be seen in the next chapter such attempts as have been made to induce conformity to the pattern of initiating family unions by formal marriage have had no effects on the situation. This suggests that the prevailing patterns of mating are of long standing and, indeed, have their roots in slavery. There are, of course, other aspects of the vexed question of the origins of West Indian family forms.[2] But considerations of these are outside the scope of the present analysis, which merely seeks to focus attention on certain social and legal factors in the history of the island that help to make meaningful the results that will emerge from the statistical analysis of Chapter 8.

Conditions of mortality decline. According to the Central Board of Health, the hospitalization services of slavery and the general policy for promoting the health of the plantations 'ceased totally or partially' on 1 August 1838, which in fact was 'a sorrowful day for

---

[1] *Report of Committee Appointed to Enquire into the Prevalence of Concubinage.* . . . 1941.

[2] The whole problem of the origin of the West Indian family forms part of the wider issue of the origin of family types among New World Negroes in general. This subject has been treated differently by Herskovits and Frazier. See for instance M. J. and F. S. Herskovits, *Trinidad Village*, New York, 1947, M. J. Herskovits, *The Myth of the Negro Past*, New York, 1941; and F. Frazier, *The Negro Family in the United States*, Chicago, 1940.

the Jamaica medical practitioner'.[1] Many doctors left the island soon after, and despite an attempt made to regularize the practice of medicine by an act of 1840 which prescribed qualifications for practitioners and set up the college of physicians and surgeons, the numbers continued to decline, and by 1854 it was estimated that less than 100 remained in the island, while many districts were without any.[2] The Central Board viewed this decline in the number of doctors as disastrous for the health of the island. And in the opinion of Thome and Kimball, the deterioration in health conditions brought about by the collapse of the policy pursued under slavery was most visible in the case of the free children: 'The situation of the free children is often deplorable. The master feels none of that interest in them which he formerly felt in the children that were his property and consequently makes no provision for them. They are thrown entirely upon their parents, who are unable to take proper care of them. . . .'[3]

Still, as we have already seen in previous chapters, there is firm evidence that death-rates after emancipation were much lower than those prevailing during the last years of slavery. Such a decline is not unexpected. Even the Central Board admitted that the hospital services of slavery probably achieved little in the face of the failure to control sanitary conditions. It is true that this lack of adequate sanitation persisted for a considerable time after emancipation. But one profound change which probably helped to offset this was the scatter of the population that the exodus from the plantations produced. It is probable that under the insanitary conditions of the period large aggregations of people such as slave society produced would be subject to greater hazards of disease than the scattered populations that emerged after emancipation. As Governor Sir Charles Grey pointed out in his criticism of Gavin Milroy's report on the cholera, many of those who left the plantations settled on the slopes of limestone hills where the natural drainage was perfect and the danger of fever and other diseases greatly reduced.[4] It is also highly probable that the less arduous conditions of work that freedom brought conduced in some measure to the decline in mortality.

[1] *Report of the Central Board of Health*, 1852, p. 262.
[2] *Report of the Central Board of Health*, 1854.
[3] J. A. Thome and J. H. Kimball, *Emancipation in the West Indies*, New York, 1839, p. 339.
[4] Despatch from Governor Sir C. E. Grey to the Duke of Newcastle, 23 September 1853, *Report on the Cholera in Jamaica*, Appendix A.

It is doubtful whether the rudimentary health measures of the immediate post-emancipation period contributed much to the improvement in mortality noted. But a brief examination of these early and short-lived measures is essential as it emphasizes the inadequate approach to the problem of health promotion at this period. It is significant that many of these early measures had their origin in one of the greatest catastrophes ever experienced by the island, the cholera epidemic of 1850–2. It is equally important to note that the more solid measures for the foundation of medical services and the improvement of sanitary conditions witnessed later in the century owed their origin to another notable event in the island's history, the disturbances of 1865 and the inquiries into the general state of the island associated with it.

The first Act of any consequence for the advancement of health in the island in the post-emancipation era was 'An Act to provide for the Establishment of Dispensaries and to extend the Facilities of Medical Aid to the Several Parishes of the Island' passed on 31 December 1845. This was intended to put medical assistance within the reach of the poorer classes in the rural areas. The law outlined a scheme for providing medical aid at nominal fees, and the cost was to be defrayed by taxes levied by the justices and the vestries. It also provided for the maintenance of adequate supplies of smallpox vaccine at the Kingston hospital as a precaution against serious outbreaks of that disease. But the act did not prove a success, partly, it was claimed, because of the unwillingness of the rural population to make use of the services offered. Indeed, most of the rural population continued to rely on what Kingsley Davis terms 'superstitious medicine'.[1] The magico-religious devices of slavery still commanded great respect in the fight against sickness.

Though an outstanding landmark in the medical, social and demographic history of the island, the cholera epidemic of 1850–2 did not give rise to any sanitary or social reforms of lasting value. Soon after the disease made its appearance an act was passed establishing a Central Board of Health as well as local boards.[2] The former was charged with the task of drawing up within a short time 'a full and correct report of the present sanitary conditions of the island and of the state of the public health'. Its long and detailed

[1] Kingsley Davis, *Human Society*, New York, 1949, p. 581.
[2] 'An Act to establish for a limited period a Central Board of Health and for other Purposes', 1851.

report was submitted in 1852. Gavin Milroy's report appeared soon after. In effect these reports taken together constitute a sharp indictment of the social conditions of the island. In the words of the Central Board, 'The welfare of its population, . . . the most sacred duty of a government . . . has been hitherto fearfully neglected in Jamaica.' Housing, particularly of the lower classes, was considered deplorable. 'These small, dark, unventilated houses' were greatly overcrowded, lacked adequate sanitary facilities and were conducive to the spread of disease. Water supplies in all areas were defective. True, Kingston, Spanish Town and Falmouth had companies supplying water, but the supplies were generally irregular and dirty. 'The mass of the people are unfurnished with this necessary of life.' In matters of personal hygiene the population maintained a remarkably low standard. According to the Central Board, 'A large proportion of our poor population . . . are never abluted, save on crossing a river or being exposed to a heavy shower of rain.' The condition of Kingston was harshly criticized. In the absence of any systematic scavenging, 'the amount of abominations upon the surface at all times is almost incredible', claimed Milroy. The tenor of both reports was that prevailing conditions were favourable to the spread of cholera, and that only by impressive changes could future outbreaks be prevented.

Both reports advanced measures designed to have lasting benefits on the social and sanitary conditions of the island. The regulations drawn up by the Central Board of Health aimed at keeping streets clean, controlling the location of domestic animals and imposing certain obligations on the local boards in the case of outbreaks of epidemic diseases. But such measures were generally deemed of secondary importance by the authorities, who tended to resort to a mixture of curious devices. Milroy reports: 'Days of public prayer and humiliation were appointed and observed to invoke the Divine mercy. Cannon were fired off, gunpowder exploded and tar barrels burnt in the streets, while houses were fumigated with incense in the hope of neutralizing the atmospheric poison.' The basic legislation of the time was framed mainly to prevent the further spread of the epidemic rather than to assure lasting control over the sanitary state of the island. And with the expiration of these acts in 1855 the whole question of the systematic improvement of health conditions was shelved. In any event, it is doubtful whether in the absence of suitable machinery for enforcement much was to be gained by continuing these measures, the success of which depended essentially on the extent to which they could be effectively enforced.

The relative freedom from epidemic disease after 1855 caused a lull in the interest in legislation to improve health conditions. As in the case of mating relations, it was not until the discussions of the state of society at the time of the disturbances of 1865 that attention was once more turned to legislation of this nature. And it can be said that within the 20 years following 1867 the foundation of the island's medical services was firmly laid, a beginning made to provide for the effective control of the spread of disease and for some basic sanitary laws. The chief measures witnessed during this period will now be noted.

Law 29 of 1867 aimed at the control of venereal disease by making compulsory the examination and treatment of common prostitutes. The quarantine regulations, which the inquiries of the 1850's had shown to be very defective, were greatly strengthened by Law 37 of 1869, which provided for the appointment of a board and the introduction of appropriate quarantine regulations. The most important act of the period was Law 6 of 1867, which re-established a Central Board of Health and the local boards in the fourteen parishes delineated in that year. This act virtually set up a medical department. Law 29 of 1872 strengthened the earlier vaccination laws, making it compulsory for children vaccinated by a public vaccinator to be brought back for inspection to ensure that the operation was successful. It also prescribed penalties of up to 20s. for parents refusing to comply. The system of burials in towns, against which the inquiries of the 1850's directed strong criticism, was brought under control in 1875.

It was many years before the medical system established in 1867 was brought to a stage of efficiency. At first the doctors employed under the scheme 'could not be regarded as civil servants in any sense unless the receipt of a fixed annual stipend . . . could be so interpreted. They were at perfect liberty to detach themselves at or without a moment's notice from this quasi-connection with the island Government.'[1] But by 1877 the department was put on a more satisfactory basis and the doctors fully integrated into the service. Soon after efforts were made to extend the medical facilities to all sections of the population by a system of attendance by tickets. Persons unable to pay the usual medical fees could

[1] *Annual Report of the Island Medical Department*, 1876-7.

obtain at a nominal fee tickets from magistrates or clergy entitling them to medical attention. This did not prove as successful as the authorities had hoped. But the fact that by 1884 there were eighteen public hospitals and six public dispensaries in the rural areas, while each parish had its almshouse, showed that the services were expanding.[1] The establishment of a school for training dispensers at the Kingston hospital in 1878 further advanced the efficiency of the medical services. Following on this the Sale of Drugs and Poisons Law of 1881 restricted the sale of drugs and poisons to suitably qualified persons and provided for the examination and licensing of such persons.

In the 1880's the Government took over the plantation hospitals as part of a campaign to assure better social and sanitary conditions among indentured workers. And such was the progress made in the expansion of the medical department that by 1884 'the expenditure connected with this Department has increased more than that of any other department in the last 10 years'.[2] In 1886 the medical services were costing the island £45,000, equivalent to a sum of 1s. 6d. per head of the population. Though this was considerably less than the per capita expenditure on similar services in British Guiana (2s. 8d.), and in Trinidad (3s. 2d.), mortality in Jamaica was lower than in Trinidad or British Guiana, as has already been shown in Chapter 6.[3]

The foregoing constituted important steps forward in the development of the medical services. But certain sections of the population remained reluctant to make use of them, preferring instead to rely on 'superstitious medicine', as in the past. 'Quackery of every description continues to thrive'; the peasantry when sick preferred to place themselves 'at the mercy of some ignorant quack or designing old man who dispenses bush mixtures or bottles of dirty water from a foul pond to hundreds of deluded people, who readily pay their dollars and fail to recognize until too late the imposition practised upon them'.[4]

Later years witnessed numerous amendments to the basic laws noted above as well as the introduction of entirely new laws aimed

---

[1] "Report of the Royal Commission to enquire into the public revenue, expenditure, etc., of the islands of Jamaica, Grenada, St Vincent, Tobago, St Lucia and the Leewards," published in the *Jamaica Gazette*, 7 February, 1884.
[2] Ibid.
[3] *Annual Report of the Island Medical Department*, 1886–7.
[4] *Annual Reports of the Island Medical Department*, 1880–1 and 1884–5.

at the same purpose. Improvements in the medical services also continued. By the beginning of the twentieth century it could be said that the campaign to improve the island's health was broadened in two ways. In the first place greater emphasis was laid on the treatment of certain specific diseases. In the second place the improvement of sanitary conditions was given more careful attention than hitherto.

Among the diseases against which direct action was taken in the twentieth century, yaws, important since the days of slavery, gained special attention. The number of cases treated increased from 1600 in 1905 to 7300 in 1910, while the sum spent on treatment rose during the same time from £200 to £1300.[1] The Yaws Notification Law (No. 23 of 1910) was intended to make action against the disease easier. Another disease that long plagued the population, syphilis, was also given special attention. It was notorious that large proportions of the aged in almshouses were in advanced stages of syphilis. Efforts to protect the population from the spread of this disease were increased by the Central Board after 1916 when large proportions of volunteers for the army were rejected because of venereal infections. Another disease against which campaigns were launched was malaria. In 1909 a malaria commission was appointed 'to investigate and take measures to remedy the conditions that give rise to malarial fever in different parts of the island'.[2] But by far the most important action taken against a specific disease was the Hookworm Campaign which began under the auspices of the Rockefeller Foundation in 1919. At this time hookworm was prevalent, perhaps 60% of the population being affected.[3] But the aim of the campaign was wider than its name implied. It was at once an attack on several kindred diseases and a basic element of the second aspect of health policy of the period, the promotion of better sanitary conditions. It was intended 'to take hookworm as an example of a preventable disease and through the different phases of the campaign to impress upon the people the desirability and necessity of practising disease prevention in their homes and of teaching them by demonstration the benefits of keeping well'.[4] The aim was to educate the people in

[1] *Annual Report of the Island Medical Department*, 1910–11.
[2] Ibid.
[3] *Annual Reports of the Island Medical Department*, 1921 and 1924.
[4] *Annual Report of the Island Medical Department*, 1924.

the prevention of typhoid, dysentery and diarrhoea. 'Selected areas are sanitated by the Central and Parochial Boards of Health in advance of the treatment campaign. This sanitation consists of the construction at each house of a latrine which will prevent soil pollution.'[1] As the work proceeded it was demonstrated that estate labourers did better work after treatment for the disease, while the incidence of typhoid and dysentery was noticeably reduced in areas where the educational and sanitary efforts were directed. It is clear that this work, which continued for many years, played a major part in the reduction of mortality throughout the island after 1921.

Since Law 6 of 1867 gave the Central Board no power to enforce its regulations or to coerce the local boards into action, progress in the development of sanitary control was slow. This did not mean that no advances in sanitary and social conditions took place. The records show several early attempts in this direction. In 1890 a law was passed (Law 31 of 1890) for the provision of a sewerage system in Kingston. Water supplies were being extended at this time in St Andrew as well as in other rural areas. Many other sanitary measures were being gradually introduced by parochial authorities on their own initiative. For instance, in 1915, the parish of Manchester under its own by-laws forced householders to adopt the dry-earth system with buckets for their privies, while in public buildings flush-out systems were installed. Adequate penalties were laid down to ensure daily attention of these facilities, and it was claimed that as a result the town of Mandeville in this parish, hitherto regarded as 'a bed of typhoid', showed great reductions in the rate of enteric fever.[2] Moreover, the programme of the Rockefeller Foundation greatly stimulated these local efforts, providing examples in sanitary improvements which many parishes tried to follow.

Despite these advances the Central Board remained, just as it was established in 1867, an advisory body without power to ensure that its regulations were adopted by the local boards. Thus the passing of a law such as Law 35 of 1910, empowering the Central Board to draw up comprehensive regulations for sanitary control in the island, failed because it did not at the same time confer on the Board the authority to enforce those regulations. In fact, it was not until 1926 that a fully effective Public Health Act was passed,

[1] *Report of the Central Board of Health*, 1919–20.
[2] *Annual Reports of the Customs and Internal Revenue*, 1914–5 and 1915–6.

establishing firmly the authority of the Central Board as the body responsible for sanitary control throughout the island and endowing it with powers necessary to ensure that its regulations were carried out.

The foregoing account indicates that the reduction of mortality, one of the signal factors involved in the demographic transition noted after 1921, was not the outcome of any revolutionary steps taken at that time. Mortality declines represented indeed the cumulative effects of many measures, some initiated more than 60 years previously, and all aimed at improving health in the island. One writer summed up the situation in 1923 as follows: 'From the provision of medical and surgical treatment followed Government efforts to control epidemics and now departmental work to prevent disease. A beginning has been made in sanitation; vaccination against smallpox and typhoid fever has been provided and so have clinics for the control of syphilis and yaws. Local Boards of Health are becoming more responsive to the advice of the Central Board of Health and the more progressive parishes are initiating sanitary work.'[1]

Conditions of mortality decline have been dealt with here solely in terms of improvements in medical services and in sanitation throughout the island. Though these remain the fundamental instruments whereby mortality has been reduced, the control of disease must also be viewed in a wider context as associated with certain social and economic changes that have taken place since 1921. It is true that under-developed areas are often able to introduce medical and sanitary techniques more cheaply than the more advanced countries of their origin. Doubtless also there is support for the view that in some under-developed areas 'public health and medical techniques have provided for the maintenance and extension of human life without an accompanying improvement in economic conditions'.[2] It cannot be denied, however, that the interrelationship between mortality movements and changes in economic and social conditions noted in the more advanced societies are also encountered in the less advanced, even if in a less clearly defined form. However, it lies beyond the compass of this study to consider mortality movements in association with such social and economic movements.

[1] J. R. Walker, 'A survey of health conditions in Jamaica', in *Annual Report of the Island Medical Department*, 1924.

[2] *The Determinants and Consequences of Population Trends*, United Nations, New York, 1953, p. 61.

CHAPTER 8

FERTILITY, MATING AND ILLEGITIMACY

DIVERSITY OF FAMILY FORMS

Most advances in the technique of fertility analysis have stressed the importance of treating mating habits of the population. Among European populations, where mating and marriage are largely synonymous terms, the study of mating is relatively straightforward. Here such unions as have not been initiated by formal marriage, as well as the children born to them, are so insignificant in number that they can be safely ignored. Under these conditions the study of fertility wholly in terms of marriage and legitimate births is justified.

But the situation in the West Indies, largely the legacy of slavery, is entirely different. Here fertility must be analysed against a background of diverse family patterns. It is an over-simplification to consider this situation as merely one of extremely high illegitimacy, as in fact exhibiting a widespread departure from the norm of the family traditionally associated with European populations, which has as one of its chief characteristics initiation by a rite or procedure endowing it with legal sanction. The previous discussion of the West Indian family through the different stages of reproduction suggests that we are dealing with types *sui generis*. The diversity of family forms, and the extent to which some of these differ from the types traditionally associated with European communities, do not mean that a wholly chaotic family situation prevails. Such an assessment, almost universally expressed in the past, is occasionally still echoed today. But, as Herskovits rightly points out, diversity of type does not indicate 'a state of demoralization'.[1] On the other hand, it is equally inaccurate to deny altogether that instability is characteristic of many West Indian family forms. And Herskovits, in his further argument on this matter, seems to read into the existing situation a greater degree of stability than actually exists, 'Here the range of permitted behaviour in organizing as in instituting the family is simply wider than in other societies. For in the final analysis the family . . . quite successfully performs the task allotted

[1] M. J. and F. S. Herskovits, *Trinidad Village*, New York, 1947, p. 107.

to it—the propagation and rearing of the young.' Admittedly the fact that many unions, even though begun as casual relationships of the keeper type, eventually attain stability through formal marriage indicates that stability of familial relationships is an ideal often aimed at; but, especially in the younger period of the reproductive span, instability and its demographic concomitants remain a cardinal element of certain types of West Indian family. Nor is it improbable that instability may have some adverse effects on the rearing of the young. But in whatever light we view the West Indian family from the sociological standpoint there remains no doubt as to the difficult problems to which the diversity of form gives rise in any attempt to introduce mating habits into fertility analysis.

Prior to 1943 it was the practice in West Indian censuses to treat marital status on the same basis as in English censuses, that is to recognize only two categories, married and never-married (the former was often subdivided into married, widowed and divorced). However, in the Jamaica census of 1943 an attempt was made for the first time to approach the situation of family unions more realistically. The never-married category was subdivided into two, common law and single. According to the Census Report, '"Common Law" means an unmarried couple living together as husband and wife.' The single thus constitutes a residual category. The same typology was adopted at the 1946 census of the other West Indian territories.[1] From the standpoint of vital statistics the only division made is in terms of legitimate and illegitimate births.

The threefold classification recognized at recent censuses may not be completely satisfactory from a sociological standpoint. But it has the advantage of being easily handled at a census enumeration and of excluding the many complexities which, if explicitly recognized, would render demographic analysis of the situation virtually impossible. The census classification is in close accord with a four-fold classification advanced by Henriques.[2] The following discussion

[1] The exact instructions to the enumerators at the 1946 census for determining unions of this type were: 'If a man and woman are living together, though unmarried, write "C.L.", i.e., common-law husband (or wife).' The definition of this type according to the census is: '"Common-law", it will be evident, is not a legal status but rather a *de facto* condition.' See *West Indian Census 1946*, vol. I, especially Chapter v.

[2] F. H. Henriques, 'West Indian Family Organization', *American Journal of Sociology*, July, 1949 and *Family and Colour in Jamaica*, 1953. See also T. S. Simey, *Welfare and Planning in the West Indies*, Oxford, 1946 and R. T. Smith, 'Family Organization in British Guiana', *Social and Economic Studies*, vol. I, no. I.

of the types used in the census is based mainly on census and registration data and on his analysis.

The first type, the married, is the only one that enjoys full legal sanction and is the only one recognized by the church. It is probably the most stable of the three and involves the continuous cohabitation of the spouses. All unions of the upper class probably fall within this category, for marriage, as Herskovits rightly indicates, is 'a prestige-giving institution'. But marriage in the West Indies does not always signify the establishment of a family union. Many if not the great majority of marriages represent no more than the cementing of unions long in existence. In other words, couples who enter into unions of a rather unstable nature early in their reproductive life often marry as they approach the end of the child-bearing period. An important consequence of this situation is that the common practice in European populations where a woman's pregnancy induces a couple to marry is in the West Indies the exception rather than the rule. Probably this practice is followed only among the higher classes where marriage usually initiates family unions in any case. Sociologically, marriage does not necessarily indicate the commencement of a family union or in the words of Malinowski 'the licensing of parenthood'; demographically, marriage does not necessarily connote the commencement of the exposure to the risk of child-bearing.[1]

The second type, the common law, also characterized by the continuous cohabitation of spouses, is akin to the type called faithful concubinage by Henriques. Unions of this type are fully institutionalized but have no legal sanction. The term common law, it should be noted, is merely a euphemism; unions so designated have no more recognition before the law than the loosest of family associations. Unlike the married class there appears to be no way of tracing the rate of formation of such unions by procedures of registration; indeed, in so far as these unions acquire permanence over a period of years, it may prove impossible to derive annual measures of their formation, since the emergence of these unions, many begun probably as casual relationships, may represent rather a slow process and thus may not be readily amenable to evaluation in terms of annual rates. Many of these unions eventually become formal marriages but generally when the partners are advanced in years.

[1] B. Malinowski, 'Parenthood—the basis of social structure' in V. F. Calverton and S. D. Schmalhausen, *The New Generation*, New York, 1930.

The third census type, the single, is not homogeneous, and in virtue of its composite nature is not strictly comparable to the two already considered. For unlike the married and the common law, it comprises not only persons at risk—mothers (or fathers) as well as those who though childless are still at risk of child-bearing—but also persons not at all at risk, those who are in fact genuinely single. The identifying feature of the single category for census enumeration was the absence of continuous cohabitation of spouses. This has important demographic consequences; it signifies that persons in these keeper unions, as they are sometimes called, are not continuously at risk of child-bearing. And there is no means of ascertaining the rate at which women and men enter into keeper unions or the rate at which these unions are dissolved, or more accurately the rate at which couples in these unions pass into the common-law or the married status. Probably in many instances the woman is involved in a casual relationship with a man which proves of short duration and is soon succeeded by a similar relationship with another man. It may be consistent to consider a woman engaged in such a succession of relationships as of keeper status. But clearly the lack of stability implied in these frequent movements raises a fundamental problem as to the justification of treating such unions as strictly comparable to the more stable ones involving the continuous association of two given mates.

An analysis of the distribution of these three family types in the 1943 Jamaica population given in Table 63 is instructive. We consider first the proportions of all persons over 15 in the three types. In the case of the males, the single type is the most important, accounting for 51% of all males over 15. Second in importance is the ever-married type, 30% of all males over 15 being so classified. Only 19% of the males over 15 are included in the common law type, which numerically is thus the least important. In the case of the females, the distribution is similar. The proportion returned as single is again the highest (57%). Once more the ever-married appears as the second most numerous, accounting for 28% of all adult females. The proportion returned as common law (15%) is again the lowest.

More significant than the overall distribution of the three family types is the distribution of each type by age groups, as reflected in the proportions entered in Table 63. Dealing first with the males, we note that the proportions returned as single decrease from 99% in the age-group 15–19 to 87% in the succeeding quinquennial group. A sharp fall then develops and by age 25–34 only half the males remain in the single category. After age 45, however, the proportions single tend to remain relatively unchanged, between 23% and 19%. As is to be expected, an entirely different

Table 63. *Proportional distribution (%) of the population by family types 1943*

Age	Male			Female		
	Single	Common law	Ever married	Single	Common law	Ever married
15–19	99·5	0·3	0·2	93·0	5·3	1·7
20–24	86·6	3·3	10·1	65·6	23·6	10·8
25–34	49·9	29·2	21·0	42·3	29·0	28·7
35–44	27·6	30·5	41·9	33·6	22·0	44·4
45–54	22·7	22·3	55·0	33·9	11·1	55·0
55–64	20·6	13·1	66·3	33·1	4·4	62·5
65 +	19·0	6·2	74·7	32·2	1·1	66·7
Total over 15	50·9	19·1	30·0	56·6	14·9	28·5

distribution is seen for the ever-married type, one which does not differ fundamentally from corresponding distributions for European populations. The proportions ever-married increase steadily with advancing age from 10% for the age–group 20–24 to 75% for males over 65. An entirely different pattern underlies the distribution of the common law type. Common law proportions increase rapidly between ages 20 and 34, from 3% to 29%; they decline after 44 and at ages over 65 only 6% of the males are returned as common law.

Females show age distributions of family types broadly similar to those for the males. In the case of the single type the proportion decreases sharply from 93% in the 15–19 age-group to 34% in the 35–44 group, after which it tends to remain unchanged. But the single proportion over 65 (32%) is appreciably higher than the corresponding male value. As in the case of the males, the ever-married proportions increase continuously with advancing age, though the proportion over age 65 (67%) is appreciably lower than in the case of the males. Again in the case of the common law we are faced with an age distribution in which the maximum frequency occurs near the age interval of greatest fertility, after which the proportions within the category decline steeply. In fact, only a negligible proportion of women (1%) is returned as common law at ages over 65.

These distributions indicate a constant movement from one type to another as couples pass through the child-bearing span. The distribution of the single type attests to the depletion of this group as persons marry or enter into common law relationships. Evidently, however, after age 35 in the case of the females and age 40 in the case of the males, movement from the single category is negligible. The most interesting feature about the distribution of the common law type is that it rises to a maximum and then declines, which means that there is a steady movement from the common law to the married. Indeed the common law appears to be only a transitional state; for after age 55 virtually no females remain in this category, while the proportions remaining in the case of the males are but negligible. Though the proportion ever-married is very low by European standards, it is important that it continues to increase appreciably after age 55, thus emphasizing a significant feature of marriage in Jamaica, the high average age at which it tends to occur. It is this movement from one type to another in the course of the child-bearing period that constitutes one of the main limitations of measures of fertility differentials by family type.

The existence of various family types, many of which exhibit peculiarities that definitely demarcate them from the type usually identified with European society, gives rise to many problems in the study of fertility, some of which are briefly noted here. Perhaps the most important of these is the linking up of mating and fertility. Since marriage does not always connote the commencement of the exposure to the risk of child-bearing, and since there are no measures of the rate of formation of other types of union, the combination of these two elements, in terms of vital statistics and census data, becomes virtually impossible. Moreover, the constant passage of couples from one type of union to another adds to the difficulty, and indeed raises the question whether it is not wholly impossible to establish conclusively the levels of completed fertility for each of the three types. Again there is the problem whether the transition from one type to another is associated with a change in fertility pattern. Another methodological issue underlying the movement from one type to another is whether fertility differentials in terms of social status can be disentangled from differentials associated with family forms. Many of these problems will emerge with greater clarity in the course of this chapter.[1]

BIRTH-RATES

Estimates of births during the intercensal years prior to the establishment of effective civil registration in the island which were given in Chapter 2 make possible the presentation of a broad picture of fertility movements over the past century. A summary of the fertility position since 1844 appears in Table 64. Birth-rates for the period of registration, it should be noted, are based on revised population estimates given in the Registrar General's Annual Reports of 1945 and later years.

Table 64. *Average births and average birth-rates*

Period	Average annual births	Average birth-rate	Index of birth-rate
1844–61	16,200	40	100·0
1861–71	18,500	39	97·5
1871–81	20,800	38	95·0
1881–5	22,040	37·4	93·5
1886–90	22,660	36·9	92·2
1891–5	25,740	38·4	96·0
1896–1900	28,260	38·8	97·0
1901–5	30,840	39·2	98·0
1906–10	31,370	38·3	95·7
1911–15	32,300	37·9	94·7
1916–20	31,190	36·4	91·0
1921–5	32,330	36·5	91·2
1926–30	34,890	36·4	91·0
1931–5	35,540	33·4	83·5
1936–40	36,760	32·1	80·2
1941–5	39,420	31·8	79·5
1946–50	43,030	31·8	79·5

Note. Values for 1844–81 are estimates; those for later years are from Annual Reports of the Registrar General's Department.

During the intercensal period 1844–61 the annual number of births is estimated at just over 16,000. It increases to over 18,000 in the following intercensal interval and exceeds 20,000 in 1871–81. The annual number of births continues to rise steadily in each quinquennium after 1881; the only period which does not show an

[1] For further discussion of these problems see *West Indian Census 1946*, vol. 1, Chapter v.

increase being the highly unfavourable years 1916–20. These years, it will be recalled, witnessed not only the influenza pandemic, but also a series of disastrous hurricanes and many hardships connected with the First World War. By 1946–50 the average annual number of births (43,000) is 2·6 times the number of a century earlier and nearly twice the number at the commencement of registration.

When allowance is made for the probable deficiencies in birth registration during the first ten years following 1878, there remains no evidence of any appreciable change in the level of fertility for the seventy-five years after 1844. On the basis of the estimated numbers of births between 1844 and 1881, birth-rates for these years appear to lie between 38 and 40. Doubtless the slightly lower rates (37) recorded for the first decade of registration are under-estimates; there probably was a measure of under-registration during these years. From 1891, by which time registration had probably attained a reasonable measure of completeness, until about 1915 the 5-year average birth-rates remain very steady, being between 38 and 39. The decline to 36 in 1916–20 is associated with the markedly unfavourable conditions then prevailing. The downward trend in fertility emerging after 1921 is unmistakable. Between 1921–5 and 1941–5 each quinquennium witnesses a decline in fertility; and within 20 years the birth-rate has declined from 36·5 to 31·8 or by 13%. From the latest available data, however, the picture at present is not one of continued decline but rather one of stabilization at a level appreciably lower than that prevailing a century ago. The birth-rate of 31·8 for 1946–50 is 20% lower than that estimated for a century earlier. The latest available birth-rates, 33·9 and 33·3 for 1951 and 1952 respectively, emphasize that no pronounced declines are imminent. But at the same time the declines that have taken place give rise to the question whether Jamaica can still be truly described as an area of very high fertility. The movement of the birth-rate since the introduction of registration is depicted in Fig. 11.

It is interesting to compare birth-rates for Jamaica with those for a few of the other West Indian populations. In Table 65 birth-rates for Jamaica are compared with rates for three larger populations, Trinidad, British Guiana and Barbados, and for one of the smaller islands, Grenada. It is clear that in both British Guiana and Trinidad, where population growth has been overwhelmingly dominated by immigration, both of indentured workers from foreign territories and of persons from neighbouring islands, birth-rates have in the past been lower than those for Jamaica. Up to 1919–23 Jamaica has shown rates appreciably higher than those for British Guiana and Trinidad. The rise in fertility in these two populations has however drastically altered the whole balance of fertility in the West Indies. Because of the growing proportion of the East Indians and the increase in fertility among this racial group, British Guiana

Fig. 11. Birth-rates for Jamaica, 1879–1952

and Trinidad now show extremely high fertility rates. On the other hand recent declines in fertility place Jamaica at a lower level than any of the other West Indian territories. Fertility in the small island of Grenada, for which a long series of rates is available, also shows a decline. The very high rate of the past, between 40 and 42, has given place to a rate in recent years very close to that of Jamaica (32·6). There is also a fall in fertility in Barbados between 1921 and 1946—from 35·3 to 32·1 or by 9%. It thus appears that Jamaica is not the only West Indian population that has experienced a reduction in its birth-rate in recent years.

The West Indies can now be divided into two areas so far as the level of fertility is concerned. The first consists of British Guiana and Trinidad. Here fertility, dominated by the phenomenal rates that now characterize their East Indian components, has increased over the past fifty years and now stands at a very high level. The second consists of other territories with negligible proportions of

Table 65. *Birth-rates for five West Indian populations, 1869–1948*

Period	Jamaica	Trinidad	British Guiana	Barbados	Grenada
1869–73	—	—	36·0	—	42·1
1879–83	36·7	—	34·7	—	41·8
1889–93	38·3	35·5	29·8	33·9	42·8
1899–1903	—	36·1	—	—	40·9
1909–13	38·7	35·0	30·5	35·4	38·7
1919–23	37·7	32·7	30·5	35·3	35·9
1929–33	—	30·5	33·0	—	—
1944–8	31·9	39·7	37·6	32·1	32·6

Note. These rates are based on the births registered during the 5 years centred on each census year and the appropriate census population. In the case of British Guiana, the aborigines are excluded.

East Indians. Here fertility is not only lower than among the first group but, in strong contrast to the latter, has shown some decline. Jamaica now stands at the lowest level of this second group. However, it must once more be stressed that the evidence is not of continued declines in fertility in the second group, but rather of a stabilization at a level lower than that prevailing in 1921.

AGE SPECIFIC FERTILITY RATES AND
REPRODUCTION RATES

When the age distribution of mothers in Jamaica is compared with that for Puerto Rico, an acknowledged area of very high fertility, a small though significant difference emerges. The former shows a relatively high concentration of mothers under 20. In fact, the proportion in the age-group 15–19 (16%) is appreciably higher than the corresponding proportion for Puerto Rican women (11%). On the other hand, the proportion of mothers between the ages 20 and 30 is correspondingly lower in Jamaica than in Puerto Rico. These differences, though not very large, are significant as they are associated with certain specific features of mating already considered.

Age specific fertility rates for Jamaica (total births per 1000 females) for the year nearest the last census for which the necessary data are available (1946) appear in Table 66, which also gives comparable rates for females in another West Indian population (Grenada, 1952) as well as for the United Kingdom (1947) and for Puerto Rico (1944) (see also Fig. 12). The comparison between the

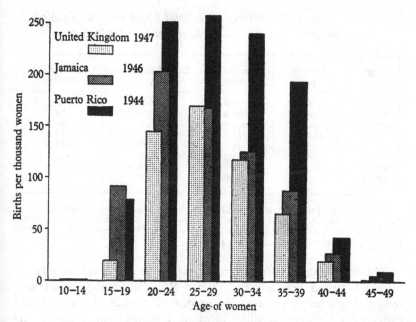

Fig. 12. Age specific birth-rates for Jamaica, Puerto Rico and the United Kingdom

Jamaica rates and those for the United Kingdom reveals two important differences. In the first place, fertility under 25 is much higher among Jamaican females. At higher ages, however, the differences between the two schedules of fertility are very small. It is thus the fertility of women under 25 that in the main results in the marked over all fertility differentials between the two populations. In the second place, the maximum fertility rate among United Kingdom females (170) occurs in the age-group 25–29, that is 5 years higher than in the case of Jamaica. More striking are the differences between the rates for Jamaica and those for Puerto Rico. Though in the case of women aged 15–19 fertility is slightly higher for Jamaica than for Puerto Rico—92 as against 79 per 1000—at advanced ages rates for Puerto Rico are by far the higher. Indeed for the age-groups 30–34 and 35–39 the Puerto Rico rates exceed those for Jamaica by 91% and 75% respectively. Again it emerges that the maximum rate for Jamaica occurs five years lower than the maximum for Puerto Rico.

Table 66. *Age specific fertility rates (total births per 1000 females) for Jamaica, Grenada, United Kingdom and Puerto Rico*

Age	Jamaica, 1946	Grenada, 1952	United Kingdom, 1947	Puerto Rico, 1944
10–14	0·6	0·8		0·3
15–19	92·4	120·2	19·2	79·0
20–24	204·2	255·1	146·0	250·6
25–29	168·2	223·3	169·7	259·4
30–34	125·9	155·7	117·6	240·3
35–39	88·5	113·3	65·6	154·5
40–44	27·5	37·6	19·4	41·1
45–49	5·4	6·9	1·4	9·7

Note. Rates for Jamaica and Grenada are calculated from Registrar Generals' data; those for United Kingdom are from *The U.N. Demographic Yearbook*, 1953; and those for Puerto Rico are from J. W. Combs and Kingsley Davis, 'The pattern of Puerto Rican Fertility', *Population Studies*, vol. IV, no. 4.

The fact that such considerable differences in levels of fertility obtain between Jamaica and Puerto Rico, a country of acknow-ledged high fertility, taken in conjunction with the very small differences between fertility rates of Jamaica and those of the United Kingdom at ages over 25, again raises the question whether Jamaica can at present be accurately described as an area of high fertility. The second feature of Jamaican fertility noted here, the maximum at five years lower than that of United Kingdom and Puerto Rico, seems to follow as a consequence of the greater concentration of births to the never-married women under 20 in the case of Jamaica, which in turn is associated with the special fertility pattern among the never-married females, to which attention will be directed at a later stage. In this connexion it is relevant that though, in general, fertility in Grenada is much higher than in Jamaica, the fertility schedules of both populations show maxima at the age-group 20–24, and at higher ages fertility rates fall off much more rapidly than in the case of Puerto Rico. Indeed the maximum at age 20–24 shown by West Indian populations contrasts strongly with the experiences of several other populations, which show maximum rates at the age-group 25–29.[1]

Recent studies have stressed the importance of considering fertility rates for males as well as those for females.[2] The great discrepancies that often arise between rates based on the two sexes indicate the limitations of rates based on a single sex either as an indication of fertility trends or as an index of the absolute level of fertility at a particular time. And in view of the changing balance between the sexes in Jamaica, occasioned by appreciable sex-selective emigration, these considerations are particularly relevant. At the same time, there is another factor to be weighed in this context. For the particular patterns of mating lead one to doubt whether the assumption on which most demographic analysis is predicated, namely, that the society is a wholly monogamous one, can justifiably be applied to the Jamaican population. In effect, the present analysis postulates complete monogamy, whereas the fluid state of mating relations in certain elements of the population suggests that conditions verging on promiscuity have to be reckoned with.

Ignoring, therefore, the elements of seeming promiscuity, we have calculated fertility rates based on males and on females. These rates have been obtained from data made available for the first time in 1951. These provide tabulations of births by age of parents and by sex, so that female births can be related to the female population and the male births to the male population. This procedure is straightforward in the case of the females because the data on age of mother at the birth of her child are easily obtained at registration. In the case of the males, however, a major difficulty is encountered.[3] For although the age of fathers of legitimate children is also easily obtained at registration, this can be secured for only a small fraction of the illegitimate births. Thus in 1951 of a total of 33,992 illegitimate births, the age of father was secured in only 928 cases.

[1] See the rates given in the *United Nations Demographic Yearbook*, 1953.

[2] See, for instance, P. H. Karmel, 'The relation between male and female Reproduction Rates', *Population Studies*, vol. I, no. 3, and *Papers of the Royal Commission on Population*, vol. II.

[3] The difficulties to be faced in securing ages of fathers of illegitimate children are discussed in recent reports of the Registrar General of Jamaica. In accordance with the registration law in England, the law of Jamaica does not permit the entry of particulars of fathers of illegitimate children except both parents sign a request that this be done.

The assumption made here to overcome this, that the age distribution of all unmarried fathers follows that shown by the small number for whom the information is available, probably involves some error. The average age of the unmarried fathers for whom ages are available is 31·0 years, as compared with 36·3 for the married fathers. As the difference (5·3 years) is close to that between the average ages of the married and the unmarried females (5·6 years) the error is probably small. The age specific fertility rates appearing in Table 67 show marked differences between those based on males and those based on females. Male rates imply a higher level of fertility than female rates; the former shows a gross reproduction rate of 2·60, which is 29% higher than that shown by the females (2·02). In view of this it therefore seems expedient to consider gross reproduction rates based on each sex as indices of fertility trends over the past.

Table 67. *Age specific fertility rates for Jamaica, 1951, based on males, on females and on both sexes*

Age	Rates for males	Rates for females	Rates for both sexes
10–14	—	0·3	0·0
15–19	3·2	54·9	28·8
20–24	66·8	117·4	92·3
25–29	118·5	100·2	108·8
30–34	108·6	68·4	86·9
35–39	79·0	42·8	60·0
40–44	74·0	16·7	44·3
45–49	37·3	2·5	19·7
50–54	23·8	0·2	11·8
55–59	5·3	—	2·5
60–64	1·9	—	0·8
65–69	1·1	—	0·5

Note. In calculating these rates male births are related to the male population, female births to the female population, and births of both sexes to the population of both sexes.

Age specific fertility rates for each sex for 1951 are therefore applied to the populations of each sex from 1891 in order to obtain the gross reproduction rates shown in Table 68. It is clear that male reproduction rates are much higher than female rates, the excess ranging from 65% in 1921 to 33% in 1943. It is also evident that each sex reveals its own distinctive fertility trend, though both show reductions after 1921. It has been argued elsewhere that the difference between male and female reproduction rates in the West Indies, both as measures of fertility levels at particular points of time and as indications of the trends of fertility, is closely linked with the imbalance between the sexes in the reproductive age span.[1] The experience of a number of West Indian populations reveals a coefficient of correlation of − 0·93 between the ratios male G.R.R./female G.R.R. and the ratios males aged 20–54/females aged 15–49. In fact the higher rates for the males in the case of Jamaica appear to be a direct result of the excess of females in the age span of maximum fertility. The desirability of seeking a measure somewhere between the upper limit set by the male rates and the lower limit set by the female rates is manifest.

Table 68.   *Gross reproduction rates based on males and females 1891–1943*

Period	Male		Female	
	G.R.R.	Index	G.R.R.	Index
1889–93	3·25	100·0	2·12	100·0
1909–13	3·20	98·5	2·14	100·9
1919–23	3·39	104·3	2·05	96·7
1941–5	2·37	72·9	1·78	84·0

Such intermediate measures are the joint reproduction rates, based on the fertility experience of both sexes. In one respect at least this approach seems especially appropriate under prevailing conditions of mating. For it ignores the seemingly intractable problem of linking up mating habits with fertility and seeks merely to determine the extent to which each person, regardless of sex and familial status, is replaced by another person, again regardless of sex. Making use of the 1951 data, we can compute for that year a joint age specific schedule of fertility, by relating the sum of the births of both sexes by age of parents to the relevant population of both sexes. The application of these rates to census populations, with the corrections based on the births registered after 1878 and on the estimated numbers of births at earlier dates, yields a series of joint gross reproduction rates for the whole period 1844–1951. The registered births used are the averages for the 5-year periods

[1] G. W. Roberts, 'Some Aspects of Mating and Fertility in the West Indies', *Population Studies*, vol. 8, part 3.

centred on the census dates; in calculating the most recent rate however the years 1950–2 are involved. For periods prior to registration the estimates are intended to refer to whole intercensal intervals. The joint age specific fertility rates can also be employed to compute a series of joint net reproduction rates from 1881. For these computations use is made of joint stationary populations, the joint radices of which are assumed to be composed of males and females in the same proportion as those shown by the sex ratios at birth for the 5-year periods covering the census dates (for 1950–2 a 3-year period is involved). The joint gross and net reproduction rates for Jamaica appear in Table 69. The joint age specific fertility rates for Jamaica are used to derive comparable gross reproduction rates for four other West Indian populations and these also are given in Table 69.

Table 69. *Estimated joint reproduction rates for five West Indian populations*

	Jamaica		Gross reproduction rates			
Period	G.R.R.	N.R.R.	Trinidad	British Guiana	Barbados	Grenada
1844–61	2·57	—	—	—	—	—
1861–71	2·67	—	—	—	—	—
1871–81	2·55	—	—	—	—	—
1879–83	2·45	1·40	—	1·79	—	3·08
1889–93	2·63	1·46	2·01	1·65	2·39	3·05
1899–1903	—	—	2·10	—	—	2·98
1909–13	2·63	1·55	2·07	1·76	2·66	2·89
1919–23	2·64	1·45	2·04	1·88	2·57	2·76
1929–33	—	—	1·92	2·17	—	—
1941–5	2·08	1·59	2·63*	2·64*	2·18*	2·73*
1950–2	2·28	1·85	2·56	3·09	2·17	2·96

* For the period 1945–8.

Except for the year 1881, when birth registration was probably defective, the gross reproduction rates for Jamaica between 1844 and 1921 show very little variation. During these years there is no evidence of any upward or downward movement, the rates being within the range 2·5 to 2·7 throughout. This confirms the picture yielded by the birth-rates; fertility between 1844 and 1921 has remained almost unchanged. Between 1921 and 1946, however, a sharp fall has been witnessed, the gross reproduction rate for both sexes declining from 2·64 to 2·08, that is by 21%. Again in conformity with the birth-rates, however, the evidence is not of continuing fertility declines, but rather of a stabilization at a level somewhat lower than that prevailing before 1921. In fact, the recovery witnessed since 1943 is sizeable, though the reproduction rate of 2·28 for 1951 remains 11% lower than that which probably obtained a century earlier.

Such irregularities as appear in the joint net reproduction rates for Jamaica between 1881 and 1921 are due mainly to slight variations in mortality. Indeed the reproductive capacity of the population in terms of the joint rate has remained fairly stable up to 1921, net reproduction rates being between 1·4 and 1·5 throughout. However, the marked mortality decline witnessed since 1921 has completely changed the situation. Despite the appreciable declines in fertility between 1921 and 1943 the net reproduction rate has still increased over these years from 1·45 to 1·59. The recovery in fertility after 1943, coupled with the continued declines in mortality, especially in the reproductive age span, has brought about a considerable rise in the net reproduction rate. The latest joint N.R.R. (1·85) is more than 30% higher than that prevailing at the commencement of registration.

Table 69 also presents a more accurate comparison between the fertility levels of Jamaica, Trinidad, British Guiana, Barbados and Grenada. In terms of the joint G.R.R., many of the irregularities noted in the birth-rates during the past are removed, but the patterns disclosed by the birth-rates remain largely unaltered. The fertility decline shown by Jamaica is all the more arresting in view of the fact that, by comparison with other West Indian populations, it shows very low literacy level. The contrast between the high fertility areas, of British Guiana and Trinidad on the one hand and the relatively low fertility areas, Jamaica and Barbados, on the other, is much more clearly drawn. Grenada, it is interesting to note, appears on the basis of reproduction rates, at a much higher fertility level than the birth-rates indicate. The great change in the relative positions between Jamaica and the two territories with large East Indian populations is shown by the fact that in 1946 Trinidad and British Guiana stood at a level exactly equal to that prevailing in Jamaica before 1921.

FERTILITY BY PARISH

The study of fertility differentials by parish and of movements in fertility levels in the several parishes assumes special relevance because of the general decline in fertility that has taken place between 1921 and 1943. Precise analyses along these lines call for fertility measures based on births corrected for usual place of residence. Such corrections have been made in Jamaica only since 1947. Data made available for that year show that only in Kingston and St Andrew are the discrepancies between the corrected and the uncorrected births of a magnitude sufficient to introduce sizeable errors into fertility rates calculated from the uncorrected births.[1] And the discrepancy virtually disappears when the two urban areas are combined. The present analysis is therefore carried out in terms of thirteen divisions of the island, a composite urban area and twelve rural areas. Probably at earlier periods when births in institutions in the Kingston–St Andrew area were small in number the discrepancies between actual and registered births were, for the whole island, negligible. But to ensure strict comparability in respect of all periods the combination of Kingston and St Andrew is maintained throughout.

In view of the limitations of the birth-rate as a measure of fertility in conditions of rapidly changing sex and age composition, such as have characterized many of the parishes between 1911 and 1943, the analysis of parish fertility is carried out in terms of joint reproduction rates, constructed from the joint fertility rates secured for 1951. The great advantage of reproduction rates in the present context is the more specific terms in which they portray many aspects of fertility differentials which are only dimly discernible from a study of parish birth-rates. Joint G.R.R.'s for 5-year periods centred on the census dates are given in Table 70.

Variations in levels of fertility among the parishes are at once evident from this table, the most outstanding feature being the relatively low level of the urban area, which, except for the year 1881, has consistently shown the lowest reproduction rates. It is possible that the marked rural-urban differential was not fully

---

[1] The proportions (%) of births registered in Kingston to the total births throughout the island accredited to that parish is 158·6, and in St Andrew the corresponding figure is 55·7. Elsewhere the proportions are very near to 100%, ranging from 99·0% to 100·7%. When Kingston and St Andrew are combined the proportion amounts to 101·2%.

established until 1891, though this cannot be firmly asserted in the face of the known inadequacies of the 1881 fertility data. From 1891 the Kingston–St Andrew area has been at a level appreciably lower than the other parishes. The urban rates for 1891, 1911 and 1921

Table 70. *Joint gross reproduction rates by parish, 1881–1943*

Parish	Gross reproduction rates				
	1881	1891	1911	1921	1943
Kingston–St Andrew	2·21	1·99	1·99	2·15	1·62
St Thomas	2·37	2·53	2·72	2·43	1·63
Portland	2·64	2·64	2·31	2·49	1·84
St Mary	2·43	2·61	2·31	2·52	1·82
St Ann	2·61	2·93	3·21	3·24	2·50
Trelawny	2·15	2·56	2·71	2·82	2·51
St James	2·09	2·15	2·60	2·70	2·28
Hanover	2·49	2·49	3·03	2·83	2·41
Westmoreland	2·50	2·60	2·79	2·75	2·38
St Elizabeth	3·04	3·13	3·07	3·09	2·84
Manchester	2·94	3·22	3·06	2·94	2·33
Clarendon	2·24	2·80	2·81	2·82	2·22
St Catherine	2·16	2·79	2·60	2·39	2·01

stood at 1·99, 1·99 and 2·15 respectively, whereas the highest rural rates for the corresponding years were 3·22, 3·21 and 3·24 respectively. The differential was even more marked in 1943; the urban rate amounted to 1·62, while the highest rural rate was 2·84. The more important aspects of the parish fertility variation will be described in detail presently, with special reference to the division of the island into areas of different levels of fertility. But it is convenient at this stage to summarize the variations at the census dates after 1881. Over the censuses 1881–1921 the coefficients of variation were 12·2, 13·0, 13·0 and 11·2 respectively; the rise to 17·1 in 1943 has, as will be argued, special connexions with the overall fertility declines between 1921 and 1943.

Despite the general stability of the island-wide fertility rates prior to 1921 some changes in the levels of parish fertility appear from the G.R.R.'s. Thus between 1891 and 1921 the parish of St James showed a rise of 26%, while in three others increases of over 10% were recorded. Such upward movements in rural areas at once suggest growing completeness of registration; but this interpretation is not pressed here since in other rural parishes declines and not increases mark the period 1891 to 1921. The possibility that the apparent rise in fertility in certain northern parishes might reflect the adverse economic conditions experienced in these areas around 1891 cannot be ignored. These parishes were heavily involved in the emigration to Panama, both through the numbers who emigrated and through the lucrative trade in food-stuffs which these parishes maintained with the Canal Zone while work was in progress. The cessation of work in 1881 led to the return of many destitute persons and to the failure of the trade with the Canal Zone, and a period of depression for these parishes ensued, a factor which might conceivably have contributed to the low fertility around 1891.

Whatever the full significance or validity of the fertility rise in certain parishes between 1891 and 1921, there can be no doubt as to the importance of the declines after 1921. A study of these declines assists in dividing the island into regions of comparatively high and comparatively low fertility. While the decline is general, it is clear that the high fertility parishes of the west of the island show on the whole a much smaller decline than the parishes on the east, that is those nearest to the urban centre of Kingston–St Andrew. The parishes of St Thomas, Portland, and St Mary show particularly sharp declines in fertility, 33%, 26% and 27% respectively. On the other hand, parishes which have consistently shown high fertility, notably St Elizabeth and Manchester, have experienced lower declines.

As a result of these shifts in levels of fertility the rankings of the thirteen divisions of the island have changed appreciably between 1891 and 1943. In general, the seven eastern parishes show lower positions in the rankings of the G.R.R.'s in 1943 than in 1891. This is most marked in the case of St Mary, which in 1891 stands seventh in position but which in 1943 is lower than all other parishes with the exception of the Kingston–St Andrew area and St Thomas. On the other hand, the western parishes have in general improved their positions on the rankings. This is perhaps most noted in the case of Trelawny, which in 1891 occupies fifth position but moves up to second position by 1943.

Reproduction rates for the parishes prove of greatest use in demarcating the island into areas of relatively high and relatively low fertility. A convenient scale for considering the levels of fertility is: under 2·60, 2·60–2·79, 2·80–2·99, and over 3·00. The fertility levels of the parishes in these terms are illustrated by maps in Fig. 13. In 1891 the Kingston–St Andrew area has by far the lowest reproduction rate; but, though there is evidence that in the nearby parishes of St Thomas, Portland and St Mary rates are comparatively low, the clear-cut division of the island into broad areas of comparatively high and comparatively low fertility is as yet not apparent. Still it is evident that St Elizabeth and Manchester constitute the high fertility areas of the island; these alone have gross reproduction rates in excess of 3·00. A division of the island into broadly similar areas of fertility can, however, be made by 1911. At this time the eastern area of St Thomas, Portland, St Mary, St Catherine, and Kingston–St Andrew comprises a section of relatively low fertility, with gross reproduction rates ranging from 1·99 to 2·72. On the other hand, the western parishes form an area of relatively high fertility. Indeed, with the exception of St James and Trelawny, where the gross reproduction rates are 2·60 and 2·71 respectively, all of these western parishes show rates well above the levels prevailing in the eastern section; and in four of them, St Elizabeth, Manchester, St Ann and Hanover, reproduction rates exceed 3·00. The division of the island into two areas of differing fertility is more clearly defined in 1921. Throughout the eastern parishes reproduction rates are at this date under 2·60, ranging from 2·15 (Kingston–St Andrew) to 2·52 (St Mary). All the western parishes show rates much greater than those observed in the east, ranging from 3·24 (St Ann) to 2·70 (St James).

The marked change in the fertility position of the island by 1943 appears most distinctly from the maps. There is no longer any parish with a gross reproduction rate in excess of 3·00. In fact, the only area standing out as one of high fertility is St Elizabeth, which has consistently shown a high level of fertility, and here the gross reproduction rate (2·84) is about equal to the fourth highest rate of 1921. And the difference between St Elizabeth and the rest of the island is striking; all other parishes indeed show gross reproduction rates below 2·60. Nevertheless the division of the island into an eastern portion of comparatively low fertility and a western portion of high fertility can still be made with assurance. Moreover it is clear that to the five divisions which since 1911 have revealed comparatively low rates must be added the parish of Clarendon. Gross reproduction rates in these parishes range from 1·62 for the largely urban area of Kingston–St Andrew to 2·22 for Clarendon.

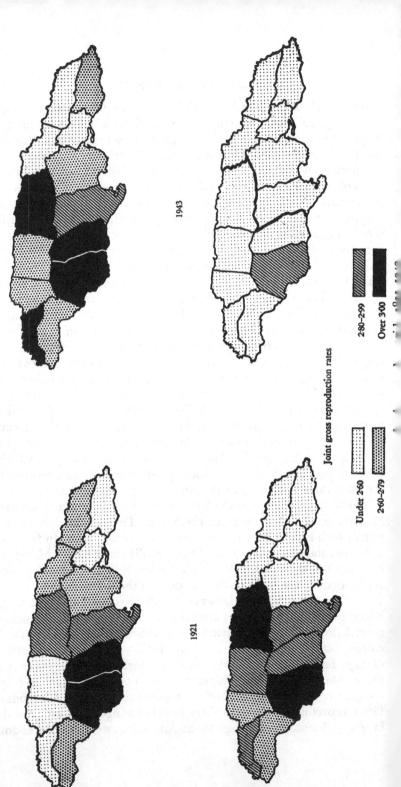

1891

1911

1921

1943

Joint gross reproduction rates

Under 2·60

2·60–2·79

2·80–2·99

Over 3·00

In the western group fertility ranges from 2·28 (St James) to 2·84 (St Elizabeth) ; and it is significant that the lower limit of the range here exceeds the higher limit of the range for the eastern group.

The reason for the marked rise in parish variation in 1943 now appears more clearly. It arises from the definite separation of the island into contrasting areas of fertility. In 1943 the gross reproduction rate for the western parishes considered as a whole amounts to 2·48 or 36% higher than the corresponding rate for the eastern parishes (1·82). In 1891 there was a difference of only 12% between these two groups of parishes, the gross reproduction rate for the eastern parishes being 2·48 as compared with 2·79 for the western. Whereas in the western parishes fertility, as measured by the joint reproduction rates, declined by 11%, in the eastern parishes the decline was much greater (27%). In 1943 the coefficient of variation for the western parishes of high fertility is 7·52 while that for the eastern group is 12·37, thus emphasizing that the greater dispersion is in fact due to the separation of the island into two areas of differing fertility.

From the various changes in fertility levels, particularly between 1921 and 1943, it is clear that declines have been most rapid in areas contiguous to the urban centre and least rapid in parishes more removed from this centre. In fact, the seven parishes of comparatively low fertility, comprising in 1943 56% of the island population, form a rough semi-circle, the centre of which is Kingston. There is, therefore, some ground for arguing that a recent development in the fertility position of the island has been a diffusion of low fertility patterns outward from the urban centre. As these patterns spread more rapidly in the parishes nearest to Kingston, fertility in these areas is now much lower than in the more distant parishes, which though themselves experiencing appreciable reductions in fertility still show rates in excess of those that now characterize the eastern area of low fertility.

But the examination of fertility differentials must be carried further in order to determine whether parish fertility, besides revealing so markedly the impact of urban contacts on levels of fertility, also discloses some general association between fertility and any socio-economic attributes. Census data do not furnish many satisfactory measures of such attributes. Possibly the best measure of this type, and the one which is employed here, is the proportion of literacy in each parish. As has been shown in Chapter 3, not only have literacy proportions been greatest in the urban districts and in the parishes adjoining it, but in general the farther away a parish is from the urban centre the smaller has been the gain in literacy in recent years. In short, the distribution and movements of literacy proportions seem to parallel the distribution and movements of fertility levels. Limiting the definition of literacy to the proportions able to read and write in the age span 15–59, that is to approximately the same span used to measure fertility, we find a high negative correlation between literacy proportions thus defined and joint G.R.R.'s ($r = -0.72$, which is significant at the 1% level). In view of the distinct division of the island into two regions of contrasting fertility and literacy more marked associations between these two variables are to be expected in each of the regions, and this indeed appears. Thus among the eastern parishes surrounding the capital fertility is lowest where literacy is highest, that is in the Kingston–St Andrew area, with a G.R.R. of 1·62 and a literacy proportion of 90%, while Clarendon, which has the lowest literacy rate (66%), also shows the highest reproduction rate (2·22). The correlation between literacy proportions and reproduction rates in these eastern parishes is high ($r = -0.75$). Again, of the western group of parishes, St James, the only area except the main urban centre showing any marked evidence of urban influence in the form of an appreciable attraction of migrants from neighbouring parishes, has at once the lowest reproduction rate (2·28) and the highest literacy proportion (73%). On the other hand, St Elizabeth, a parish long noted for its high fertility and low literacy status, is at the opposite end of the scale, with a G.R.R. of 2·84 and a literacy proportion of 64%. These western parishes show a very high correlation between literacy and fertility ($r = -0.80$).

What makes this strong association in the two regions of the island between literacy and fertility levels especially significant is its recent origin. Before the decline in fertility there was indeed some evidence of a relationship between levels of fertility and literacy proportions over the whole island. Thus in 1921 there was a small negative correlation ($r = -0.58$) over the thirteen areas between G.R.R.'s and literacy proportions measured on the basis of the population over 5. But there was no evidence of such associations within the eastern and western parishes; indeed, as we have already shown, the division of the island into such regions could be only roughly drawn at this time.

The association between literacy and fertility in 1943 strongly suggests that the declines so far witnessed derive in the main from controls exercised over fertility, as a consequence of the spread of low fertility ideals and of the effective practice of birth control. Thus two significant and interrelated factors are involved in the general decline in fertility. In the first place, the direct urban contacts have created a centre of relatively low fertility. But over and above this urban influence there is a fundamental relationship between the level of fertility and literacy, an important socio-economic attribute. And as both urbanization and social advancement in general are certain to continue, the prospects of further decline in fertility are high, though the influence of other factors associated with the differing levels of fertility among the several family types cannot be ignored, as will be argued further on.

The widespread phenomenon of association between fertility levels and urbanization, so evident in Jamaica, is also a prominent feature of fertility in other West Indian populations. This has already been pointed out by Lampe, who using children/women ratios as a basis of fertility measures, also showed that fertility tended to rise as we proceed outward from the main urban centre in many West Indian populations.[1]

MATING AND ILLEGITIMACY

The existence of several family types makes the distribution of births among them hardly less important than their rates of formation. But just as prevailing laws and conditions limit the study of such rates of union formation to marriage rates, so they also limit the study of the distribution of births by family type to the usual legitimacy-illegitimacy dichotomy. Indeed the only way in which the association between entry into unions and the division of births among the several family types can be fruitfully considered is in terms of marriage rates and illegitimacy rates.

From Table 71 it can be seen that marriage rates for Jamaica are very low compared with marriage rates for European populations. On only three occasions have the 5-year marriage rates for Jamaica exceeded 5 per 1000. It is significant to note that the highest marriage rates ever recorded followed one of the greatest

[1] P. H. J. Lampe, 'Human fertility in the British Caribbean Territories', *Caribbean Economic Review*, vol. III, nos. 1 and 2.

catastrophes that ever visited the island, the destruction of Kingston by earthquake and fire in 1907. In the year 1906–7, 5507 marriages were celebrated and of these 1228 took place within a month of the disaster. The number in the following year was even greater—6251 —or 7·71 per 1000 of the population, the highest rate ever recorded in the island.[1] There is no evidence of any long-term upward or downward movement in the marriage rate. Indeed the latest marriage rate, 4·66 in 1946–50, is very close to the rate prevailing at the time of the commencement of effective registration. Apart from the high rate of 1906–8, the most interesting departures from the almost unbroken level of rates of between 4 and 5 per 1000 are the declines to under 4 during 1911–25, a period marked by unfavourable economic conditions and comparatively heavy emigration, and to 3·8 during the period of economic depression, 1931–5.

Table 71.   *Marriage and illegitimacy rates*

Year	Marriage rates	Illegitimacy rates	Index of Marriage rates	Index of Illegitimacy rates
1881–5	4·53	58·9	100·0	100·0
1886–90	5·04	60·38	111·3	102·5
1891–5	5·18	60·64	114·3	103·0
1896–1900	4·41	62·62	97·3	106·3
1901–5	4·17	64·47	92·1	109·5
1906–10	5·43	62·44	119·9	106·0
1911–15	3·50	65·34	77·3	110·9
1916–20	3·76	69·23	83·0	117·5
1921–5	3·75	71·63	82·8	121·6
1926–30	4·43	72·14	97·8	122·5
1931–5	3·77	71·85	83·2	122·0
1936–40	4·44	70·76	98·0	120·1
1941–5	4·81	69·52	106·2	118·0
1946–50	4·66	68·61	102·9	116·5

High illegitimacy rates may appear to be no more than the obverse of low marriage rates. Undoubtedly illegitimacy rates are by world standards very high. As can be seen from Table 71, illegitimacy rates have been in general in excess of 60% and at times exceed 70%; in fact, throughout the period 1921–40 they lie

[1] This affords suggestive material for speculation on the part of those who view low marriage rates and high illegitimacy rates as evidence of the 'immoral' aspects of social life in the West Indies.

between 71 and 72%. But rates so far in excess of those that characterize European populations suggest that the comparatively low level of marriage is only part of the explanation of the large proportion of births to the never-married population. There is in fact no evidence that marriage rates and illegitimacy rates are in any way associated in movements over time. Nor is this surprising. For inasmuch as the number of illegitimate births in any given year is determined not by the marriages in that year but by the accumulated effects of past marriages, the basic relationship between marriage and illegitimacy must be sought in the association between proportions ever-married in the child-bearing age and illegitimacy rates.

As the 1943 census is the first to tabulate marital status by age, the association between proportions ever-married and illegitimacy cannot be investigated over any length of time. However such an analysis can be carried out on the parish data for 1943; these appear in Table 72. Here only a small correlation between marriage rates and illegitimacy levels is revealed ($r = -0.36$). But, as is to be expected, illegitimacy rates and proportions of never-married mothers show an almost perfect correlation ($r = -0.97$). These never-married proportions, it must be emphasized, reflect a feature of marriage just as important as its generally low level, the high age at which, on the average, it occurs.[1]

Vital statistics underline a significant aspect of illegitimacy, its decreasing importance with rising age of mother. This is shown in Table 73, which, in order to illustrate the similarity of experience throughout most of the populations of the West Indies, also gives comparable data for Grenada and St Vincent. All births to women under 15 years are illegitimate. In the case of births to women aged 15–19 illegitimacy rates, though somewhat reduced, still remain very high, between 92 and 95%. A considerable reduction in illegitimacy rates takes place between ages 20 and 25 and by age 30 just over half the births are to unmarried women. Towards the end of the child-bearing span, illegitimacy for Jamaica is about 32%, or less than half the rate for all births registered. In summary it can be said that illegitimacy rates decline on the average by about 10%

[1] This very close association between proportions never married and rates of illegitimacy is evidence of a high degree of consistency between registration and census data on the allied question of marriage and illegitimacy. Previous census data on marital status in Jamaica have been from many angles unsatisfactory.

for every five years' advance in age of mother. Put in another way, these rates imply that the greater proportion of illegitimate births are to young women; in fact, 81% of the total illegitimate births

Table 72. *Proportions of illegitimate births and never-married females and marriage rates, by parish*

Parish	% Illegitimate births, 1941–5	% Never-married females, 15–44, 1943	Marriage rates, 1941–5
Kingston–St Andrew	61·1	60·1	8·58
St Thomas	79·9	74·2	3·47
Portland	71·9	66·1	4·46
St Mary	74·6	70·4	3·63
St Ann	64·3	58·2	4·21
Trelawny	71·8	65·5	3·89
St James	76·6	70·8	4·71
Hanover	74·4	68·4	4·29
Westmoreland	68·1	61·8	3·99
St Elizabeth	71·5	66·4	3·67
Manchester	62·8	56·7	4·16
Clarendon	70·3	63·4	3·40
St Catherine	77·7	73·1	3·61

are to women under 30. On the other hand, a much smaller proportion of illegitimate births are to younger women; only 50% of all legitimate births are to women under 30. Two significant facts emerge from these data: firstly, unmarried women commence their

Table 73. *Illegitimacy rates by age of mother for Jamaica, Grenada and St Vincent*

Age	Jamaica 1951	Grenada 1952	St Vincent 1952
10–14	100·0	100·0	100·0
15–19	95·1	93·3	91·7
20–24	82·4	80·6	81·0
25–29	64·2	63·6	69·5
30–34	52·7	49·2	57·8
35–39	43·8	38·0	55·0
40–44	37·4	35·1	46·5
45–49	31·6	38·5	40·0
All ages	70·2	70·2	73·8

reproductive life in general earlier than the married women; secondly, fertility is to a large degree independent of nuptiality.

Detailed study of marriage habits of a population generally involves the computation of age specific marriage rates, where the marriages may denote first marriages only or all marriages and the never-married populations constitute the numbers at risk of marrying. This can be done in the case of Jamaica, but evidently the rates thus obtained are in no way comparable to similar rates for populations where formal marriage is the only means of entry into a family union. It is nevertheless possible to derive estimates of never-married females in some way comparable to the never-married in European populations. The isolation of such groups requires the exclusion from the total never-married females of all common law women, all single mothers and a small proportion of single non-mothers taken to indicate those who, though at a given time are childless, are cohabiting with men and are therefore at risk of child-bearing.[1] Thus we can distinguish between gross and net never-married women: the former signifying all who have not been married and the latter a specific section of the never-married, namely those not at risk of child-bearing. A comparison between the gross and net never-married populations for Jamaica is given in Table 74. Clearly, marriage rates based on net never-married should be more comparable to marriage rates for European populations. Again, marriage rates based on the gross never-married populations and on total marriages must always tend to be comparatively low. The fact that distributions of never-married females, similar in some respects to never-married females of European populations, can be derived, does not mean that comparable marriage rates can also be derived. This could be done only if the marriages taking place were divided at registration into two classes—those signifying the formation of unions for the first time and those indicating processes taken by couples to give full legal sanction to unions long in existence.

Despite the limitations of the age specific marriage rates for Jamaica, gross nuptiality tables constructed from them serve to emphasize much more adequately than the crude marriage-rate the

---

[1] The proportions of the single non-mothers assumed to be at risk are: 5% for the age group 15–19, 8% for the age group 20–4, and 10% for all higher age groups. These are based on the proportion of first births to never-married women aged 15–29.

Table 74.   *'Net' never-married women as a percentage of 'gross' never-married women, 1943*

Age	'Net' never-married as % 'gross' never-married
15–19	80·5
20–24	43·0
25–29	28·4
30–34	22·7
35–44	19·1
45–54	18·1
55–64	23·0

special features of marriage in the island. Gross probabilities of marriages yielded by these rates are shown in Table 75, together with comparable values for the United Kingdom population. Strong contrasts between the marriage experience of the two populations emerge. The chances of marriage for Jamaican males below 35 are only a fraction of the corresponding chances for United Kingdom males. However, between the ages 35 and 45 the

Table 75.   *Gross probabilities of persons marrying within five-year age intervals for Jamaica and the United Kingdom*

Age interval	Male		Female	
	Jamaica 1945–7	U.K. 1942–7	Jamaica 1945–7	U.K. 1942–7
15–20	0·0024	0·0457	0·0425	0·1772
20–25	0·0790	0·3749	0·1669	0·5745
25–30	0·1908	0·5546	0·1706	0·5144
30–35	0·2014	0·4234	0·1493	0·2359
35–40	0·1885	0·3016	0·1220	0·1432
40–45	0·1391	0·2075	0·0860	0·0764
45–50	0·1316	0·1226	0·0652	0·0632
50–55	0·1030	0·0390	0·0412	0·0540
55–60	0·0851	0·0336	—	—

Note. The values for the U.K. are taken from J. Hajnal, *Papers of the Royal Commission on Population*, vol. II, p. 412.

difference narrows, and above age 45 the Jamaican males have a small advantage over their United Kingdom counterparts. Similar differences appear in the case of the females.

The fact that at advanced ages the chances of marriage in

Jamaica are either in excess of, or very little below, the corresponding chances of the United Kingdom population, is only one aspect
of the major characteristic of marriage in the West Indies, the
comparatively high average age at which it takes place. In fact the
average age at marriage in Jamaica is, according to the nuptiality
tables, 34·1 years for the males and 28·6 for the females. These
values underline the fact that marriage, occurring so late in the
child-bearing period for so large a section of the population, has
little bearing on levels of fertility. Illustrating the same aspect of
marriage are the proportions of those marrying who do so at
comparatively high ages; 23% of the males marrying are over 40,
while 20% of the females marrying are over 35.

The large proportions of the population remaining unmarried at
the end of the child-bearing period are fully reflected in the
survivors columns of the marriage tables. Only 68% of the males are
married by age 55, while at age 45, 55% of the females remain
unmarried.

Though the information available on the formation of family
unions in Jamaica is limited to marriage rates, a rough idea of the
wider aspects of mating can still be secured from an analysis made
for Barbados on the basis of the 1946 census. There are indeed some
differences among West Indian populations, but, so far as the
formation of family unions is concerned, the conditions prevailing
in Barbados are probably sufficiently representative of West Indian
conditions as a whole (with the exception of the East Indian components) to give a rough indication of the process of family formation in Jamaica, in terms of the threefold census typology.[1] The
close similarity between Jamaica and Barbados is well illustrated
by the average age at marriage. In the case of Barbados, this stands
at 31·7 years for males and 27·1 for females, values differing only
slightly from those already given for Jamaica.

It is not possible to estimate directly from census data the age at
entry into keeper unions. But if we assume that the data secured for
the Negro elements of the population approximate the experience
of the keeper unions, then a rough estimate of this age may be
obtained. The average age of motherhood for the Negro women
(1941) for Barbados amounts to 20·7 years. And if we assume that
about one year elapses between the commencement of a keeper
relationship and the birth of the first live-born child, it can be

[1] See G. W. Roberts, op. cit. and *West Indian Census 1946*, vol. 1, Chapter v.

taken that the average age at entry into a keeper union is between
19 and 20.[1] Some of the women in the keeper unions marry at
higher ages, though it is not possible to estimate the average age at
which this occurs, as the marriage experience of this type is included
in the total marriage experience of persons other than those of the
common law type. The average age at which marriage among these
types takes place is 29·8 years for the males and 26·1 for the females.
It is presumably the continued keeper unions which ultimately
attain stability in the form of common law unions. The average age
at which this stability is attained is 36·4 for the males and 29·9
for the females. The majority of unions that attain common law
status eventually become formal marriages, though at high ages—
on the average 41·9 years for the males and 34·1 for females. From
census data at least, the common law type seems to be merely a
transitional one, in general a brief stage between a comparatively
loose keeper attachment and a legally sanctioned union. But it
would be an over-simplification to read into the existing situation
a definite continuum in the family cycle, opening with a keeper
relationship and ending in formal marriage. This doubtless is a
pattern traced by some unions. But it is equally clear that many
unions begin as formal marriages and are in every respect similar
to married unions among European populations. It must, however,
be emphasized that this is merely a provisional analysis, based on
data which were not primarily tabulated to illuminate these
questions. Further study of the problem should assure more
satisfactory estimates of the rates of formation of the different types
of unions and clearer indications of the ways in which they are
initiated.

FERTILITY DIFFERENTIALS BY FAMILY TYPE

The difficulties in the way of measuring fertility differentials by
family types have already been underlined. It is evident that though
the acknowledgement of the existence of three types at recent
censuses constitutes a big step forward, the relevant information
provided remains meagre, permitting only the roughest estimates of
the fertility position of the three types. So far as vital statistics are
concerned, all that can be gleaned from them is a breakdown of

---

[1] This may be an overstatement, for the study of Judith Blake indicates that
the interval between the entry into a keeper relationship and the birth of the
first child is longer than one year.

births in terms of legitimate and illegitimate; this in fact is all that prevailing registration and marriage laws make possible. The present limited discussion of these differentials will be carried out in four stages. The first is based on registration data; the second, and by far the most convincing, is based on census data; the third seeks to combine both classes of data in order to secure fertility rates for married and never-married women; finally, an attempt is made to determine, by an indirect approach, whether the differentials of 1943 were in evidence at earlier census periods.

The illegitimacy rates by age of mother already given indicate that illegitimacy is highest at young ages, or, in other words, that the average age of unmarried mothers is lower than the average of married mothers. In fact, the average age of married women to whom children were born in 1951 stands at 30·5 years as compared with 24·9 for the unmarried. Half the legitimate births are to women under 30 years, while in the case of the illegitimate births half are to mothers under 23·8 years. Similar age differentials are to be observed in the case of the males. The average age of married men who fathered children in 1951 stands at 36·3 years, as compared with 31·0 for the unmarried. Median ages also show the same differences as in the case of the females. One-half of the married fathers are less than 35·9 years, whereas in the case of the unmarried the median age is 29·5 years. From these data it is clear that the never-married, both males and females, begin reproduction at an earlier age than the married.

Two sets of average ages from the census data throw light on differential fertility among the three census types, single (keeper), common law and married. The first averages refer to women of reproductive age (15–44). These show that married mothers have the highest average age (34·1), while second come the common law (30·8) and lowest is the single (29·5). Again, therefore, the evidence is that the never-married begin reproduction at an earlier age than the married. A somewhat different pattern is shown by the second set of average ages, those of mothers of all ages. Again the highest value is that for the married mothers (43·1 years); but the position of the common law and the single mothers is reversed. Single mothers show an average age of 37·9, which is nearly five years higher than the average age of common law mothers (33·2). This is evidently due to the movement of common law women to the married state, which is relatively important at advanced ages. This movement leaves the general distribution of common law mothers more heavily weighted with young mothers than is the case among the single category and thus forces down the average age among the former.

A variety of fertility measures for the three family types can be derived from census data. It is true that in view of the constant passage of women from one type to another many of these measures call for qualification. A further limitation of the analysis is that the experience of males cannot be treated, as there are no data on this aspect of the problem in the census. Nevertheless the information at hand does illuminate several basic aspects of the fertility position of the three family types.

The first of these to be discussed appears in Table 76 which shows the proportions of mothers of the three types with 1, 2, 3 . . . 6 and more children. Clearly the size of family is largest among married mothers and smallest among the single. Whereas one-third of the single mothers had one child, the proportion of married women with the same family size is less than half of this (15%), while falling between these two are the common law mothers, 23% of whom have one child. Further, whereas 66% of the single women have less than four children, the corresponding proportion is considerably lower among the married mothers (40%) and lower also among the common law (56%). On the other hand, the proportion of mothers with large families is much higher for the married mothers. Thus 34% of these have six children or more, which is twice as high as the corresponding proportion for the common law and much higher than that for the single mothers (13%).

This same differential appears in the three other measures of fertility given in Table 77. These have been discussed for the West

Table 76. *Proportion (%) of mothers of three family types, 1943, with families of various sizes*

Mothers with	Single	Common law	Married
1 child	32·7	23·4	15·2
2 children	20·3	18·7	13·3
3 children	12·8	14·2	11·1
4 children	9·3	11·1	9·7
5 children	6·8	8·6	8·4
6 children	5·3	6·7	7·9
Over 6 children	12·8	17·3	34·4

Table 77. *Fertility measures by family type, 1943*

Family type	Children per woman over 45		Children per mother over 45		Total fertility of mothers	
	Number	Index	Number	Index	Rate	Index
Married	5·88	100·0	6·64	100·0	4·60	100·0
Common law	4·76	81·0	5·60	84·3	3·87	84·1
Single	3·32	56·5	4·74	71·4	3·23	70·2

Indies as a whole elsewhere.[1] The data for Jamaica parallel quite closely those for the other populations. The first of these, a measure of completed fertility, shows that there are 5·9 children per married woman over 45 as compared with 3·3 in the case of the single woman, and 4·8 in the case of the common law woman. But such measures do not afford strict comparability between the three types; for in the case of the single women many are genuinely single and therefore not at risk of child-bearing, whereas by definition women of the two other types comprise only those at risk of child-bearing. The second measure of completed fertility, the number of children per mother over 45, assures full comparability among the three types, as it utilizes only women who have already borne children. In terms of this measure, the difference between the single and the married levels of fertility is considerably reduced but the essential pattern of the differentials remains. Whereas the average married mother over 45 has 6·6 children, the number in the case of the keeper type is only 4·7 or 29% less. Occupying a position midway between these two is the common law mother with an average of 5·6 children, that is 16% lower than the married. In addition to measures of the stock type, it is also necessary to consider those of the flow type, rough estimates of which can also be secured from the census data. By using the first differences of the numbers of children born to mothers of various ages, summary indices can be made of current fertility performance of the three family types. The calculation is based on mothers rather than on all women because, as has already been pointed out, the use of all women in the case of the single type introduces many women not at risk of child-bearing. The measures computed, termed here total fertility rates of mothers, again show that fertility among the

[1] G. W. Roberts, op. cit.

married women is highest. From these estimates the total fertility of married mothers (4·6) is 43% higher than that shown by single mothers (3·2) and 19% higher than that shown by the common law mothers (3·9).

Still another aspect of these fertility differentials emerges when we consider the average age of mothers under 45 with 1, 2, 3 . . . 10 children for the three family types. As can be seen from Table 78, the average age of single mothers with one child stands at 26·3 years, which is 2·1 years lower than the corresponding value for the common law mothers and 5·5 years lower than the value for the ever-married mothers. Similarly, the average age of single (keeper) mothers with two children (28·7 years) is slightly lower than that for the common law (28·9) and markedly lower than that for the ever married (32·2). These figures indicate what has already been revealed by registration data that reproduction commences much earlier among the single women than among the married. A further important consequence of these data, one primarily associated with the fertility differentials noted, is the much longer spacing between successive births among the single mothers than among the married. They seem to demonstrate convincingly that the difference in age between women with $n + 1$ children and those with n children is greatest in the case of the single mothers and least in the case of the married. Thus the difference between the average age of single mothers with one child and single mothers with two children (2·4 years) is many times greater than the corresponding values for the common law (0·5 years) and the married (0·4 years). Similarly the difference between the average age of mothers with three children and those with two children is considerably higher among the single (2·1 years) than among the common law (0·9) and the married (0·4). In fact, up to women with six children, these differences are greatest among the single and least among the married.

The foregoing analysis emphasizes the importance of three factors in the fertility position of the three family types. In the first place, the level of fertility is, from all available data, lowest among the keeper unions and highest among the married. Secondly, reproduction begins earlier in the case of the single than in the case of the married. Thirdly, the length of time between successive births is generally much longer for the single mothers than for the married. This latter is indeed the crucial aspect to be borne in mind when considering the seeming paradox yielded by the experience of the three family types: namely, that in terms of these family types the level of fertility tends to vary inversely with the total length of the period of exposure to the risk of child-bearing.

Table 78. *Average age of mother under 45 according to size of family (no. of children), 1943*

No. of children	Average age in years		
	Single mothers	Common law mothers	Married mothers
1	26·3	28·4	31·8
2	28·7	28·9	32·2
3	30·8	29·8	32·6
4	32·7	30·9	33·2
5	34·2	32·4	34·3
6	35·4	33·5	35·2
7	36·5	34·8	36·0
8	37·2	35·7	37·1
9	38·1	36·9	37·9
10	38·4	37·5	38·4

Combination of registration and census data cannot, of course, be made on the basis of the three family types. But a consideration of the data in terms of married and never-married still adds to our knowledge of the fertility differentials. As a first approximation, two sets of age specific fertility rates can be computed by relating illegitimate births to the unmarried population and legitimate births to the married. Rates of this sort have been described by demographers as legitimate and illegitimate age specific fertility rates, and sets of these for the male and female population of Jamaica appear in Table 79. Fertility rates for both sexes show the married type as much higher than the unmarried. In the case of the females the highest fertility rate for the never-married (183 per 1000 for the age-group 20–24) is less than half that for the married females of the same age (375). Throughout the reproductive span never-married rates fall well below those for the married. The differences in the case of the males are much smaller; in fact, at advanced ages fertility is higher among the never-married.

Suggestive as these rates are, they remain of very limited use and actually grossly overstate the differences in fertility between the married and the never-married. For the two sets of rates are basically dissimilar, as in fact the extremely large differences between the female rates indicate. The use of married women in the calculation of the legitimate rates means that only women at risk are involved. On the other hand, not all the never-married women are at risk of child-bearing. Accordingly an attempt is made to refine the never-married rates for the females (this cannot be done

Table 79. *Age specific fertility rates for ever-married and never-married populations (per 1000), 1946*

Age	Male		Female	
	Ever-married	Never-married	Ever-married	Never-married
10–14	—	—	—	0·6
15–19	172·7	5·8	370·9	87·5
20–24	439·9	125·6	375·2	183·4
25–29	321·2	181·4	260·9	139·2
30–34	226·4	161·2	177·4	98·8
35–39	195·0	142·7	122·3	64·9
40–44	125·7	101·2	35·3	20·1
45–49	70·5	60·3	6·9	3·7
50–54	28·8	57·5	—	—
55–59	10·4	29·7	—	—
60–64	6·5	6·5	—	—

for the males) by re-calculating them on the basis of women deemed to be at risk rather than on all never-married women. Approximate estimates of never-married women at risk may be taken as the difference between the gross and net never-married categories already dealt with. In other words, the illegitimate births are related to unmarried women of the following classes: all common law women, all single mothers, and a small proportion of single non-mothers deemed to be at risk though still childless. Rates thus derived probably present a more realistic pattern of never-married fertility, and are shown in Table 80. These rates place the fertility of the never-married women on a higher level than the unrefined rates. They show the highest rate (449) within the age-group 15–19 whereas the highest rate shown by the unrefined rates (183) occurs in the age-group 20–24. The overall differences between the fertility of the married and that of the never-married are much reduced when the latter is measured in terms of the refined rates. Also it should be noted that in the age-group 15–19 the unmarried females show a fertility rate 21% higher than that of the married, though at all ages over 20 the advantage lies in favour of the married. Whether there is any justification for pushing this analysis to its logical conclusion and obtaining reproduction rates for the two schedules of fertility is debatable. But if this is done the married women appear at a level of fertility 13% higher than the never-married. For the reproduction rate for the former amounts to 3·32, as against 2·95 for the latter. This difference is smaller than that shown by the total fertilities of mothers obtained from the census data; these put the married mothers at a level 31% higher than the single and common law.

Table 80. *Age specific fertility rates based on never-married women at risk, 1946*

Age	Rates per 1000
15–19	449
20–24	322
25–29	194
30–34	128
35–39	78
40–44	26
45–49	5

Though exploration of the fertility differentials along the foregoing lines cannot be made for earlier periods, indirect study of the problem, that is linking illegitimacy rates with joint gross reproduction rates for the several parishes, furnishes some further confirmation of the basic findings. In limiting the analysis to legitimate and illegitimate births we are in effect treating births to the common law and the keeper unions together. On the basis of the previous findings we should expect fertility rates to be highest where illegitimacy rates are lowest, that in short a negative correlation should exist between illegitimacy rates and reproduction rates. Utilizing the thirteen divisions of the island already identified, we observe the small negative correlations between the two variables shown in Table 81. But, with the exception of 1879–83, these are too low to lend much support to the differentials already established. Indeed the period 1909–13, which was disturbed by the marriage boom that followed the destruction of Kingston in 1907, actually showed a positive correlation of 0·23. However, closer examination of the data suggests a reason for the low correlation. It appears that special factors operate in the urban area to reduce fertility in general to a level lower than would be expected on the showing of illegitimacy. In other words under urban conditions factors other than the relative importance of married women are the prime causal elements determining overall fertility levels. It has already been shown for instance that the level of literacy is strongly correlated with the level of fertility. According to the 1943 census data, all family types within the urban areas show fertility rates much lower than those of the rural areas. Thus the number of children per mother over 45 in Kingston–St Andrew is 3·60 for the single, 4·24 for the common law and 4·94 for the married, all substantially lower than the island-wide levels. If, therefore, on the ground that special characteristics obscure the relationship between illegitimacy and fertility in urban areas, we exclude Kingston–St Andrew from the correlations, the coefficients are more impressive. With the exception of 1911, when evidently as a result of the high marriage rates experienced in 1907–9 the relationship seems masked, the correlation coefficients from 1881 to 1921, ranging from −0·76 to −0·69, are significant at the 5% level. That for 1943, though not significant at this level, is still considerable (−0·52). These negative correlations between illegitimacy and fertility not only support the more direct results derived from the 1943 census, but, more important, indicate that the basic differentials are of long standing. There is also some suggestion that the strength of the relation has declined, especially after 1921.[1]

The much higher fertility rates for the married women in Jamaica parallel a differential noted for Puerto Rico by P. K. Hatt.[2]

[1] Though the foregoing approach strongly supports the hypothesis that the differentials in fertility by family type are of long standing, it must be emphasized that to some degree the relationship between levels of fertility and of illegitimacy is here over-simplified. In effect a linear relationship between the two variables has been secured by excluding the urban area on the ground that factors other than illegitimacy are primarily responsible for the comparatively low fertility in the Kingston–St Andrew area. But the fact that when the urban area is included the linear relationship is largely obscured, especially in the case of 1943, suggests that no simple linear relationship between the two variables obtains. In fact it seems that the inclusion of the urban area gives rise to a J-shaped relationship. Presumably an analysis of the relationship in terms of smaller (and more numerous) geographical divisions would make this J-shaped relationship more pronounced.

[2] P. K. Hatt, *Backgrounds of Human Fertility in Puerto Rica*, 1952, p. 315.

Table 81.   *Correlation between illegitimacy rates and joint gross repro-*
duction rates by parish

Period	Correlation coefficients	
	Including Kingston and St Andrew	Excluding Kingston and St Andrew
1879–83	– 0·71	– 0·76
1889–93	– 0·40	– 0·73
1909–13	+ 0·23	– 0·06
1919–23	– 0·35	– 0·69
1941–5	– 0·16	– 0·52

Here fertility among the married unions is higher than fertility among the consensual unions, which correspond broadly to the common law category of Jamaica. These differentials, in form and origin so much unlike those usually observed in European populations, seem to be associated with certain basic features of the three types involved. The fact that fertility is lowest among the keeper type despite the earlier age at which these women begin child-bearing suggests that important elements of fertility control are in operation. These, it would seem, are the products of the special institutional patterns of mating and not in any way akin to the motivations at work in the control of fertility differentials in European populations, to which modern Western civilization has given rise. There is no evidence to suggest that the fertility differentials in Jamaica, and in fact in the West Indies at large, are of recent origin. And if they are the products of conditions of mating they may date back to the time of slavery.

As we have already seen, an important characteristic of the keeper union, indeed in recent censuses it is the only feature by which this type is identified, is the absence of the male partner from the domestic unit. This means that the female partner of such a union is not continuously at risk of child-bearing and the chances of her bearing a child are to that extent reduced. On the other hand, the continuous cohabitation of spouses among the married type indicates that in this case the women are continuously at risk of child-bearing and thus have much higher chances of bearing children than the single. The common law type would seem to partake in some degree of the characteristics of both the keeper and the married types. In terms of census definitions, common law unions involve the continuous cohabitation of spouses and to this extent the intensity of exposure to the risk of child-bearing should be much greater than in the case of the keeper unions. There is, moreover, evidence that the birth of a child to an unmarried woman tends to affect her association with the father of the child and with men in general to an extent that she may take steps to minimize the risk of future pregnancy.[1] Discussing the fertility differentials between the married and the consensual unions in Puerto Rico, Hatt suggests that marriage may be in part at least a function of fertility, that in short the birth of a child might lead to the legalization of the union. Indeed it may be that as the number of children to unmarried parents increases forces develop which tend to make formal marriage both desirable and advantageous. The fact that those women who begin their reproductive life in the unmarried categories marry on the average at advanced ages lends support to this. However, it must be repeated that the constant passage of couples from one type of union to another during the child-bearing period and the fact that the formally married union is the last one entered, create special difficulties in interpreting the differentials, especially as it is not definitely known whether any change in fertility pattern accompanies such shifts.

While available evidence strongly supports the foregoing sociological interpretation of fertility differentials by family type, the possibility that factors other than the differential rates of exposure to the risk of fertility may be involved must be acknowledged. One such factor may be the differential incidence of venereal disease. If, for instance, the incidence of venereal disease is markedly higher among unions of the keeper and common law type than among those of the married type, this may tend to impair the reproductive performance of the former by promoting large-scale sterility and stimulating spontaneous abortion. It is conceivable that other medical and social factors also contribute to the establishment of the fertility differential. However, the assessment of the relative importance of various elements as causal factors in the fertility pattern of the population must await more detailed analysis of fertility in the island.

[1] This and kindred factors associated with fertility among the never-married in Jamaica will be dealt with by Judith Blake in her study of fertility in the island.

Inevitably one question that will arise from a consideration of the fertility differentials by family types revealed is what relation, if any, they bear to differentials by social status. Presumably if some reliable index of social status for the three types could be constructed this relation could be clarified. Unfortunately the limitations of the census data preclude the calculation of any satisfactory index of this nature. As has already been indicated, a basic characteristic of the common law type is its transient nature. It, in short, represents a transitional state between keeper and married unions and the constant passage to the married type, especially at advanced ages, results in a preponderance of women of young age in this type. Unless, therefore, an index of social status can be effectively standardized for age, it remains of little use. And since none of the census data which could be used for this purpose provides age breakdowns, no reliable indications of social status can be derived. Two measures which could give some indication of social status—the average floor space available per head of the household and the average wage of the wage earner—are given in Table 82. Both of these show that the married type is higher than the other two; but the extremely unfavourable light in which the common law type appears derives largely from the great differences in age distributions involved.

Table 82.   *Floor space per head of household and earnings per person, 1943*

Family type	Floor space per head of household (square feet)		Earnings per person (shillings per week)	
	Male	Female	Male	Female
Single	136	144	17·6	12·0
Married	285	248	40·7	16·8
Common law	131	140	16·3	7·4

It still remains clear that if we take the two never-married types together the general fertility differential in the West Indies seems to be the reverse of what is usually encountered in European society. For even though, as is evident from the very high proportion of illegitimate births, the population continues to be recruited largely from births to the never-married, fertility remains highest among the married, which from census data constitutes the highest social class. It will also be recalled that infant mortality is also much higher among the never-married. If we are disposed to equate the higher social status of the married type with inherently superior biological traits it may well be asked whether the long persistence of these fertility and mortality differentials has not been wholly beneficial for the population and indeed whether they do not constitute an essentially eugenic process. Of necessity, these aspects of reproduction, in the light of our present knowledge of fertility differentials, must remain on the plane of speculation.

In conclusion we may note that the fertility differentials prevailing effectively refute a view widely held in the West Indies, that fertility is high because illegitimacy is high.[1] It is indeed tempting to speculate whether this all too frequent assertion is in some obscure way linked up with the realization of the widespread social phenomenon that high fertility rates tend to be associated with lower social classes. With the scanty data available it would be hazardous to make the inference that a decline in fertility would of necessity follow a decline in the marriage rate, or more precisely a decline in the proportions ever-married in the reproductive age span. But it remains clear that no assessment of the possible future trends of fertility in Jamaica can ignore the probability that future changes in mating habits may affect fertility in a way wholly at variance with what would be expected on the basis of the experience of European populations. This argument will be developed in Chapter 9.

[1] For instance the *Report of the Committee appointed to Enquire into the Prevalence of Concubinage* . . . (1941) states that one of the social effects of high illegitimacy is 'a rapid increase in the birth rate with the present enormous unemployment rate whereby many children are born without any prospect of employment or living a decent life'.

CHAPTER 9
GROWTH PROSPECTS

As the island completes its third century of British rule it stands poised, apparently, for population growth on a scale greater than anything it has experienced in the past. The third century, unlike the one preceding it, has been characterized by unbroken population growth, which, though varying slightly from one intercensal interval to another, has remained, by comparison with other West Indian colonies, very stable. The annual rate of growth during the first intercensal interval, 1844–61, amounted to 0·9%, and this increased to 1·4% during the period 1861–81. A decline to under 1% was witnessed between 1881 and 1891, but in the succeeding intercensal interval the rate rose once more, to 1·3%. It was only during 1911–21, when emigration was on a relatively heavy scale and mortality conditions unfavourable, that a very low rate of growth prevailed (0·3%). However, the appreciable declines in mortality after 1921 and the cessation of emigration resulted in a considerably higher rate of growth, even in the face of some declines in fertility. The annual rate of growth during 1921–43 was the highest ever recorded (1·67%). Present rates of growth are much higher than any experienced in the past. Indeed since 1950 the island has consistently shown rates of natural increase in excess of 2% which, despite some emigration, results in a rate of growth nearly twice that prevailing in the middle of the nineteenth century.

It is easy to show that the continuation of a rate of increase of 2% over even a moderate length of time will yield massive additions to the population. Growth at such a rate would mean a population of 3·8 million within 50 years—that is more than the present population of the whole British Caribbean—while within a century a population of over 10 million would be massed in the island. Even a continuation of the more modest intercensal rate of 1921–43 (1·7%) would, if continued, yield totals hardly less impressive. Within 50 years the island population would amount to 3·3 million and within a century to 7·5 million. Densities of the order implied in such growth of the population—nearly 2000 persons per square mile within a century—cannot be contemplated with equanimity. But attempts to project populations into the remote future, on the basis of assumed constant rates of growth, are very unrealistic. For it is impossible to say with certainty what course the vital rates will trace in the remote future, especially if, as seems probable, considerable economic and social changes take place. Indeed it is doubtful whether rates of the order of 2% can persist indefinitely; almost inevitably the resulting population growth will give rise to conditions making for their reduction.

In this chapter we shall follow the more realistic course of estimating what changes in size and composition of the island's population may take place, within a period of 20 years, on the basis of certain simple assumptions concerning mortality, fertility and migration. These assumptions are developed mainly on the basis of the past experience of the population. Two projections are made. The first is based on constant fertility, declining mortality and no migration. The second seeks to consider the implications of two apparently reasonable steps to curb population growth in the near future. It assumes declining mortality, declining fertility and small annual emigrations. In both cases the initial population is the mid-year estimate for 1951 prepared by the Registrar General. A summary of the several aspects of fertility and mortality in the island enables us to assess more realistically the probable lines of their future development.

THE COURSE OF MORTALITY

Professor Notestein has urged that the prime objective of a population policy is the securing of better health for the people.[1] There is no reason to suppose that Jamaica, now firmly committed to the furtherance of this policy, will relinquish it in the future. And unless some shattering social disaster supervenes it is highly unlikely that any checks to the processes of mortality control will appear as a curb to population growth in the future. Mortality declines seem indeed to be inevitable, though the rates and patterns of decline are not fixed, and this is assumed in both of the projections made here. In seeking to evaluate the probable rates of mortality decline, it is useful to draw together the threads of foregoing discussions on mortality movements throughout the island's history.

[1] F. W. Notestein in P. K. Hatt, *World Population and Future Resources*, 1951, p. 57.

Attempts to control mortality in Jamaica date back to the late eighteenth century, when, during the pro-natalist era of slavery, every effort was being made to stimulate population growth among the slaves. Furthermore, the evidence, scanty though it is, suggests that these early attempts were not altogether without success. For the rates of natural decrease revealed by the slave registers are considerably lower than any that can be gleaned from the fragmentary data on earlier slave populations, with which historians have provided us. Emancipation definitely brought about a further reduction in mortality, probably as a result of the less arduous conditions of work and the reduced hazards of disease, following the scatter of the population over the rural areas. The foundations of the island's medical services and the basis of its sanitary controls, laid in the years after 1867, added still further to the control of mortality. The expansion of the medical services and the improved sanitation which the twentieth century witnessed, made possible the transition from the 'pre-industrial' to the 'expanding' phase of demographic development, which, we have already seen, can be conveniently fixed at 1921.

The control over mortality effected in Jamaica has generally been greater than that shown elsewhere in the British Caribbean. Even the slave registers and the analysis of Tulloch and Marshall suggest that death-rates prevailing during slavery were lower in Jamaica than in most of the other West Indian populations. This advantage has continued up to recent times, though clearly the widespread decline in mortality shown in the West Indies has resulted in a general convergence to a more or less common level of mortality. Doubtless in the past Jamaica had certain advantages over other Caribbean colonies which conferred this relatively favourable mortality. Uninfluenced by the ravages of malaria to the same extent as British Guiana, unsaddled by such extremely high infant mortality rates as characterized Barbados, with a racially homogeneous population, and with a density much lower than most of the eastern islands, Jamaica has apparently, throughout most of its history, been able to maintain this relatively favourable mortality position vis-à-vis the other West Indian populations. But the most recent indications are that the differences in mortality levels between the various populations are disappearing. This, together with the decline in fertility, means that growth rates in Jamaica are no longer higher than those of many other West Indian populations;

but the fact remains that declining mortality in Jamaica will suffice to ensure very considerable increases in the near future.

Though mortality falls have been witnessed in most age-groups, the rates of decline vary throughout the life span. The analysis of cause of death statistics suggests that in some cases at least the relatively low rates of decline achieved are due to resistant cores of causes at particular ages. An important core of this nature appears in the case of infant deaths. For though infant mortality has declined appreciably since 1921, largely as a result of the control of diseases of the digestive system, and diseases of the respiratory system, progress in the reduction of mortality from congenital debility, malformations, birth injuries and similar causes associated with infancy has been negligible during the period 1946–51. It is also probable that other resistant causes have been operating to reduce the rates of decline in mortality under five in general, which have been much lower after 1946 than before 1946. When inroads into these mortalities are made, further drastic reductions in infant and child mortality are to be expected. Furthermore, the growing proportion of deaths from cancer and diseases of the circulatory system, particularly among females, points to another though smaller core of deaths, one certainly much less susceptible of reduction than the core of deaths in infancy.

Desirable though it might be to evaluate the rates of future mortality decline in terms of growing control over the various major causes of death, this line of approach cannot be pursued because of the insufficiency of the data. The most fruitful approach seems to be on the basis of some assumed rates derived from past mortality experience. We have already seen that mortality declined strikingly between 1946 and 1951. The death-rate for 1945–7 stood at 14·1 and this declined to 11·8 by 1950–2. This reduction of 16% within five years becomes all the more impressive when the annual rates of decline among the several age groups are examined. The annual rates of decline during these years were for a considerable portion of the life span in excess of 6%, and in one age-group reached nearly 10%. The continuation of rates of decline of this order would rapidly transform the mortality pattern of Jamaica, and indeed endow it with a level comparable to that now shown by European populations. Should these rates continue for 10 years, the average length of life for the males would amount to 63 years. Though it is highly probable that the success that has hitherto attended the efforts to reduce mortality in the island will continue, it is unlikely that the extremely favourable rates of decline achieved during 1946–51 can be indefinitely maintained. Accordingly it is here assumed that the rates of decline that prevailed over the whole period of mortality control will obtain. Straight lines have been fitted, by least squares, to the logarithms of the $_nq_x$ values for 1921, 1946 and 1951, and these are used to project mortality to 1971. The annual rates of decline thus secured are in excess of 2% under age 50, and reach a maximum of 4·9% for the age interval 25–30 in the case of the males. The rates for the females are in general slightly lower, and the maximum (4·3%) occurs at a much lower age, in the age interval 10–15. The levels of mortality assured on these assumptions can be effectively summarized by the average length of life at quinquennial intervals from 1956 to 1971. These are shown in Table 83. At 1971 the assumed average length of life (63·3 years for males and 67·3 for females) will still be much lower than the levels now prevailing among most European populations, thus suggesting that the rates of decline are not unreasonable.

Table 83. *Average length of life in years according to projected mortality*

Year	Male	Female
1956	57·4	61·0
1961	59·5	63·3
1966	61·5	65·4
1971	63·3	67·3

THE COURSE OF FERTILITY

The hazards that attend any attempt to extrapolate fertility rates into the future are many and obvious. And in view of the complex elements determining fertility levels and differentials in Jamaica, these hazards are all the more numerous. Still, any basis used to estimate future fertility levels must bear some relationship to past rates, and a study of the movements might help to make possible the development of more realistic estimates of its future course. It is therefore pertinent to summarize briefly the main factors involved in the course of fertility since 1921. Two aspects of the subject can be conveniently distinguished. The first turns on the elements of social change in the island that can be associated with declining fertility: these fall into two broad categories, the considerable shifts in occupational status and the progressive urbanization. The second aspect, the peculiar system of mating and the diverse forms of family, together with their different levels of fertility, constitutes by far the more difficult to integrate into any discussion of the possible future course of fertility. As we have suggested, these differentials long antedated the decline of fertility, but in what manner they were involved in the changes witnessed since 1921 cannot be readily ascertained.

Occupational shifts constitute a striking feature of the population after 1921. The experience of the two sexes is discussed separately. The changing pattern in the case of the males reveals a growing absorption of workers into spheres of activity other than agriculture. Prior to 1921 there is no evidence of any considerable movements of males away from agriculture. But after 1921 male participation diminished drastically. This is not to say that agricultural production in the island declined. On the contrary, with the exception of bananas, considerable expansion of agricultural production was witnessed, especially in the case of sugar. This expansion was not achieved by the employment of more labour but by more capital and more efficient use of available land. The workers released from agriculture, following such increases in the efficiency of labour as were witnessed, could be absorbed in other sectors of the economy which were expanding after 1921. Although by 1943 the island, with 57% of its gainfully employed males still in agriculture, remained largely dependent on its major crops, the increase in the numbers employed in other spheres was impressive. This was particularly marked in the case of the so-called industrial class. Whereas the proportion of males in agriculture declined from 71% in 1921 to 57% in 1943, the proportion in the industrial class rose from 16 to 27%. The manufacture of textile goods, leather and chemical products, of negligible importance in 1921, was by 1943 engaging sizeable proportions of the male population, as was also building and construction and communications. Thus, despite the agricultural background of the island, a core of skilled and semi-skilled personnel was in the making.

Even more significant from the standpoint of social change and ultimate effects on levels of fertility have been the striking shifts in the pattern of female occupation since 1921. We have seen that this was dominated by sharp withdrawals from agriculture. Indeed the withdrawal of females from agriculture has been in progress since 1891, but was most pronounced after 1921. Whereas in 1921 57% of the gainfully occupied females were in agriculture, the proportion was by 1943 reduced to 28%. Doubtless some of the reduction may be accounted for by alterations in definitions of the gainfully employed, but even allowing for this the magnitude of the change remains. Concurrently with this withdrawal from agriculture has been witnessed an increased participation in domestic (personal) service and commercial occupations. Whereas most of the males displaced from agriculture have gone into other spheres of activity, it is clear that, in the case of females, displacement from agriculture has not meant shifts to new kinds of activity to the same extent. In fact the proportion of females over 10 employed has fallen from 60% in 1921 to 35% in 1943. This development has been particularly marked in the rural areas, where the withdrawal from agriculture on the scale witnessed has meant that the gainfully occupied females declined from 186,300 in 1921 to 131,800 in 1943.

Closely associated with these occupational shifts has been the appreciable migration to the urban centre. This movement, complementary to the widespread decline in employment in agriculture, was, as we have seen, inevitable in view of the cessation of emigration. Though nothing bearing on the social and demographic characteristics of these migrants is available, it is probable that the movement involved a measure of selectivity, that the majority of the migrants were young, eager to advance their earning power, and less concerned with the establishment of a family than those who remained in the rural areas. The prevailing patterns of family life seem most fitted to promote such migration. The lack of legal ties in so many unions makes migration easy. Moreover, the custom whereby a woman is often able to leave her (illegitimate) children in the country, under the care of her mother or some other relative, while she herself migrates to the city, also facilitates the migration of individuals on a wide scale.

All the foregoing changes are such as to conduce to fertility declines. The shift from agriculture and the growing urbanization have aided materially in eroding the high fertility patterns, which despite the special familial features of the population predominated in rural areas. Inevitably, urban concentration and growing contacts with the urban centre, as a result of improved transportation, aided powerfully in the reduction of fertility throughout the island. For although methods of fertility control were known and practised as early as the slave days, it is among the elite of the population that we should expect such practices to be most widely and effectively used, and these classes were largely concentrated in Kingston. As the fertility rates indicate, the city has always shown comparatively low levels.

Although the urban concentration and the changes in occupational patterns in Jamaica have been such that, so far as their effects on fertility are concerned, they conform to the experience of the technologically advanced societies, we have still to face some baffling questions on the interpretation of the course of fertility after 1921. How have the fertility differentials by family type been connected with the fertility declines witnessed throughout the island? Have the institutional elements of fertility control inherent in the special systems of mating conduced in any way to this decline and if so in what way? These and cognate questions are not easily answered. The island-wide fertility decline is fully consistent with growing control in a particular family type, since all types are equally spread throughout the island. For instance it is conceivable that a reduction of fertility among the upper classes of the population, those who in the main compose the married type, has been the prime cause of the general fall in fertility. Indeed the fact that illegitimacy rates after 1921 are higher than those prior to 1921, is in harmony with such an argument. For this could mean that increased fertility control among the married type has led to a smaller proportion of the total births being legitimate. But despite its seeming plausability such a hypothesis cannot be sustained, since there is no evidence of any relation between annual movements of illegitimacy rates and birth-rates between 1921 and 1943. On the other hand the general decline in fertility is equally consistent with declines in all types of the family. This indeed is suggested by the spread of fertility control outward from the urban centre, the effects of which appear as the main determinant of fertility declines.

Unfortunately the only data on geographical distribution of family types which could be linked to fertility in 1921 are illegitimacy rates. These offer no satisfactory clue as to the part played by fertility among the various types in the course of fertility throughout the island after 1921. The areas showing the greatest decrease, those near the urban centre, show no distinctive change between 1921 and 1943 which can be causally linked with fertility movements. At both dates they show slightly higher rates of illegitimacy than the other areas, but this is to be expected; it is indeed merely evidence of the negative correlation between fertility and illegitimacy and tells nothing about the relationship between fertility declines and patterns of fertility among the several family types. The crucial weakness of the available data is that there are virtually no indices of fertility by family type prior to 1943.

Though fertility differentials by family type cannot, on the basis of our present knowledge of the subject, be linked up with the decline in fertility in any meaningful way, the possible implications of such phenomena in an era of social change cannot be ignored. Even though our discussions of such implications must, under the circumstances, verge on the speculative, we shall attempt to assess the possible interrelationships between industrialization of the island and the movements of fertility.

Since most areas which have long evolved into what W. S. Thompson calls the 'stationary' phase of demographic development have experienced declines in fertility of varying magnitude, it is generally urged by students of population that industrialization of the so-called underdeveloped areas might well lead, in the long run, to similar fertility declines, even though the short term consequences, with the growing control of mortality, will actually greatly augment population growth. For ultimately industrialization will mean declining dependence on agriculture, growing urbanization, increased education and a general rise in the standard of living, all of which inevitably tend to make for growing control over fertility. Let us suppose that some of the plans for the rapid industrialization of Jamaica materialize in time, so that the spread of urbanization continues, that new urban centres spring up and existing ones develop still further, that the decline in agricultural employment continues, that more and more industrial and commercial and service occupations dominate the occupational patterns of the population; what are the prospects that these changes will induce fertility declines on a larger scale than that hitherto witnessed? Will the movements of fertility in Jamaica under such circumstances parallel those which the technologically advanced societies have experienced? The declines in the urban area and its immediate surrounding that have accompanied the relatively slight changes since 1921 at once suggest that industrialization will lead to still further declines in fertility. On the other hand the fertility differentials by family type may tend to offset this decline, if not indeed to induce a rise in fertility. For presumably with the rising standard of living, greater urbanization and general social improvement of the population that would probably follow rapid industrialization, the idea of marriage as a binding element of family life may spread, and the general stability of family life may be enhanced. In view of the basic characteristics of the existing family unions, which, we have already urged, are causally associated with the levels of fertility, this development, far from inducing a fall in fertility, may actually lead to a rise. For marriage on a wide scale would convert many unions now no more than casual associations where there is no continuous cohabitation of spouses into stable unions in which both partners share a common home and where consequently the risk of child-bearing would be greatly increased. The existing pattern of partial or sporadic exposure to child-bearing would thus disappear and in its place would appear the spread of unions with both partners sharing a common household and consequently a greater degree of intensity of exposure to the risk of fertility. Admittedly this is speculative, but the possibility that the fading of the institutional checks on fertility will attend rising industrialization and offset any fertility declines that may be expected to develop cannot be ignored.

From the foregoing the extreme difficulty of projecting with any certainty the future course of fertility in the island is obvious. The very fact that some decline has been witnessed in the past 35 years strongly suggests that the orderly course of events will not engender an increase in the near future, though the possibility that rapid industrialization may add entirely new factors to the situation cannot, as we have already argued, be entirely ruled out. On the other hand the present indication is not one of continuing declines in fertility, but rather one of stabilization at a rate admittedly lower than that prevailing prior to 1921, but which still shows a birth-rate above 30. It thus seems possible that for some time at least this fairly high level of fertility will continue and this is one of the assumptions on which the future population of the island will be estimated. Later, when the problem of controlled growth is taken up, we shall introduce an assumption of declining fertility, following as a consequence of measures designed specifically to curb population growth. In Jamaica, as in so many areas of rapid population growth, the tracing of what R. P. Vance terms the demographic gap is severely hampered by uncertainty as to the precise course of fertility in the future.[1]

PROJECTION I

On the assumption of constant fertility, declining mortality and no migration, a steady and considerable growth of the island's population up to 1971 will result, as can be seen from the summary in Table 84. An increase of 25%, from 1·4 million to 1·8 million is estimated between 1951 and 1961. And by 1971 the population will stand at 2·3 million, showing an increase of 58% within 20 years. These increases are much higher than any experienced in the past. The magnitude of these changes can best be appreciated by noting that the addition to the population between 1951 and 1971 (822,000) is not far short of the total increase to the island's population during the century 1844–1943 (860,000). This projection discloses much greater increases than a projection based on constant fertility of 1946 and mortality declining according to the rates of 1911, 1921 and 1946, and using the population of 1946 as the starting point.[2]

It is instructive to analyse in more detail the estimated growth in four broad age intervals shown in Table 84. In terms of this projection, the population under 5 will increase from 197,000 in 1951 to 314,000 in 1971, or by 60%. As a consequence of the greater rates of mortality decline assumed for the population aged 5–14, a more considerable increase for this age-group is indicated between 1951 and 1971, from 320,000 to 525,000 or by 64%. Estimates of the future population for these two age-groups, based on projected rates both of fertility and mortality, are probably

---

[1] R. P. Vance, 'The demographic gap: dilemma of modernization programs', in *Approaches to Problems of High Fertility in Agrarian Societies*, Milbank Memorial Fund, New York, 1952.

[2] The estimated populations according to the previous projection are 1,552,700 for 1956 and 1,700,800 for 1961 (see G. W. Roberts, 'Population trends in the British Caribbean', *Caribbean Economic Review*, vol. IV, no. 2) thus differing from the present estimates by 47,700 and 91,400 respectively. These differences are largely due to the different fertility assumptions involved. According to the present estimates the number of births amount to 50,700 per year during 1951–6 and to 55,300 during 1956–61, whereas the corresponding estimates of the earlier projection are 43,700 and 46,500 respectively. The difference in regard to deaths however is very small despite the difference in the assumptions taken. The present assumptions lead to an average number of deaths of 16,600 in 1951–6 and of 16,900 in 1956–61, as compared with 17,500 and 16,900 respectively for the earlier projection.

Table 84. *Summary of population estimates according to Projection I*

Age Group	1951	1956	1961	1966	1971
			Male		
0–4	98,900	115,600	127,800	141,200	157,900
5–14	159,900	179,500	207,800	236,900	262,900
15–64	414,200	461,700	512,300	570,000	638,800
65 +	22,100	24,300	29,300	37,000	47,700
Total	695,100	781,100	877,200	985,100	1,107,300
			Female		
0–4	98,200	114,900	126,900	140,100	156,500
5–14	159,700	179,200	207,300	236,200	261,900
15–64	444,500	489,300	538,400	595,000	661,500
65 +	32,300	35,900	42,400	51,300	64,800
Total	734,700	819,300	915,000	1,022,600	1,144,700
			Both sexes		
0–4	197,100	230,500	254,700	281,300	314,400
5–14	319,600	358,700	415,100	473,100	524,800
15–64	858,700	951,000	1,050,700	1,165,000	1,300,300
65 +	54,400	60,200	71,700	88,300	112,500
Total	1,429,800	1,600,400	1,792,200	2,007,700	2,252,000

subject to greater error than estimates for the higher age-groups, the validity of which must depend only on mortality (assuming of course no migration takes place). Large increases to the population aged 15–64, those of child-bearing and working age, are indicated. This group should increase by one half between 1951 and 1971, from 859,000 to 1·3 million. But the most striking increases are those shown by the population over 65, those past working age. This group, it will be recalled, has since 1921 been the most rapidly growing sector of the population, a consequence of the considerable declines in mortality throughout most ages of the life span. This rapid growth, according to this projection, should continue in the future and by 1971 the numbers over 65 will be twice as large as in 1951, increasing from 54,000 to 112,000. As a consequence of this considerable expansion, the proportion of the population over 65 rises from 3·8% in 1951 to 5% in 1971. This proportion of aged dependents is small by comparison with European populations, but the absolute increases will necessitate the provision of much greater sums than hitherto for their support.

Mainly because of the difficulty of foretelling the direction and extent of future internal migration, estimates of the growth of the urban centre can at best be tenuous. Between 1921 and 1943 the proportion of the population in the Kingston–St Andrew area rose from 14 to 19%. Obviously this proportion will continue to rise in the future. But it is also highly likely that other urban centres will develop and that some of those now existing will attain greater prominence as a consequence of industrial development throughout the island. In this event the Kingston–St Andrew area may in the future grow much more slowly than in the past. If the wholly arbitrary assumption is made that concentration in the main urban centre will continue and by 1971 will account for 25% of the island's population then this area may support about 560,000, thus more than doubling its size between 1943 and 1971.

One of the chief consequences of recent population growth in Jamaica (and in the West Indies at large) has been the considerable increments to the school population it has produced. The necessity for increased expenditure on education, even at such levels as prevail, has posed weighty financial and social problems. Clearly the growth envisaged here indicates still further additions to this category. On the assumption that the proportions of the population aged 5–14 at school in 1943 remain unchanged up to 1971, the estimated numbers for whom facilities will have to be provided are as follows:

1956	237,000
1961	274,000
1966	314,000
1971	349,000

It should also be recalled that the proportions attending school in 1943 (50% in the age-group 5–9 and 85% in the group 10–14) are lower than in many other West Indian populations, so that an increase in the general level of schooling would involve the provision of facilities on a wider scale than these estimates indicate. Moreover, these estimates take no account of requirements for school populations over 14 years of age. In fact the estimates given here are on several counts very conservative. However, it should be stressed that the school population of the island increased very greatly between 1921 and 1943—from 130,000 to 207,000—so that the increase envisaged here between 1943 and 1971 is not markedly greater than that within the transitional period 1921–43.

Though a multitude of factors combine to determine the numbers employed at any given time, estimates of the population at least provide a rough means of securing indications of the numbers who may be available for work in the future on the assumption that some stated rates of participation prevail. The assumption that the rates of 1943 continue to 1971 is made here. This seems defensible in the case of the males. For though the slight rise in the average age of male workers noted between 1891 and 1943 will doubtless continue into the future, this projection does not indicate any shifts in age distribution of a magnitude which will introduce any drastic alterations in the structure of the working population. Less secure is the assumption of constant worker rates for the females. It is true that in the future increasing absorption of females in such activities as commercial occupations, professional services and occupations especially connected with urban centres may tend to offset the continuing decline of female employment in agriculture and thus tend to maintain the general level of female employment. But numerous other factors will help to determine the proportions of females employed in future, probably foremost among them being changes in family forms in general. For these reasons the assumption of the persistence of the rates of female participation of 1943 may be less firmly grounded and the estimates of the number of females employed at future dates are of necessity less firmly based than those for the males.

On the assumption of constant worker rates for both sexes estimates can be derived of the numbers gainfully employed from 1951 to 1971. These are shown in Table 85. In the case of the males the total increases from 418,000 in 1956 to 585,000 in 1971. In fact, if the present assumptions are not seriously disproved, the numbers for whom work will have to be provided will increase by 265,000

Table 85. *Estimated numbers of gainfully employed, according to Projection I*

Year	Male	Female
1956	418,000	209,000
1961	466,000	230,000
1966	520,000	254,000
1971	585,000	283,000

between 1943 and 1971. This increase is much greater than that noted between 1891 and 1943. During those years, largely as a result of emigration, the number of males employed increased by only 137,000. The much more uncertain estimates for the females indicate an increase to 283,000 by 1971, a figure which is 73% higher than the number employed at 1943.

CONTROLLED POPULATION GROWTH, PROJECTION II

The quest for means to curb population growth in densely settled areas experiencing rapid population increases is now being actively pursued in many quarters. Indeed it can be said that the very magnitude of population growth in certain areas has created or is tending to create situations which necessitate the serious considera- tion of the introduction of measures to control this growth. To that extent therefore it seems valid to argue that under certain conditions massive increments to populations themselves produce situations which make the search for measures to restrict further growth advantageous. In fine it is becoming increasingly unlikely that extremely high rates of population growth will continue indefinitely into the future; almost certainly measures aimed at restricting growth will be applied. This of course does not constitute the inevitable solution of the problem of the progressive massing of people in a given area. As Kingsley Davis has argued, 'the possibility that sudden and widespread increases in mortality will occur' has always to be reckoned with.[1] However, in view of the growing awareness of the potentials of population growth implicit in the present demographic situation in Jamaica, it seems safer to dwell on the possibility that the promotion of measures to control population growth may ultimately constitute a segment of social policy in the island. Consequently we shall discuss two possible lines such control may take, emigration and fertility control. Our main concern here, it must be stressed, is not with the propriety or inevitability of such turns in social policy, but merely with some of their main demographic implications, should they in fact be pursued.

Prospects for large scale emigration from Jamaica are uncertain. In view of the world-wide immigration restrictions now in force, such places as the United States, Cuba and other Latin American

[1] Kingsley Davis, in P. K. Hatt, *World Population and Future Resources*, p. 25.

countries, which in the past absorbed appreciable numbers of Jamaicans as permanent settlers, must now be completely ruled out as possible emigration outlets. There has been within recent years some movement out of the island, the total net emigration during 1950–4 amounting to nearly 23,000. Probably a large proportion of this represents emigration to the United Kingdom, but available data furnish no reliable estimates of this phase of recent emigration. This movement to the United Kingdom is receiving special attention, largely it seems, because of the implications of the cultural contacts resulting from it.[1] It is impossible to tell whether this movement will continue on a scale sufficient to have marked and lasting effects on the growth and composition of the population of Jamaica, and consequently assumptions based on its continuance are not directly involved in the second projection.

As there has been recent discussion of emigration in terms of a transfer of population from the densely settled islands of the West Indies to the sparsely settled mainland territories, it is instructive to glance briefly at some aspects of such movements in the past.[2] The two British territories which in the past were areas of strong immigration were British Guiana and Trinidad. Largely as a result of immigration, the latter has shown the most rapid rate of population growth in the West Indies and population density there (330 per square mile) now exceeds that of Jamaica. In view of its extremely high fertility and density Trinidad must be completely ruled out as a possible outlet for Jamaican emigration. Indeed the only area which could conceivably absorb immigrants from neighbouring territories are British Guiana and British Honduras. British Guiana, long an area of immigration, was also the only Caribbean territory to promote government sponsored immigration from West Indian territories, though this was, admittedly, on a very limited scale. A brief consideration suffices to show that the conditions which made immigration into British Guiana in the past possible, if not inevitable, no longer exist. Demographic and other conditions favouring past immigration into that territory have drastically altered. Immigration into British Guiana from neighbouring colonies, supplementing the influx of indentured labourers

[1] See for instance Michael Banton, 'Recent migration from West Africa and the West Indies to the United Kingdom', *Population Studies*, vol. 7, no. 2.

[2] See *Report of the British Guiana and British Honduras Settlement Commission*, Cmd. 7533, 1948 and W. A. Lewis, 'The industrialization of the British West Indies', *Caribbean Economic Review*, vol. II, no. 1.

from India, Africa and China, was essential because under prevailing conditions of high mortality this was the only way of maintaining the labour force. Towards the end of the nineteenth century for instance only about 10% of all males who attained age 15 could expect to reach age 65. Even as late as 1911 the proportion who could expect to reach age 65 was only 15%. Without recruitment from overseas the plantation economy, with its demand for a cheap and abundant source of labour, would have been endangered. In British Guiana at least the planter's claim for a constant addition of foreign labour seemed therefore justified. But such conditions are now radically altered. Mortality has declined drastically. Thus the proportion of males aged 15 who can expect to reach age 65 is, according to the mortality of 1950–2, 45%. This, together with the improved morbidity conditions, means that immigration would no longer constitute replacements of wastage occasioned by high mortality, but genuine additions to the population. Moreover, the sugar industry is no longer based on such large supplies of cheap labour as in the past. Here as in Jamaica agricultural densities have been declining steeply since 1921.

Another change in terms of which inter-Caribbean migration must be viewed is the altered mechanism of migration. All that state sponsored immigration into British Guiana consisted of was a bounty of $4 per head to captains of schooners bringing immigrants from neighbouring colonies. In fact under the low scales of passages then obtaining immigrants preferred to pay their own way; the attractions of the immigrant colony sufficed to assure a ready supply of immigrants. Trinidad relied solely on the attractions it had to offer immigrants and, unlike British Guiana, never experimented with state sponsored immigration of West Indians.[1] As we have shown in Chapter 4, emigration of Jamaicans to Panama developed under similar conditions. The Canal Commission rightly assumed the remunerative work available at the Canal was a sufficient inducement to Jamaicans to emigrate. However, later legislation changed the conditions of inter-Caribbean migration fundamentally, imposing many curbs on the movement. Indeed, it is doubtful whether, even if basic changes in the political relationships among the Caribbean territories develop, the halcyon days of unrestricted,

---

[1] These aspects of inter-Caribbean migration are discussed in G. W. Roberts, 'Emigration from the Island of Barbados', *Social and Economic Studies*, vol 4, no 3.

inexpensive migration, enabling individuals to move with ease from one part of the Caribbean to another in response to demands for labour, will ever return. Probably the only large scale migration that may reappear in the future is some form of rigidly controlled movement entirely financed by the state.

The Commission which examined migration between West Indian territories favoured the promotion of government sponsored migration, expressing the view that the two mainland territories could 'absorb about 100,000 men, women and children, including some 25,000 adult workers'.[1] This however would have to be conditional on the 'vigorous development of their latent resources'; indeed immigration was envisaged only as part of a wide programme of economic development urged by the Commission. At present the chances of any such scheme of migration materializing seem dim, and discussions of this nature must remain merely hypothetical.

Undoubtedly formidable difficulties would have to be overcome in any such migration; schemes of this sort would moreover prove immeasurably more costly than indenture immigration, the only government sponsored migration ever developed in the West Indies on an appreciable scale. Further, many have questioned whether the absorptive capacity of the mainland territories is such as to make even this comparatively small scale immigration successful. But despite the vastly altered context in which inter-Caribbean migration must now be viewed, there is nothing absurd in a discussion of the possible demographic implications of a scheme having as its aim the transfer of population from an island which within fifty years' time may face a density of 700 per square mile to a mainland area, vast by comparison and which even at rates of growth in excess of the phenomenal ones now exhibited could not within the next century attain a density even approaching that which Jamaica supported in 1943. However, as the advocacy of such a migration policy forms no part of the present treatment, direct considerations of its possible implications will not be attempted. All that is done here is to incorporate into the second population projection an assumption of a small net emigration from Jamaica over a moderate period of time. In effect a series of computations is made to show the possible effects on population growth and composition of such emigration, irrespective of the direction of the movement. Complementary computations showing

[1] *Report of the British Guiana and British Honduras Settlement Commission*, p. 189.

the effects of such a movement on the receiving country can of course be easily made.

Clearly emigration on a scale which would completely wipe out the annual increase now shown by Jamaica, even if desirable, would be wholly impracticable. In the past several of the eastern Caribbean colonies experienced emigration on a scale sufficient to induce population declines. Of these Barbados is an outstanding example, an area which was literally saved from disastrous overcrowding by the prolonged emigration of relatively large numbers of its population. Here net emigration at an annual rate of 1% of the island's population was maintained for 30 years (1891–1921) and this induced rates of decline of up to 1% a year.[1] This was possible only because of the high mortality then prevailing and the consequent low natural increase. With a birth-rate of 33 and a death-rate of 11, Jamaica would have to promote emigration at a rate of 2% of its population annually to maintain an unchanging population. But we are not concerned with the futile speculation of entirely halting population growth in the island, by emigration or by any other means. Indeed in view of the more pressing problems of population pressure on resources in many of the eastern islands of the Caribbean, emigration to achieve such an unrealistic end is unthinkable. In any serious consideration of Caribbean migration the claims of the eastern islands might probably outweigh those of Jamaica. In the present context therefore we consider a modest level of emigration of about 1% of the island's population annually for a period of 15 years (1956–71). Even this is larger than anything experienced in the past. During the 40 years when emigration was in progress, the total net emigration amounted to only 146,000, whereas we are assuming a total net emigration of 180,000 in 15 years. An equal number of emigrants is supposed to be selected each year and to consist of 8000 adults equally divided between the sexes and 4000 children also equally divided between the sexes. Male adults are assumed to be selected in accordance with the age distribution of the gainfully occupied between the ages of 20 and 49 at 1943. It is further assumed that each male is accompanied by a mate 5 years his junior. According to the last census each woman aged 15–44 had on the average 2 children. But on the assumption that not all children accompany their parents and that in fact only

[1] See G. W. Roberts, op. cit.

their younger children are involved only 4000 emigrant children are introduced and these are distributed as follows: 50% under 5, 40% aged 5–9, and 10% aged 10–14.

The declines in fertility already witnessed and the marked association between the levels of fertility and of literacy indicate that contraceptive devices and practices have been more widely and effectively used since 1921, especially in and around the urban areas and generally among the more socially advanced sections of the population. Undoubtedly therefore further declines may be expected as more people enter the urban areas, as educational standards rise and as social development in general proceeds. It is highly probable that the active promotion of methods of fertility control, a line of public policy by no means unlikely in view of the growing awareness of the problem of mounting population pressure on resources, may accelerate this decline. The particular framework of the Jamaica family, with its institutional patterns seemingly suited to the restriction of fertility may prove most receptive to such plans, and indeed the inauguration of such a policy may result in declines much more impressive than those hitherto witnessed. Actually the assumptions made here do not envisage any drastic reductions, but merely a fall of 17% within 20 years. In effect, it is assumed that a general spread of urban patterns of reproduction, coupled with effects due to the postulated public policy, will force down the island fertility to a level roughly equivalent to the 1951 level of the urban areas. The joint gross reproduction rates assumed for the various years are as follows:

1951	2·28
1956	2·18
1961	2·08
1966	1·99
1971	1·90

Table 86 summarizes the results of Projection II, based on declining mortality, declining fertility and a constant emigration from 1956 onwards. The modest measures of control assumed here result in a rate of growth appreciably lower than that shown by the first projection, but still the indications are that within 20 years (1951–71) 496,000 will be added to the population, an increase of 35%. Though less than the increase shown by the first projection (822,000), it still exceeds the total increase between 1911 and 1943

{406,000). As the measures of control are assumed to run from 1956 to 1971 it is appropriate to examine the growth during these years more closely. During this period the increase in the population, according to the present projection, amounts to 330,000, as compared with 652,000 shown by the first projection. Though the net emigration assumed is larger than any previously experienced its effects on population growth are much less than the effects of emigration during 1911–21.

Table 86. *Summary of population estimates according to Projection II*

Age group	1956	1961	1966	1971
		Male		
0–4	112,900	111,100	112,000	115,500
5–14	179,500	200,200	209,100	209,100
15–64	461,700	492,300	529,500	571,900
65 +	24,300	29,300	37,000	47,700
Total	778,400	832,900	887,600	944,200
		Female		
0–4	112,300	110,300	111,100	114,400
5–14	179,200	199,700	208,400	208,100
15–64	489,300	518,400	554,300	594,100
65 +	35,900	42,400	51,300	64,800
Total	816,700	870,800	925,100	981,400
		Both sexes		
0–4	225,200	221,400	223,100	229,900
5–14	358,700	399,900	417,500	417,200
15–64	951,000	1,010,700	1,083,800	1,166,000
65 +	60,200	71,700	88,300	112,500
Total	1,595,100	1,703,700	1,812,700	1,925,600

Note. The values for 1951 are not shown as these are the same as in Table 84.

A convenient way of relating emigration to population increase is to express the net emigration as a percentage of the natural increase between two census dates. These percentages for 1956–71 can be compared with those previously recorded:

1881–91	30%
1891–1911	19%
1911–21	74%
1956–71	36%

Thus when account is taken of the much more favourable mortality conditions almost certain to develop after 1956, the level of emigration assumed, though more effective as a controlling factor in population growth than emigration between 1881 and 1911, is much less effective than that which took place during 1911–21.

As a result of the assumed decline in fertility and the small emigration of children, the estimated increase in the population under 5 between 1951 and 1971 amounts to 17%, which is less than one-third of that shown by the first projection. Whereas the effects of the assumed reduction in fertility fall mostly on the population under 15, the main effects of the emigration assumed are on the number of people aged 15–65, that is within the child-bearing and working age. Here the numbers increase from 859,000 in 1951 to 1·2 million in 1971, or by 36%, as compared with 51% in the first projection. The movements in the case of those over 65 remain unchanged. As a consequence these constitute a relatively greater proportion of the total population by 1971 than that shown by the first projection. The proportion over 65, according to the second projection, amounts to nearly 6% by 1971. This, though small by European standards, is much higher than any likely to be found in the other West Indian populations for a long time to come.

The assumptions of emigration and declining fertility yield much lower increases in the estimates of school population between 1951 and 1971. Again on the assumption that the proportions attending school at 1943 continue unchanged up to 1971 the numbers at school amount to the following:

1956	237,000
1961	265,000
1966	281,000
1971	280,000

Thus there is evidence that a continuation of emigration and declines in fertility may ultimately result in only small annual increments to the population of school age. In fact between 1966 and 1971 the indications are that there will be no increase in the school population. However these estimates must be qualified. For probably the proportions attending school may be appreciably higher than those here assumed. In fact an overall rise will be inevitable if the low proportions attending school in the rural areas are to be increased. In terms of this projection therefore much of the increase in educational facilities will be necessitated not so much by the annual increments to the population, but by efforts to raise the educational status of the population as a whole.

Emigration affects sensibly the future estimates of the working population. As can be seen from Table 87, the males should, on the assumptions taken, increase to 524,000 by 1971, or by 64% between 1943 and 1971. This, though proportionally much less than the increase indicated by the first projection, is none the less appreciable in terms of absolute numbers. For the increase of over 200,000 within a period of 28 years is much more than the increase in the number of males employed between 1891 and 1943 (137,000). The increase in the case of the females, which is subject to much wider margins of error, is from 163,900 in 1943 to 255,000 by 1971: in other words by 1971 there may be 55% more females occupied than in 1943.

Table 87. *Estimated numbers gainfully employed, according to Projection II*

Year	Male	Female
1956	418,000	209,000
1961	447,000	222,000
1966	482,000	236,000
1971	524,000	255,000

Thus both projections emphasize that the major problem of population growth in Jamaica is the growing numbers of potential workers. It is true that the numbers of children are increasing rapidly and that this presages growing claims on the island's financial resources for greatly expanded educational facilities. It is also true that the considerable increases in the numbers over 65 mean increased economic provision for aged dependents. But the burden of the economically unproductive, both young and old, on the economically active elements of the population is a secondary matter compared with the central problem of finding work for the growing number of potential workers in the island.

APPENDIX I

Populations of British Caribbean Territories

Territory	1841–4	1851	1861	1871	1881
					Year of
Barbados	122,198	135,939	152,727	162,042	171,860
British Guiana	98,133	*125,692*	*148,026*	*193,491*	*244,478*
British Honduras	(*10,000*)	—	25,635	24,710	27,452
Jamaica	377,433	—	441,255	506,154	580,804
Leeward Islands					
Antigua	36,687	*37,757*	37,125	35,137	34,964
Montserrat	7,365	7,053	7,645	8,693	10,083
St Kitts–Nevis	*32,748*	20,741	*34,125*	42,576	44,220
Virgin Islands	6,689	—	—	6,651	5,287
Trinidad and Tobago	*73,023*	82,978	99,848	126,692	171,179
Windward Islands					
Dominica	*22,469*	—	25,065	27,178	28,211
Grenada	*28,923*	32,671	31,900	37,684	42,403
St Lucia	*21,001*	*24,318*	26,674	31,610	38,551
St Vincent	27,248	30,128	31,755	35,688	40,548
Total British Caribbean	863,917	—	—	1,238,326	1,440,040
No. of censuses at each date	13	9	12	13	13

EXPLANATORY NOTES

In this Table are collected populations given in censuses known to have been taken in the British West Indies since 1841–44. For reasons given below some of these differ slightly from the corresponding totals given in the 1946 Census Report, vol. 1, Table 1. All figures in italics in the Table represent slight departures from those adopted in the 1946 Census Report. These notes merely indicate the nature of the departures and supplement the notes given in the 1946 Census (volume 1). The discrepancies between the two series in respect of 1841–4 call for special comment. Such is the variation in presentation that many versions of population totals at this date are possible. The procedure followed here is to use the versions given in the summary table accompanying the

330

census						No. of censuses in each colony
1891	1901	1911	1921	1931	1946	
182,867	—	172,337	156,774	—	192,800	9
270,865	—	289,140	288,541	302,585	359,379	10
31,471	37,479	40,458	45,317	51,347	59,220	10
639,491	—	831,383	858,118	—	1,321,054	8
36,819	34,971	32,269	29,767	—	41,757	10
11,762	12,215	12,196	12,120	—	14,333	10
47,662	46,446	43,303	38,214	—	46,243	10
4,639	4,908	5,562	5,082	—	6,505	8
218,381	273,899	333,552	365,913	412,783	557,970	11
26,841	28,894	33,863	37,059	—	47,624	9
53,209	63,438	66,750	66,302	—	72,387	10
42,220	49,883	48,637	51,505	—	70,113	10
41,054	—	41,877	44,447	47,961	61,647	10
,607,218	—	1,951,327	1,999,159	—	2,851,032	
13	9	13	13	4	13	125

censuses in the *P.P.* 1845. But the compiler of that summary followed no systematic plan and some of the totals adopted in the 1946 Census seem more defensible.

Barbados. The populations in the Table are identical with those given in the 1946 Census; both versions include shipping. But there has been in the past no uniformity of procedure in the treatment of shipping. In some censuses this group is excluded from the population on which the tabulations are based; in others it is treated as an integral part of the population.

British Guiana. Except for 1841, the figures used here are lower than those given in the 1946 census because they exclude all aborigines and, for the years 1861–81, small numbers covering shipping and garrisons. The exclusion of aborigines is justifiable on the ground that complete coverage of this group has never been achieved and the enumerated populations vary widely from census to census. Also it appears that, in some instances at least, the Registrar General has based his annual population estimates on census totals that do not include garrison populations.

British Honduras. The estimate of 10,000 made by the Governor for 1844 (despatch from Elgin to Stanley, 6 July 1844) is entered here in order to complete the record for the years 1841–4. Actually the first general census of this colony was not taken until 1861.

Jamaica. These population totals are for Jamaica only, that is they exclude the dependencies of Turks and Cayman Islands. Census populations for the dependencies that have been located for 1851 to 1921 are:

	1851	1861	1871	1881	1891	1901	1911	1921
Turks	3,250	4,372	4,723	4,732	4,744	5,287 (?)	5,475 (?)	5,612
Caymans	—	—	—	3,066	4,322	—	5,564	5,253

The Jamaican census of 1943 shows an island total of 1,237,063. On the basis of the Registrar General's end of year estimates for 1945 and 1946, an estimate of the population at 7th April 1946 (the date of the census for the other colonies) has been made.

Antigua. The 1851 population given here is taken from the census of that year and differs slightly from the figure appearing in the 1946 Census Report. It should also be noted that a census covering Antigua alone i.e. excluding Barbuda was taken in 1856, giving a population of 35,408.

Montserrat. The only point calling for comment is that the population given for 1851 is really from a census of 1855. There appears to have been no census taken in 1851.

St Kitts–Nevis. The population for 1844 is that given in the summary in the *P.P.*, 1845. It is lower than that given in the 1946 Census because the latter includes an estimate of the population of Anguilla. Available data do not indicate clearly whether or not the 1844 census excluded Anguilla. The population entered in the Table for 1851 is for St Kitts alone, and follows from a census of 1855; no census was taken in 1851. The population of St Kitts alone for 1861 is 24,303, and not 24,440 as given in footnote (m) to the compilation in the 1946 Census, while that for Nevis is 9,802; but apparently the census did not cover Anguilla. The population for 1871 used here is that given in the St Kitts–Nevis Census Report, 1891, which shows the following returns:

St Kitts   28,169
Nevis   11,703
Anguilla   2,704

But according to the St Kitts–Nevis Census of 1911, the population of Anguilla in 1871 was 'not ascertainable'. Access was not had to the original 1871 census report.

Virgin Islands. The only comment called for here is that there are slight discrepancies in the census tables of 1871. In fact three versions of the total are possible: 6651, 6641 and 6626.

Trinidad and Tobago. The correct population of these two islands for 1844 is 73,023. Up to 1891 separate censuses were taken in each island and it was not until 1901 that the two were treated as one unit for the purpose of enumeration. Throughout, the populations given here are for the two islands combined.

Dominica. Following the summary in the *P.P.*, 1845, we have included a group of 269, said to be 'off the island' at the time of the census. The population given in the 1946 census summary excludes this group.

Grenada. Again the only difference between the two series is the population of 1844. The population in the Table includes certain groups which, though incorporated in the 1845 summary, can legitimately be excluded, as was done in the compilation in the 1946 Census. These groups are:

Persons in St Georges Gaol	20
Crews of vessels belonging to the colony	59
Crews of vessels not belonging to the colony	244
Garrisons Fort George and Richmond Hill	404

St Lucia. The population adopted here (21,001) is that of the 1845 summary. No census was taken in St Lucia in 1844. A census was taken in August 1843, and this apparently formed the basis of the population given for 1844. The Governor issued instructions 'to have births and deaths ascertained to the 3rd June next. . . . This together with the census will give a correct statement of the whole population'. The 1851 figure in the Table, which differs from that given in the 1946 Report, is taken from a despatch from Colebrook to Newcastle, dated 7 June 1853. The population given here is 24,318, composed of 11,763 males and 12,527 females but clearly there is a discrepancy here. The figure of 24,185 appears in the St Lucia Census Report, 1901. The Blue Book of 1862 gives a population of 27,480, apparently taken from the census of 1861. But the Census Reports of 1871 and later dates give the 1861 population as 26,674.

St Vincent. The populations entered here agree throughout with those given in the 1946 census compilation.

APPENDIX II

INDENTURE IMMIGRATION INTO JAMAICA

Year or season	East Indians	Africans	Europeans	Others	Total
1834	—	—	2	—	2
1835	—	—	864	24	888
1836	—	—	1,145	67	1,212
1837	—	(772)	360	—	360
1838	—	(446)	—	—	—
1839	—	(170)	—	—	—
1840	—	—	—	71	71
1841	—	686	1,684	186	2,556
1842	—	1,006	19	3	1,028
1843	—	301	—	133	434
1844	—	540	—	—	540
1845	261	297	13	35	606
1846	1,851	619	—	—	2,470
1847	2,439	—	—	—	2,439
1848	—	1,940	—	—	1,940
1849	—	1,080	—	—	1,080
1850	—	468	—	—	468
1851	—	808	—	—	808
1852	—	16	—	—	16
1853	—	32	—	167	199
1854	—	—	—	472	472
1855	—	—	—	212	212
1857	—	362	—	—	362
1859–60	598	47	—	—	645
1860–1	1,522	649	—	—	2,171
1861–2	1,982	608	—	—	2,590
1862–3	542	533	—	—	1,075
1866–7	1,625	11	—	—	1,636
1868–9	1,393	—	—	—	1,393
1869–70	906	—	—	—	906
1870–1	1,354	—	—	—	1,354
1871–2	1,207	—	—	—	1,207
1872–3	1,518	—	—	—	1,518
1873–4	1,356	—	—	—	1,356
1874–5	1,250	—	—	—	1,250
1875–6	748	—	—	—	748
1877–8	895	—	—	—	895
1878–9	167	—	—	—	167
1879–80	747	—	—	—	747
1880–1	504	—	—	—	504
1882–3	396	—	—	—	396
1884–5	570	—	—	680	1,250
1890–1	2,135	—	—	—	2,135

Year or season	East Indians	Africans	Europeans	Others	Total
1892–3	484	—	—	—	484
1894–5	1,167	—	—	—	1,167
1899–1900	1,276	—	—	—	1,276
1902–3	659	—	—	—	659
1905–6	812	—	—	—	812
1906–7	814	—	—	—	814
1907–8	609	—	—	—	609
1908–9	414	—	—	—	414
1909–10	1,118	—	—	—	1,118
1911–12	813	—	—	—	813
1912–13	1,985	—	—	—	1,985
1913–14	293	—	—	—	293

Total indenture immigration, 1834–1914

East Indians	36,410
Africans	10,003
Europeans	4,087
Others	2,050
Total	52,550

INDENTURE IMMIGRANTS KNOWN TO HAVE RETURNED TO INDIA

Year or season	East Indians returning	Year or season	East Indians returning
1853	1,125	1883–4	78
1854	395	1884–5	471
1858	126	1885–6	161
1870–1	925	1888–9	553
1871–2	420	1890–1	567
1874–5	356	1892–3	486
1875–6	251	1894–5	348
1876–7	316	1902–3	1,126
1877–8	237	1904–5	318
1878–9	416	1905–6	680
1879–80	376	1909–10	111
1880–1	403	1910–11	171
1881–2	448	1912–13	331
1882–3	415	1915–16	270
		Total	11,880

NOTE ON COMPILATION OF ESTIMATES OF INDENTURE IMMIGRATION INTO JAMAICA

The estimates of indenture migration given here are not intended to be more 'exact' than any previously compiled (e.g. those in I. Ferenczi, *International Migration*, vol. 1, 1931). They constitute merely a further attempt to consider the movement in the light of sources available in Jamaica and elsewhere. The nature of these sources and of the migration itself makes it impossible to derive any exact and definitive measures of the movement. For instance, in the late nineteenth century East Indians entering the West Indies were generally classified as indenture immigrants, as passengers or as casuals. Admittedly the last two categories were only a very small proportion of the total entries, especially in the case of Jamaica. But clearly estimates of the magnitude of the movement must to some degree vary according to whether or not all these classes of immigrants are used. (All East Indians recorded as entering Jamaica are used here.) Again not all East Indians leaving Jamaica were persons returning to India on the completion of their industrial residence. Many left on their own initiative and were not entered as emigrants. In general it appears that East Indians about to leave any West Indian colony at their own initiative and expense were obliged to secure passports and the numbers of passports so issued were recorded in Reports of the Agent General of Immigration. In the case of Jamaica the number of passports recorded was negligible and the returning migrants listed here are restricted to those going to India in exercise of their right to the return passage. Despite these and other limitations of the original source material it is possible to secure fairly accurate measures of the magnitude of indenture migration.

The following sources, which are hereafter referred to by the numbers affixed to them, have been either directly used or consulted in compiling the records of indenture immigration (of East Indians and others) and of immigrants who returned to India:

(1) 'Returns showing the Number of Free Emigrants into Jamaica, British Guiana, Trinidad . . . for each Year since the Abolition of Slavery . . .', signed by S. Walcott, Secretary, Colonial Land and Emigration Commission, and dated 11 June 1847, *P.P.*, 1847.

(2) Later compilations made by the Colonial Land and Emigration Commission. Some of these were published in the *P.P.*, but the main data on migration for later years appear in the appendices to the *General Reports of the Emigration Commission*.

(3) Entries in the *V.H.A.J.*

(4) When the *Votes* were no longer printed the *Reports of the Agent General for Immigration*, included in the *Administrative Reports*, formed the only source, as the Colonial Land and Emigration Commission ceased to supervise indenture immigration after 1873, publishing no annual reports after that year.

(5) *Annual Reports on Emigration from the Port of Calcutta to British and Foreign Territories*, published at Calcutta. These Reports give detailed statistics of emigrants leaving under indenture from 1875 onwards, as well as detailed statistics of those returning after completing their industrial residence.

(6) D. W. D. Comins, *Note on Emigration from the East Indies to Jamaica*, Calcutta, 1893.

(7) J. Geoghegan, *Note on Emigration from India*, Calcutta, 1873.

(8) Despatch from Metcalfe to Lord Stanley, dated 22 March 1832, *P.P.*, 1842.

Africans. The sources and methods of compilation of African immigration are discussed in G. W. Roberts, 'Immigration of Africans into the British Caribbean', *Population Studies*, vol. VII, no. 3. As indicated in Chapter 4, the figures given in this paper for immigrants returning to Africa are erroneous. The entries of immigrants for 1837, 1838 and 1839 represent liberated Africans landed in Jamaica from captured slavers before the official sanction of African immigration into the West Indies and are not included in the total African immigration.

East Indians. The sources used in compiling the numbers of East Indian immigrants are (1), (2), (3) and (4). Where these sources cover the same period they are in agreement except for two seasons, 1861–2 and 1871–2. For the former the figure appearing in (2) is 1982 and this is used here. The *Votes* give 2123, the difference being accounted for, apparently, by the inclusion in this source of some Africans who came over in one of the ships bringing East Indians. The difference between (3) and (2) in respect of 1871–2 is small. The former gives 1188 as compared with 1207 given in (2). The figure in the *Votes* is not used here because this source shows some discrepancies in the returns published for this season. For years after 1872 the only source of arrivals into Jamaica is (4). Returns from this source have been checked against the records of emigrants leaving India given in (5). When allowance is made for mortality experienced on the voyage, and for births to the emigrants during the voyage, there is almost perfect agreement between the two sources. The present compilation is also very close to that prepared by Comins in (6) for the period up to 1891; the chief differences occur in the first three years of the movement, for which period probably Comins did not have access to the original sources. An earlier compilation by Geoghegan for years up to 1870 differs somewhat from the present one.

To some extent this may be explained by the fact that the entries in (7) refer to calendar years and not to seasons as in the present case. Whatever the reasons for the differences (7) remains the most comprehensive account ever written of the course of indenture migration up to 1870.

Estimates of migrants returning to India are obtained from (3) and (2) for years up to 1872. The entries for 1853 and 1854 are from (3) and these differ from those in (2) by 76, because the latter includes infants, presumably born in Jamaica. It is not clear whether children born to immigrants in Jamaica were always excluded from the numbers given as returning to India. Returns from (4) are the only source of numbers leaving Jamaica after 1872. These have been checked against the returns in (5) and when allowance is made for mortality during the voyage, the two sources are in almost perfect agreement.

Others. Returns of European indenture immigrants are given in (1) and (3). These are in agreement for all years. However the very detailed statement in (8) of immigration into Jamaica in 1841 is used here as the basis of the present compilation. This is 351 more than the figure given in (1) and (3). For immigration from Canada, and the United States the source is (1), which also gives small numbers coming from other areas.

APPENDIX III

ESTIMATES OF NET EMIGRATION TO UNITED STATES, PANAMA, CUBA AND OTHER AREAS, 1881–1921

The following outlines the methods of estimating the net emigration from Jamaica entered in Table 31.

United States. According to the *Fourteenth Census of the United States,* the numbers of persons born in the West Indies (exclusive of Puerto Rico and Cuba) were:

1920	64,090
1910	32,502
1900	14,354
1890	23,256 (including Cuba)
1880	9,484

It is safe to assume that the majority of these came from Jamaica. It seems also safe to assume, from the discussion in the text, that the importance of the United States as an absorber of Jamaican emigrants increased with the cessation of emigration to Panama. Consequently it is arbitrarily taken here that the proportions of the American population of Jamaican birth were 60% in 1900, 70% in 1910 and 80% in 1920. The differences between the various census enumerations of West Indian born populations suggest the following numbers emigrating from Jamaica to the United States:

Before 1910	14,000
1910–20	28,000

These are minimum estimates, as they do not include emigrants who died in the United States. Therefore they are arbitrarily increased to 16,000 and 30,000 respectively. The estimate for years prior to 1910 is to some extent supported by the fact that between 1899 and 1910 the number of West Indians (excluding those from Cuba) who entered the United States amounted to 11,569 (*see Statistical Review of Immigration, 1820–1910,* Washington, 1911).

Panama. Published figures on migration for the period 1881–91 (exclusive of indenture migration) emphasize that emigration to Panama was important, but the fact that they disclose virtually no net emigration to this area indicates their uselessness. It is here arbitrarily

339

assumed that 70% of the net emigration during this period was to Panama (17,000).

Kuczynski (op. cit. vol. II, pp. 7–9) has suggested that deaths by place of birth registered in Panama can serve as a rough guide to the numbers of West Indian immigrants into Panama. On the assumption that the proportion of Jamaican born among the total deaths in Panama gives a rough estimate of the numbers of Jamaican born in the population of Panama, it appears that the numbers of Jamaicans resident in Panama in 1911 and 1921 were 33,000 and 29,000 respectively. Furthermore, as the number of deaths between 1911 and 1920 was 6000, the total Jamaican born population in 1921 would have been 35,000 had there been no deaths. Therefore a rough estimate of net immigration during 1911–21 is 35,000 – 33,000 = 2000. Again, the number of deaths during 1906–10 was 3000, so that the Jamaican born population of 1911 would have been 36,000, had there been no deaths during 1906–10. It can therefore be taken that total net immigration of Jamaicans into Panama prior to 1911 was 36,000. If we further assume that about 10,000 of the original net immigration during 1881–91 were still alive, it appears that net immigration between 1891 and 1911 was about 26,000.

Cuba. Kuczynski suggests that there was not much emigration to Cuba before 1916, but migration between Jamaica and Cuba was in operation since the nineteenth century. According to *Censo de la Republica de Cuba, Año de 1919*, Habana, 1924, the number of persons born in Jamaica amounted to 18,539. It is here arbitrarily assumed that allowing for mortality among these immigrants the total net emigration to Cuba between 1911 and 1921 totalled 22,000.

Other Areas. These entries represent the differences between total net emigration estimates taken from Table 7 and the estimates of emigration to the United States, Panama and Cuba.

INDEX OF NAMES

INDEX OF SUBJECTS

Printed in the United States
By Bookmasters